The Cretaceous World

The rich geological record of the Cretaceous Period reveals a world that expe[rienced] climatic warmth and significantly higher global sea-levels than today. Elevate[d] atmospheric carbon dioxide have been implicated in these conditions. It thus provides a natural case study of the Earth in 'greenhouse' climatic mode, which this interdisciplinary textbook analyses from the perspective of Earth System Science. With mounting concerns over global warming today, an understanding of how the Earth system operates when in greenhouse mode is very relevant to studies of future climatic change.

Part 1 (Chapters 1–5) surveys what the Cretaceous world was like, covering the evidence for the major changes in palaeogeography, sea-levels, life and climates that took place during the period, and especially the remarkable responses to climatic conditions of high-latitude vegetation and the shallow marine biota at low latitudes. Part 2 (Chapters 6–9) explores the interactions between the physical, chemical and biological processes, both within the Earth and at its surface, that together controlled conditions on the Cretaceous Earth, and highlights how they differed from those of our present world. Comparison is made between the global carbon cycles of then and now, with particular attention to the geological sources (especially volcanism) and sinks (organic carbon on land and carbonate plus organic carbon in the sea). Other biogeochemical cycles are also discussed. The results from computer modelling of climates are also critically reviewed. Part 3 (Chapters 10–13) investigates the infamous mass extinction that terminated the period, and its causation. Finally, a short Epilogue considers broader issues arising from this case study of the Cretaceous world.

Designed for use on undergraduate and graduate courses, this textbook includes many features that will aid tutors and students alike, including full-colour figures, boxed summaries of supplementary and background information, chapter summaries, and bulleted questions and answers. The book is supported by a website hosting sample pages, selected illustrations to download, and worked exercises: **http://publishing.cambridge.org/resources/0521831121**

All the authors are based at The Open University, UK. They have shared interests in Earth System Science and the Cretaceous, with complementary areas of specialist expertise, in each case internationally recognized from numerous publications. **Peter Skelton**'s research concerns the marine sedimentary and fossil record, especially that of the giant carbonate platforms which characterized low latitudes during the Cretaceous. **Robert Spicer** uses palaeobotanical evidence to test computer-based climate models, with a particular interest in the Cretaceous flora of high latitudes, where the strongest climate signals can be detected. **Simon Kelley** applies high-precision radiometric dating methods to a variety of geological issues, such as the ages of meteorite impacts and large igneous provinces, which were major features of the Cretaceous world, and the provenance of sediments (in time as well as space). **Iain Gilmour** uses stable isotope geochemistry to trace the origins of organic compounds, and has specifically deployed this approach to investigate impact-related effects at the Cretaceous/ Tertiary ('K/T') boundary. The nature of the Cretaceous world is thus a common theme for the major research groups at The Open University to which the authors variously belong — 'Environmental Change in Earth History', 'Isotope Geochemistry and Earth Systems' and 'Volcano Dynamics' — ensuring both the broadly interdisciplinary character of this book and its incorporation of some of the most recent research results. The authors presented an earlier version of this text to Open University students in 2002. Three of them have also collaborated previously in producing acclaimed Open University teaching texts, including those on Evolution (Skelton and Gilmour) and Earth System Science (Skelton, Spicer and Gilmour), while Kelley has contributed to several books on radiometric dating and noble gases.

Cover photograph   Limestone that formed on a shallow marine platform in Early Albian times (c. 110 Ma ago), exposed in fields outside the town of Teloloapan, Guerrero State, SW Mexico, and showing classic karstic ('limestone pavement') weathering, yielding thin, well-drained soils. Such carbonate platform deposits were a common product of the climatically warm Cretaceous world, in low latitudes, giving rise today to characteristic landscapes from Mexico to the Far East. (Peter Skelton, Open University.)

The Cretaceous World

Edited by Peter W. Skelton

Authors:

Peter W. Skelton

Robert A. Spicer

Simon P. Kelley

Iain Gilmour

PUBLISHED BY THE PRESS SYNDICATE OF THE UNIVERSITY OF CAMBRIDGE
The Pitt Building, Trumpington Street, Cambridge, United Kingdom

CAMBRIDGE UNIVERSITY PRESS

The Edinburgh Building, Cambridge CB2 2RU, UK

40 West 20th Street, New York, NY 10011-4211, USA

477 Williamstown Road, Port Melbourne, VIC 3207, Australia

Ruiz de Alarcón 13, 28014 Madrid, Spain

Dock House, The Waterfront, Cape Town 8001, South Africa

http://www.cambridge.org

This co-published edition published 2003, first published 2002.

Printed in the United Kingdom by Bath Press, Blantyre Industrial Estate, Glasgow G72 0ND, UK.

Typefaces: text in Times 11/13pt; headings in Futura.

System: Adobe Pagemaker.

A catalogue record for this book is available from the British Library.

ISBN 0 521 53843 2 paperback   ISBN 0 521 83112 1 hardback

This publication forms part of an Open University course, S369 *The Geological Record of Environmental Change*. Details of this and other Open University courses can be obtained from the Call Centre, PO Box 724, The Open University, Milton Keynes MK7 6ZS, United Kingdom. Tel. +44 (0)1908 653231, e-mail ces-gen@open.ac.uk Alternatively, you may visit the Open University website at http://www.open.ac.uk where you can learn more about the wide range of courses and packs offered at all levels by The Open University.

2.1

Contents

Preface

We live in an age of growing, and justified, concern about the impact of humans upon the Earth. And with that has come a desire for a deeper understanding of how the solid Earth and its atmosphere, oceans, and life itself, as well as extraterrestrial influences, interact with one another to yield the conditions experienced at its surface. Huge advances in recent decades both in the scope of investigative techniques, ranging from satellite imagery to mass spectrometry, and in the computing power necessary for modelling such a complex system have spawned the kind of science needed to satisfy that desire — Earth System Science. The record of the rocks shows us that the Earth is by no means a stranger to change, however, and that conditions in the geological past differed in many ways from those that we experience today. In particular, former climates were usually even warmer on average than in our current 'interglacial' state, with polar ice caps that were of only limited extent or even absent. The Cretaceous Period furnishes one of the most extreme and also accessibly recorded examples of such a contrast. Not only is that period well represented by widely exposed strata on the continents, but there still remains an extensive geological record of it on the ocean floor, which is not the case for earlier such periods because of the subduction of most pre-Cretaceous oceanic crust. By examining what the Earth was like in such a contrasting state, the Cretaceous record can provide insights into the variability of the feedbacks that govern the overall behaviour of the Earth system.

This book, then, is primarily intended as an exploration in Earth System Science, viewed through the exotic perspective of the Cretaceous world. It is this central theme that has largely determined the topics addressed by the authors. Hence, the global carbon cycle features prominently, for example, entailing coverage of the major geological sources, especially volcanism, and sinks, such as marine carbonates and high latitude coals. On the other hand, we apologize in advance for the relative lack of discussion of some of the more conventional icons of the period such as the dinosaurs, which are nevertheless only of secondary importance in this context. Besides, the latter, especially, are more than adequately treated in other publications. It would be disingenuous, however, to pretend that the actual balance of topics was wholly determined by our central theme. Inevitably, there is also some bias towards the special areas of research expertise of the authors, and we acknowledge that any other group of Cretaceous 'buffs' might well have written a somewhat different book. Also worth pointing out is that within the confines of a reasonably transportable volume it would not be possible to describe all aspects of the known Cretaceous world; the data are far too rich and extensive for that. What we have done is provide some examples of the more extreme differences between the Cretaceous world and that of the present, as well as showing how such data are culled from the rock record. Above all we would hope that any group of Cretaceous experts would stress alike the interdisciplinary nature of the enterprise, for Earth System Science is all about the connectedness of our dynamic planetary system. If our, perhaps naïve, enthusiasm for this nascent field of science succeeds at least in stimulating others to take it towards maturity, then our main objective will have been achieved.

This book is aimed at advanced undergraduates and graduates. Readers are expected to have a basic understanding of sedimentary rocks, fossils, sedimentary geochemistry and major Earth processes (such as plate tectonics and the rock cycle). Further information can also be obtained from the references cited in the text. The book was designed as a teaching text for independent study. For this reason, it includes some short bulleted questions followed directly by answers. These are designed to make the reader pause and think about fundamental points concerning the subject matter. Supplementary and/or background information, which some readers may already be familiar with, has been placed in boxes.

The Cretaceous World forms a part of an advanced undergraduate course offered by the Open University, UK, entitled S369 *The Geological Record of Environmental Change*.

The authors, February 2003.

PART 1 SURVEY OF THE CRETACEOUS WORLD

1 Introduction to the Cretaceous

Peter W. Skelton

According to current absolute dating methods, the Cretaceous Period extended from around 144, or 142, million years (Ma) ago — depending on definition — until its famously eventful termination 65 Ma ago. During that time the Earth was a remarkably different world from the one we live on today. Its strange-looking extinct life forms — especially the dinosaurs — have long been regarded in popular imagination as emblematic of a 'Lost World' (Figure 1.1). This book explores how it became what it was.

Since the conceptual revolution of plate tectonics in the 1960s, geologists have been charting the past arrangements of the world's continents and oceans. One notable contrast between Cretaceous geography and that of today was the existence of an east–west oriented ocean at low latitudes — the Tethys Ocean — separating the clustered northern and southern continents (Figure 1.2). Furthermore, the Atlantic Ocean and Indian Ocean, still in their early stages of formation, were of relatively smaller size, and the Pacific Ocean was correspondingly broader. Empirical and modelling studies, especially since the 1980s, have shown that there were some significant differences in the characteristics of the atmosphere and oceans, with warm

Figure 1.1 Cover of paperback edition of Sir Arthur Conan Doyle's *The Lost World* (1960), a classic escapist fantasy of 'derring-do' among the dinosaurs.

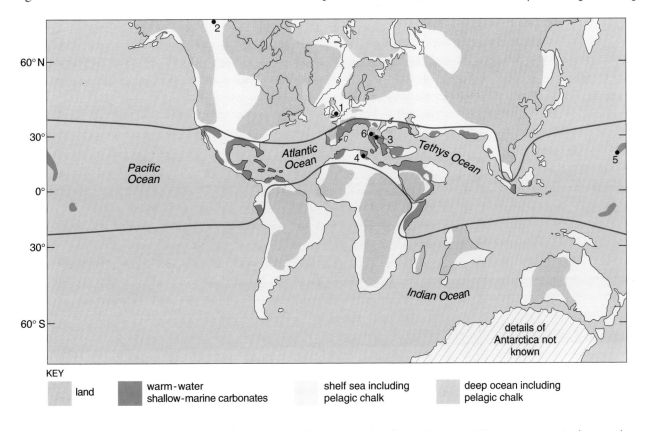

KEY

| | land | | warm-water shallow-marine carbonates | | shelf sea including pelagic chalk | | deep ocean including pelagic chalk |

Figure 1.2 Approximate reconstruction of Cretaceous palaeogeography about 98 Ma ago (Cenomanian Age), showing the former positions of the continents (the latter are shown with their present outlines, to aid recognition, though at the time their shapes would also have been somewhat different). Red lines show the belt of carbonate platform development. Numbers indicate the Tour stops referred to in Section 1.1. *(Modified from Sohl, 1987.)*

climates extending to high latitudes. Were we able to go back in time to visit the Cretaceous Earth, it would indeed have seemed very strange to us, although some familiar sights, such as flowering plants, would have reminded us that we were still on our own planet. It is this dissimilarity from today's world that we investigate in this book: the Cretaceous provides us with an excellent case history of the Earth in an extreme greenhouse state, in contrast to today's icehouse state. By viewing the Earth in such a contrasting condition, we aim to provide a fuller insight into the workings of the Earth system — the complex mesh of interactions between the solid Earth and its oceans, atmosphere and life — under varying influences, both internal and extraterrestrial. Such understanding is not merely of academic interest, given the increasing probability of our re-creating a warmer world, albeit at an unprecedented rate, which could entail alarming consequences — more literally a 'Lost World' than we might care for.

1.1   A short reconnaissance trip

By way of a preview for some of the major topics that we will be exploring in this book, let us take a brief tour of six locations around the world and reconstruct the conditions that prevailed at each during various times in the Cretaceous.

1.1.1   Tour stop 1: the Chalk Sea

We will begin our tour some 50 miles north-west of London, England. In quarries scattered along the escarpment of the Chiltern Hills, a fine-grained, somewhat powdery white limestone of Late Cretaceous age, known as the Chalk, is excavated to manufacture cement (Figure 1.3). The Chalk is one of the most characteristic deposits of the Cretaceous, and, indeed, the name of the period was derived by the French geologist J. J. Omalius d'Halloy in 1822, from the Latin word 'creta' (chalk). It has an extraordinarily broad distribution and may be seen not only in the Northern Hemisphere, both in North America (Texas) and across Northern Europe (from the United Kingdom to Russia), but also in the Southern Hemisphere (Western Australia).

Figure 1.3   A Chalk quarry on the Chiltern escarpment. *(Peter Skelton, Open University.)*

Scanning electron micrographs (Figure 1.4a) show that the Chalk largely consists of minute skeletons of single-celled, 'golden-brown' algae (coccolithophores) and isolated plates of calcite (coccoliths) derived from their breakdown, together with the remains of other, predominantly planktonic organisms. Living coccolithophores (Figure 1.4b) inhabit the photic zone (the water depth to which light penetrates — usually up to a maximum of 200 m) mainly in open oceanic waters, especially in areas of upwelling or vertical mixing of water masses. On death, their skeletons rain down through the water column to accumulate on shallower parts of the ocean floor as ooze. Calcareous ooze does not accumulate at a greater ocean depth, where the supply from above becomes matched by dissolution due to a combination of the high pressure of the water and its low temperature. The depth limit for accumulation — the carbonate compensation depth (CCD) — varies according to circumstances; in the Atlantic Ocean for example, the CCD is at a depth of about 4000 m for calcite, whereas in the colder waters around Antarctica it is at a depth of around 500 m.

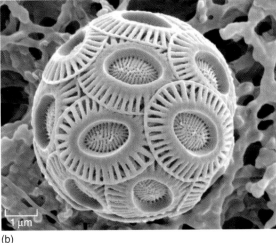

(a)   (b)

Figure 1.4 Scanning electron micrograph of: (a) the Chalk *(John Jones, Open University)*; (b) a Recent coccolithophore *(© Jeremy Young, Natural History Museum, London)*.

Today, calcareous ooze is essentially an oceanic deposit, so the accumulation of the Chalk over vast areas of continental crust in the Late Cretaceous requires an explanation.

○ How might the deposition of the Chalk on the continental crust be explained?

● The areas of continental crust in question must have been inundated by seas (Figure 1.5).

With no reason to suppose that the ecology of the Cretaceous coccolithophores and associated planktonic organisms differed significantly from that of their living relatives, their requirement for living space within the water column would imply water depths of up to 200 m. This, in turn, suggests that either those areas of continental crust had subsided commensurately in the Late Cretaceous, or that sea-levels rose by that much on a global scale. Uniform subsidence over such a vast area and on several different continents at the same time seems unlikely, so the latter explanation — a global, or eustatic change in sea-level — is more plausible. This pattern, and the effect of the major rise in sea-level, are aspects of the Cretaceous world that we will investigate in Chapter 3.

Figure 1.5 Reconstruction of life on the Late Cretaceous sea-floor in southern England, based on fossils found in the Chalk. The stalked tubular objects are sponges. *(John Watson, Open University.)*

1.1.2 Tour stop 2: the polar forests

We will now travel to the Arctic, to the northernmost part of Alaska, to an area known as the Arctic Slope. Today, this region extends to latitude 70 °N, is without trees and is covered in tundra (Figure 1.6). It has an annual average temperature of –5 °C.

Figure 1.6 Aerial photograph of Alaskan tundra with cliff exposure showing interbedded coals and mudstones. *(Bob Spicer, Open University.)*

The rocks in this area record quite a different kind of environment in the Cretaceous. They are composed of interfingering marine shales and non-marine sandstones and mudstones rich in coal (Figure 1.6). In fact, the United States Geological Survey has estimated that the Cretaceous rocks of the Arctic Slope contain approximately 2.75 trillion (10^{12}) tonnes of low ash, low sulfur coal. This is about one-third of the total United States coal reserves. Looking more closely at these non-marine sediments, we see that they represent large deltas that grew outwards into the ancient Arctic Ocean. Rooted in ancient fossil soils are the upright remains of large trees (Figure 1.7a). Scattered throughout the sediments are the fossilized remains of leaves of conifers, broad-leaved flowering plants (Figure 1.7b), ferns, and even cycads (Figure 1.7c). In clear contrast to today's vegetation, this region must have been covered in forests, and highly productive ones at that. This is because the huge quantities of coal preserved in this region represent only a fraction of the carbon captured by photosynthesis from the Cretaceous atmosphere.

(a)

(b)

(c)

Figure 1.7   (a) Upright tree stump in an Alaskan cliff face. (b) Fossilized leaves of a conifer, and broad-leaved plant. (c) Fossilized cycad leaf. *(Bob Spicer, Open University.)*

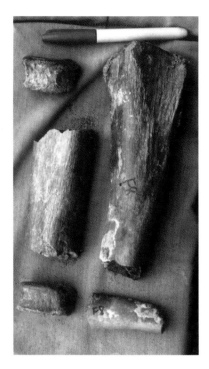

Figure 1.8 Dinosaur remains from northern Alaska. *(Bob Spicer, Open University.)*

For these luxuriant forests to have grown, they must have experienced an average annual temperature higher than that of the present-day tundra. There are two ways in which this could have been achieved: the first explanation is that Alaska might have been situated at a lower latitude in the Cretaceous; the second explanation is that the polar temperatures were higher. To distinguish which explanation is correct, we have to determine the original latitude (palaeolatitude) of the region in the Cretaceous. Palaeomagnetic data, comprising the remanent magnetism of iron-bearing minerals preserved in rocks from across the Northern Hemisphere, and even from the Arctic Slope itself, provide the answer. The declination of this faint residual magnetization records the local orientation of the Earth's magnetic field, hence the palaeolatitude, at the time of formation of the rocks. These data show that the Alaskan forests grew even further north (up to 85° latitude) than their present position. The inescapable conclusion is that the Cretaceous Arctic was considerably warmer than it is today. In addition, there must have been abundant around-the-year rainfall to have allowed the organic material to accumulate and be preserved long enough to become buried and converted to coal.

The forests were not restricted to Alaska. They stretched around the North Pole as revealed by Cretaceous leaf and wood fossils in Greenland, Canada, and across Russia. As in Alaska, there are huge Cretaceous coal reserves across Siberia and north-eastern Russia. Moreover, the Arctic forests were mirrored at high palaeolatitudes in the Southern Hemisphere, where again they were associated with coal deposition. These polar forests teemed with animal life. The remains of turtles, small mammals, toothed birds such as *Hesperornis*, and various dinosaurs, have been found (Figure 1.8).

In combination, animal and plant life in the Arctic (Figure 1.9) represents an ecosystem without parallel in today's world. This was an ecosystem existing in a warm climate that also experienced several months of continuous darkness each year — the polar light regime.

Figure 1.9 Reconstruction of northern Alaskan forests of the Late Cretaceous. *(John Watson, Open University.)*

1.1.3 Tour stop 3: the tropical carbonate factory

Our next port of call is the Istrian Peninsula, which juts out from western Croatia into the northern part of the Adriatic Sea. Much of the peninsula consists of various types of Cretaceous limestone, although these are all very different in character from the Chalk in Section 1.1.1. Prominent among these rocks are moderately well-cemented bioclastic limestones that contain abundant shelly fossils of Late Cretaceous age. This rock type makes an excellent building stone, which was much favoured, for example, by the Romans (Figure 1.10).

(a)

(b)

Figure 1.10 (a) Roman arena at Pula, Istria, built using local Cretaceous limestone. (b) Abandoned Roman quarry in similar Cretaceous limestone on Frašker Island, south-west Istria, Croatia. *(Peter Skelton, Open University.)*

Much of the bioclastic material in this location was derived from the shells of an extinct group of sessile bivalves known as rudists. In life, these highly gregarious animals were either partially implanted in the sea-floor sediment, or lay prone upon it, forming vast underwater shelly 'meadows'. Congregations of shells were sometimes buried in life position (Figure 1.11a). The fate of most of the shells, however, was to be toppled or even swept away and broken up by storm currents, so fuelling the supply of bioclastic sand and debris (Figure 1.11b). Local depressions were often filled in by such material, which was swept away from neighbouring shallow areas to accumulate in gently dipping inclined beds (clinoforms) on their flanks (Figure 1.12).

Figure 1.11 Rudist limestones dating from the beginning of the Late Cretaceous (Cenomanian; see Section 1.3) in south-west Istria, Croatia (see Tišljar *et al.*, 1998). (a) Close-up of a congregation of rudist bivalves preserved in life position on the platform top. Scale is 12 cm. (b) Beds of redistributed bioclastic sediment containing abundant transported rudist shells and fragments, shown in the cut face of a quarry. *(Peter Skelton, Open University.)*

(a)

(b)

Figure 1.12   Clinoforms (sloping down to the right) of redistributed bioclastic sediment (as in Figure 1.11b), exposed along the coast of Frašker Island, south-west Istria, Croatia. *(Peter Skelton, Open University.)*

These and other limestones of shallow marine origin accumulated over large areas in and around the Tethys Ocean and were scattered across the Pacific on shallow volcanic promontories (Figure 1.2), to form massive carbonate platforms, similar to the Bahama Banks of today, although on a far bigger scale (Figure 1.13). Their development through the Cretaceous was episodic, with extended phases of widespread platform growth repeatedly terminated over relatively short time-spans. Whereas the Chalk was the most abundant type of limestone deposited in mid-latitude seas, at least in the Late Cretaceous, carbonate platform deposits dominated at low latitudes. Together, the carbonate platforms and the Chalk constituted such a large geological reservoir for carbon (as carbonate) that their development formed an important part of the global carbon cycle of the Cretaceous — an aspect to which we will return later in this book.

Figure 1.13   Reconstruction of the carbonate platform illustrated in Figures 1.11 and 1.12. On the left, rudist 'meadows' are spread across the top of the platform, while slopes covered by their shelly debris run off to the right. *(John Watson, Open University.)*

1.1.4 Tour stop 4: the susceptible sea

Our next tour stop takes us even further south, to western Central Tunisia. Here, strata dating from the earlier part of the Late Cretaceous crop out in wooded hills near the town of Kasserine. They show a remarkably abrupt change of facies, from thick beds of well-burrowed limestone below, to thinly bedded limestone above a sharp boundary (Figure 1.14a). The latter splits quite readily into thin plates (Figure 1.14b), and, although a pale colour when weathered, is dark grey on freshly broken surfaces. Planktonic microfossils are diverse below the boundary, but impoverished and relatively small above the boundary between the two facies. Moreover, the burrows that testify to bottom-dwelling animals in the lower part disappear above the boundary, where rare ammonites are the only macrofossils encountered (Figure 1.14c).

Figure 1.14 Shelf limestones deposited in the earlier part of the Late Cretaceous (across the Cenomanian/Turonian boundary; see Section 1.3) exposed on Jebel Bireno, Kasserine, Central Tunisia. (a) Due to local tectonic folding, the beds are dipping steeply towards the right. Thick-bedded limestone (on the left) abruptly overlain by platy limestone (on the right). The geologist's right foot is on the boundary between the two facies. (b) Detail of the platy limestone above the boundary. (c) External moulds of ammonites on a bedding surface of the platy limestone. *(Peter Skelton, Open University.)*

(a)

(b)

(c)

○ From the information given above, what is the most likely explanation for the abrupt change in facies?

● A change in conditions in the water column extinguished life on the sea-floor, allowing undisturbed accumulation of laminated sediment.

Much of the overlying water column was evidently also affected, as shown by the change in the plankton, although a few pelagic animals such as ammonites survived, presumably by living close to the surface of the sea. The dark grey colour of the laminated limestone would suggest a relatively high content of organic material. The most likely circumstance in which such material would be allowed to accumulate would be in the absence of aerobically respiring bacteria, i.e. in anoxic conditions. So, the most plausible explanation for the change of facies is the onset of anoxia throughout much of the water column.

Such a change of facies in one locality alone need not seem remarkable. After all, today, anoxia occasionally afflicts many restricted marine basins, such as parts of the Gulf of Mexico offshore from the Mississippi Delta, the Black Sea and local areas of the Adriatic (although human pollution also plays a part in all these cases). The case considered here, however, is different, as a similar change can be seen at the same stratigraphical level in marine successions deposited on the continental crust at low latitudes from south-west Mexico to several other sites around the Mediterranean. It has also been documented from sites within all the ocean basins on the basis of cores drilled from the bottom sediments. Not surprisingly, the event was associated with a mass extinction of many marine organisms (especially those dwelling on the carbonate platforms), although the extent to which the extinctions can be attributed directly to the spread of the anoxia remains a matter for debate.

Such drastic events are termed Oceanic Anoxic Events (OAEs) and they occurred on a number of occasions, on varying geographical and temporal scales, during the Cretaceous, although not, it appears, thereafter. They clearly represent major perturbations in the operation of the global carbon cycle, which we will consider in greater detail later in the book.

1.1.5 Tour stop 5: the submarine volcanic plateau — Allison's story

We now take a large detour from the side of the globe where the continents are clustered, to the other side, which is dominated by the Pacific Ocean. For many years the Pacific Ocean floor was poorly explored and its form was not well understood. Since 1968, however, thanks to the Deep Sea Drilling Programme (DSDP) and Ocean Drilling Programme (ODP), we have come to appreciate better what lies on the ocean floor and to understand a great deal about the rocks below. We will focus on an area of the western Pacific called the Mid-Pacific Mountains (Figure 1.15). Although this name might sound counter-intuitive, the Mid-Pacific Mountains cover parts of the Pacific which formed during the Early Cretaceous, and are, in truth, a collection of volcanoes sitting on a plateau, which is, in turn, 1–2 km higher than the surrounding ocean floor. The western Pacific Ocean floor, is covered by thousands of volcanoes; some in chains like the Hawaiian Islands, some isolated and some in groups, such as the Mid-Pacific Mountains (Figure 1.15).

In comparison, the Cretaceous areas of the Atlantic Ocean floor are relatively flat and draped in fine-grained, deep oceanic muds.

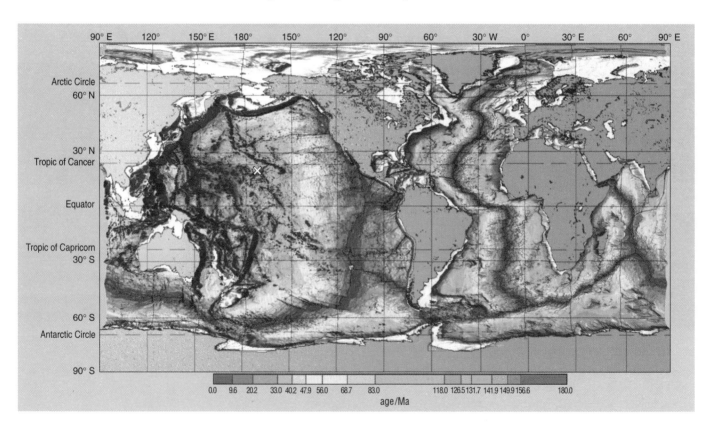

Figure 1.15 Topography of the Earth's ocean floor, colour-coded according to its age of formation (sea areas in grey show where age is undetermined). X marks the Mid-Pacific Mountains. (National Geophysical Data Center, USA.)

Many of these mid-Pacific volcanic mountains were sampled by ocean research cruises as early as the 1950s. The first studies of the Mid-Pacific Mountains showed that they were indeed volcanoes and ruled out the best previously existing hypothesis for their formation, which was that they were ancient Precambrian volcanoes flattened by erosion at sea-level.

○ Why must the idea that the Mid-Pacific Mountains might be Precambrian volcanoes be wrong, but forgivably so, in the light of what we have learnt since the 1950s about the way the Earth works?

- The elucidation of plate tectonics in the 1960s revealed that, as a consequence of sea-floor spreading, the oceans are now underlain entirely by post-Palaeozoic ocean crust, older ocean floor having been consumed down subduction zones.

In fact, by 1956 it was discovered that the Mid-Pacific Mountains were volcanoes that had been exposed at sea-level during the mid-Cretaceous, because they were encrusted by carbonate platform deposits containing abundant rudists (such as those encountered in Section 1.1.3). Only later had their summits subsided beneath sea-level. As more information became available, it emerged that almost all the volcanoes on the floor of the western Pacific formed between 125–75 Ma ago, at the same time that spreading at mid-ocean ridges had been faster than today. These eroded volcanic edifices are called guyots (flat-topped volcanoes rising from the ocean floor, usually covered by at least 200 m of water, and named after Arnold Henry Guyot (1807–1884), a Swiss–American geographer and geologist).

Let us now take a look at the Allison Guyot, which lies in the central part of the Mid-Pacific Mountains, not far from the trace of the Hawaiian volcano chain (Figure 1.16). Allison has a structure typical of the several dozen seamounts rising from a plateau which together make up the 200-km-long Mid-Pacific Mountains. Allison has a roughly flat top, 20 km across, surrounded by steep volcano walls (Figure 1.17a). It was first mapped by DSDP scientists using seismic reflection techniques and sampled by dredging in the late 1970s (dredging is simply trawling a reinforced bucket behind a research ship so it drags across the ocean floor and collects loose fragments of rock — not very sophisticated but it works well!). Like other guyots in the Mid-Pacific Mountains, the Allison Guyot is a basaltic volcano topped by carbonate platform deposits (Figure 1.17b). The volcano that forms its base erupted around 111 Ma ago. Allison was then only perhaps 500 m higher than the surrounding plateau but emerged above sea-level and was capped by sub-aerial lava flows, i.e. extruded in the air rather than underwater.

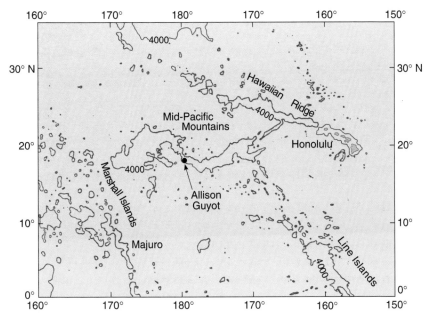

Figure 1.16 Bathymetric map of the western Pacific, showing the position of the Mid-Pacific Mountains and Allison Guyot. The 4000 m depth contour is shown and islands are indicated in yellow. (Ocean Drilling Program, Texas A. & M. University.)

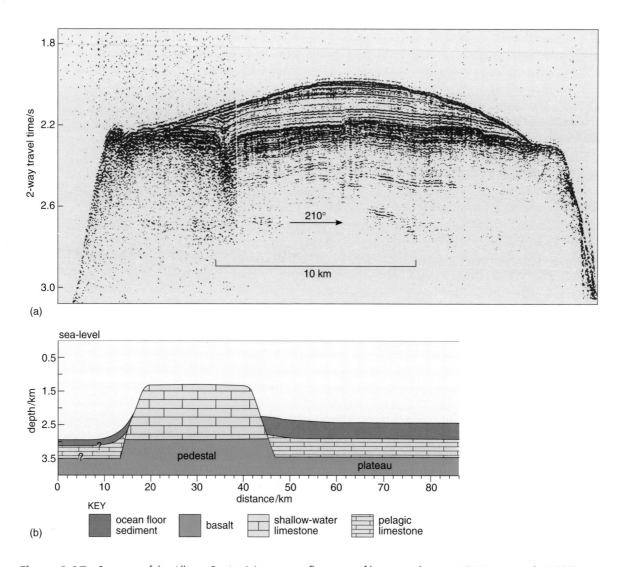

Figure 1.17   Structure of the Allison Guyot: (a) seismic reflection profile across the guyot *(Winterer et al., 1995)*; (b) diagram showing basic structure, in vertical section, as interpreted by Winterer and Sager (1995).

Allison subsided on moving away from the mid-ocean ridge, as all ocean crust does as it cools (and so becomes denser), but that was not the end. To start with, Allison sank rapidly in geological terms, at a rate of about 9 cm every 1000 years, although it sank somewhat more slowly later on. However, for the abundant life that inhabited sea-floor environments close to the ocean surface, Allison was a very stable home. Shallow submarine 'meadows' of rudists (like the examples shown in Figure 1.18) and sponges congregated around the edges of the platform in front of shifting perimeter sand shoals and storm-built islands which damped incoming waves, allowing finer-grained sediments to accumulate in shallow lagoonal areas behind. As the original volcanic relief became eroded and buried, a virtually flat-topped carbonate platform thus became established on the sinking volcano. Indeed, the carbonate platform continued to build up for over 13 Ma, rising a staggering 1330 m from the top of the basalt volcano and maintaining its surface close to the ocean surface as Allison sank deeper. During this time, the carbonates accumulated as a succession of shallowing-upward cycles 3–10 m thick, like a pile of stacked blankets, in response to small-amplitude oscillations in sea-level with a periodicity of about 100 000 years.

(a)

(b)

Finally, about 98 Ma ago, sea-level apparently fell by 100–200 m (relative to the guyot at least) and Allison was an island again with the carbonate platform exposed. All the guyots in the Mid-Pacific Mountains, as well as some further to the west, suffered the same fate, although it is unclear whether this was due to a eustatic fall in sea-level or regional uplift. Rain falling on the new islands caused erosion and started to dissolve Allison in the same manner that it dissolves exposed limestone areas today. The top surface of the carbonate platform on Allison still bears the scars of this episode, in the form of a karstic landscape (named after the Karst region of Slovenia), which includes sinkholes, caverns and cavities. In particular, the central area, where freshwater could collect and dissolve the limestone, was preferentially hollowed out, leaving a residual rim around parts of the guyot's margins that gave it an atoll-like profile (Figure 1.17a). Then sea-level rose again, but this time the carbonate platform did not become re-established, although it is unclear why not. By around 93 Ma ago, Allison was too deep to support the carbonate platform dwellers and became covered in a thick layer of ocean mud (forming the convex 'cap' in Figure 1.17a).

The Mid-Pacific Mountains illustrate the profuse nature of volcanic activity within the Pacific Ocean basin, reflecting the existence of large-scale rising plumes of hot material in the mantle during the Cretaceous. Not only did this activity result in major topographical prominences on the ocean floor, with implications for global sea-levels, but the associated release of gases as well as the convective circulation of ocean water through the young hot crust would have had significant impacts on atmospheric and oceanic chemistry. In addition, the sedimentary record of the carbonate platforms reveals long periods of small-amplitude sea-level oscillations, punctuated by infrequent larger changes. Large-scale sea-level changes in the last couple of million years, especially, have been caused by water released or taken up by continental glaciation. This mechanism cannot be invoked for most of the Cretaceous, however, as Section 1.1.2 has already shown. Later in this book, we will explore in detail the possible interaction of volcanism with the atmosphere during the Cretaceous.

Figure 1.18 Images of Pacific rudists: (a) a cluster of conjoined tubular shells from a guyot south-east of Japan; (b) section across a shell from another guyot in the Mid-Pacific Mountains, showing distinctive narrow canals within the shell wall, a feature typical of certain types of rudist. *(Peter Skelton, Open University; both specimens courtesy J.P. Masse.)*

Figure 1.19 Limestones of the uppermost Cretaceous (right) and lowermost Tertiary (left) are separated by a thin layer of clay in this exposure in the Bottaccione Gorge, near Gubbio in Italy. The clay layer contains anomalously high amounts of iridium, an element rare in the Earth's crust but more abundant in meteorites, comets and in the Earth's mantle. *(Iain Gilmour, Open University.)*

1.1.6 Tour stop 6: Apocalypse Then

Our final stop takes us to the medieval town of Gubbio in the Italian Apennines and to one of the many gorges that cut through the mountains. The rocks of this area were deposited in deep water during the Cretaceous and Early Tertiary (the Cainozoic is frequently subdivided into the Tertiary, comprising the Palaeocene to Pliocene epochs, and the Quaternary, comprising the Pleistocene and Holocene epochs). The strata are dominated by large thicknesses of pelagic deep-water limestones. In the Bottaccione Gorge, just east of Gubbio, these limestones are exposed along the side of the road as it winds its way north-east. At the western end of the gorge the rocks are of Late Cretaceous age, and as we move up through the gorge the rocks become younger, until eventually we pass into rocks of Early Tertiary age. The boundary between the two geological periods is shown in Figure 1.19 — a boundary that also marks the end of the Mesozoic Era. On the face of it, there is little remarkable in the rocks across the boundary: the pinkish-coloured limestones of the uppermost Cretaceous become whiter as we approach the boundary and are separated from the grey-coloured limestones in the lower Tertiary by a thin layer of clay. However, within these rocks is the evidence for a theory that has radically changed the way geologists interpret events at the end of the Mesozoic Era.

The Earth's geological record is punctuated by several abrupt events when many of the species living at the time became extinct, disappearing from the record over a relatively short time interval. One such event marks the boundary between the Cretaceous and Tertiary Periods, and this was the first major boundary between two eras of the geological record to be identified. These short episodes when the rate of extinction was much higher than normal are known as mass extinctions and the end-Cretaceous mass extinction is one of the biggest. The boundary marks the change from the world of the dinosaurs, ammonites and other now extinct species; above the boundary, we enter the age of mammals and the eventual emergence of *Homo sapiens*. The seemingly unremarkable rocks at Gubbio contain dramatic evidence of this change and its possible cause. Under the microscope, we can see that these limestones are composed of the carbonate shells of millions of microscopic organisms known as foraminifers (Figure 1.20). To this day, these organisms are common in the surface waters of oceans, and as they die they slowly accumulate on the sea-floor, together with other planktonic remains. The fossil foraminifers at Gubbio therefore provide us with a snapshot of life in ocean surface waters during the mass extinction that marked the end of the Cretaceous. If we compare thin sections of the limestones immediately above and below the Cretaceous/Tertiary boundary at Gubbio (Figure 1.20), a dramatic change is apparent. Above the boundary, while foraminiferan microfossils are still present, they are greatly reduced in number, diversity and in size. In fact, detailed studies of the palaeontology of the limestone of the Bottacione Gorge reveals that the majority of the foraminiferan species occurring as fossils in the latest Cretaceous limestones did not survive into the Tertiary; they became extinct right at the end of the Cretaceous.

The microfossil record from Gubbio is just one part of a global picture that geologists have reconstructed to examine the mass extinction that occurred at the Cretaceous/Tertiary boundary. However, what could have caused such a sudden and extraordinary change in the history of life on Earth? An important clue does not come from the kilometre-thicknesses of limestones that make up the Cretaceous and Tertiary rocks around Gubbio, but from the thin clay layer, barely

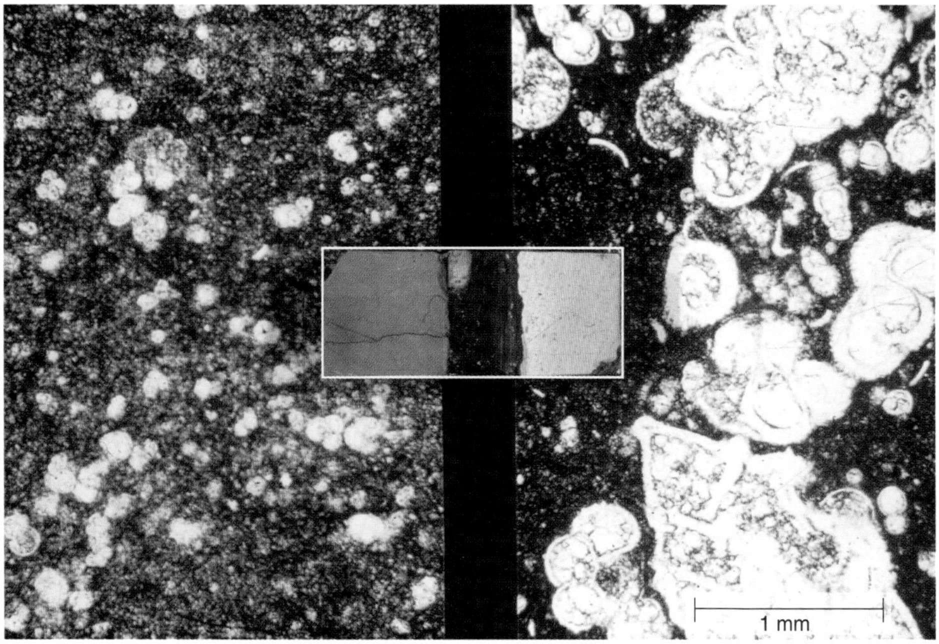

Figure 1.20 Photomicrographs (at the same scale) of foraminiferan microfossils: (a) immediately above; (b) immediately below the Cretaceous/Tertiary boundary at Gubbio in Italy. The small inset rectangle in the middle shows the clay layer between the two limestones, at natural size. *(Iain Gilmour, Open University; specimens courtesy of A. Montanari.)*

a centimetre thick, that separates the rocks of the two periods. The clay contains anomalously high amounts of the element iridium (Ir), which is rare in the Earth's crust but much more abundant in meteorites and in the Earth's mantle. This discovery was made in 1980 by a team of scientists, led by Walter Alvarez, from the University of California at Berkeley. The Alvarez team explained the Ir anomaly by the hypothesis that there had been a massive, catastrophic accretion of an Ir-rich extraterrestrial object — a giant asteroid or comet impact — and that this impact was responsible for the mass extinction at the end of the Cretaceous. We will examine the impact–extinction hypothesis, as it came to be known, in more detail in Part 3.

1.1.7 Matters arising

The major features that emerge from this brief reconnaissance of the Cretaceous world are:

- the attainment of considerably higher sea-levels than we see today;
- the development of widespread climatic warmth, extending to high latitudes;
- the repeated growth and demise of vast, shallow carbonate platforms in low latitudes;
- episodic development of widespread marine anoxia;
- extensive oceanic volcanic activity, especially in the Pacific;
- the occurrence of a catastrophic event that marked the end of the period.

These are some of the phenomena that we will be investigating in this book. Although we now have convincing evidence for all of them, we cannot yet explain with confidence how they might have been connected with one another. What we have are theories, more or less well-grounded in observation but all deserving further testing by recourse to yet more observation and/or experimental modelling, as Earth system science is still a relatively young, fast-moving field. Hence it is not so much that we wish to emphasize any currently preferred conclusions in this book — interesting, and in some cases probable, though they may be — but rather the manner of thinking that leads to them, for such is the way that our understanding will gradually improve.

The remainder of Part 1 will look in greater detail at the evidence for conditions throughout the Cretaceous. In Part 2, we will compare and contrast the workings of the Earth system today, and as it appears to have been in the Cretaceous. Finally, Part 3 will focus upon the extraordinary events that brought the period, and indeed the Mesozoic Era, to a close.

1.2 Interesting times — putting the Cretaceous into context

Before dissecting the Cretaceous world, it will be useful to outline how it fits into the broader history of changes in palaeogeography, sea-level, climate and life that occurred over the latter half of the Phanerozoic Eon.

1.2.1 Palaeogeography

The dawn of the Mesozoic Era makes a good starting point for considering the palaeogeographical framework. By that time, all the former continental masses had coalesced to form a single supercontinent, called 'Pangaea' (Figure 1.21a). This had a configuration broadly shaped like a 'less than' symbol (<), with a large wedge-shaped ocean (the Palaeotethys Ocean) intruding on its eastern side from the encircling Panthalassa Ocean (the precursor to the Pacific), which covered the rest of the globe. The succeeding periods witnessed the progressive break-up of Pangaea to form today's continents (Figure 1.21b–d).

The break-up of western Pangaea commenced early in the Jurassic with rifting between eastern North America and the conjoined northern South America and West Africa, to form the incipient Central Atlantic. Meanwhile, to the east, rifting along the north-eastern margin of the southern 'limb' of Pangaea had caused widespread block-faulting and foundering of broad carbonate platforms that had developed along the southern margins of the Palaeotethys Ocean. Further opening of this rift gave birth to the Tethyan Ocean, and the microcontinental fragments that had separated drifted northwards, eventually to 'dock' onto the Asian margin later in the Jurassic. Their collision marked the closure of the Palaeotethys, the old ocean floor of which was consumed by subduction alongside the Asian margin. By Late Jurassic times, if not earlier, the opening of the Central Atlantic gulf had extended eastwards to join up with the Tethys, so creating a narrow seaway between the northern and southern continental masses (Figure 1.21b). Hence, Pangaea split into two discrete supercontinents, 'Laurasia' in the north (comprising North America and most of Eurasia) and 'Gondwana' in the south (consisting of South America, Africa, Arabia, India, Madagascar, Antarctica and Australia). A hallmark of the Cretaceous, then, was the presence of a continuous equatorial seaway between Laurasia and Gondwana, which allowed ocean currents to flow between the two. However, some obstruction to this flow was offered, especially in the constricted western part of the Tethys, by the constellation of continental fragments and platforms (Figure 1.21c) that are now incorporated in a tectonic collage running around the northern part of the Mediterranean and across the Middle East.

While the Tethyan story unfolded in low latitudes, further south Gondwana began to break up in the Early Cretaceous with the opening of the South Atlantic between South America and Southern Africa. Later in the period, the nascent ocean joined the Central Atlantic, while to the north, the North Atlantic also began to open up between North America and Western Europe. Hence by the start of the Late Cretaceous, there existed a continuous, although locally constricted, Atlantic Ocean. Likewise, the drift of India and Madagascar from an original position nestled between Africa and Antarctica gave rise to the Indian Ocean during the Cretaceous.

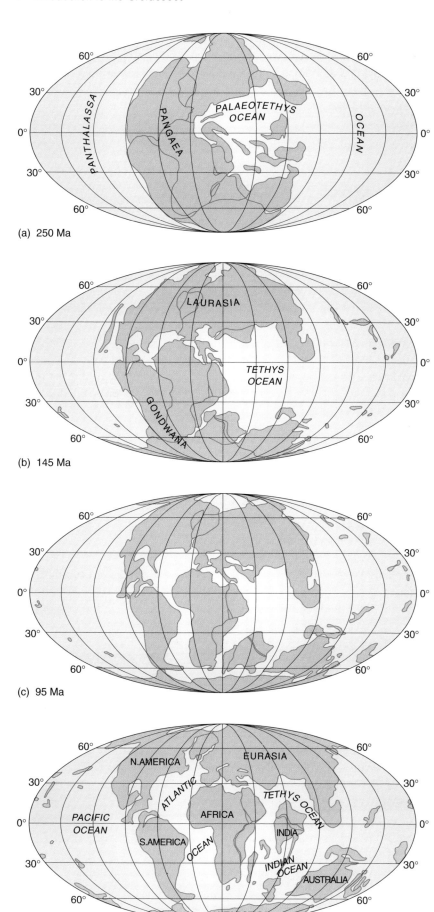

(a) 250 Ma

(b) 145 Ma

(c) 95 Ma

(d) 55 Ma

Figure 1.21   Palaeogeographical maps showing continental positions for: (a) around the Permian/Triassic boundary; (b) the Late Jurassic; (c) the mid-Cretaceous; and (d) the Early Tertiary. The outlines of today's continents are indicated by thin lines. Note that shallow seas on the continents are not indicated. *(Based on (a) University of Chicago 'Paleogeographic Atlas Project', and (b–d) ODSN Plate Tectonic Reconstruction Service.)*

In the Southern Hemisphere, large parts of southern Gondwana were situated at or near the pole, but in the Arctic there was a small, almost totally enclosed, Arctic Ocean. The connections between this northern polar sea and the rest of the world's oceans were shallow and narrow. From time to time these connections widened and shrank, as the sea-level rose and fell, but at all times deep-water connections were absent. This had profound implications for ocean circulation, the poleward transport of heat and global climate.

As the Atlantic Ocean continued to widen through the Late Cretaceous and Tertiary, Africa, together with then conjoined Arabia, first rotated anticlockwise, pivoting around Iberia, and then drifted northwards relative to Eurasia, leading to the progressive closure of the western Tethys (Figure 1.21d). The resulting crush of continental promontories and microcontinental fragments between the jaw-like masses of Africa/Arabia and Eurasia created an array of mountain chains from the Pyrenees to the Zagros Mountains of western Iran. The first effects of the collisions were felt towards the end of the Cretaceous, but the main period of mountain building ensued in the Tertiary. Only parts of the Mediterranean basin remained as constricted relics of the old ocean, while slivers of ocean crust became caught up in the collision zones as ophiolites. Likewise, the collision of India with Asia during the Tertiary gave birth to the Himalayas, while Australia separated from Antarctica to begin its own northward drift. Many of the distinctive wrinkles on the face of today's globe, due to continental collisions, are thus essentially Tertiary features.

1.2.2 Global sea-level

Turning now to eustasy, post-Palaeozoic sea-levels underwent large, long-term ('first order') oscillations (Figure 1.22a) according to changes in the capacity of, and hence displacement of water from, the ocean basins. The main variable in this respect was the total volume of mid-ocean ridges. As oceans opened and closed, the cumulative length of mid-ocean ridges increased or decreased, respectively. Moreover, given the constant proportional rate of subsidence of cooling oceanic crust as it moves away from a ridge, changes in the rate of sea-floor spreading affected the relative widths of mid-ocean ridges, such that more rapid spreading created relatively wider profiles (Figure 1.22b).

○ Study the eustatic sea-level curve in Figure 1.22a, together with the palaeogeographical map for the Permian/Triassic boundary (Figure 1.21a), and decide why global sea-level appears to have been relatively low at that time, compared with later Mesozoic times.

● With the assembly of the supercontinent Pangaea, only the mid-ocean ridge systems of the old Panthalassa Ocean and Palaeotethys Ocean remained to displace water from the ocean basins. Continental break-up thereafter gave birth to new oceans, each with their own mid-ocean ridge systems. These mid-ocean ridges cumulatively displaced more and more water, thus leading to higher sea-levels.

Despite several pronounced oscillations of global sea-level (probably largely related to changes in sea-floor spreading rates together with the subduction of the Palaeotethys ridge system), an overall long-term rise ensued during the Mesozoic, reaching a peak in the Late Cretaceous. Following further oscillations during the Early Tertiary, a pattern of overall decline then set in from the Oligocene to the Quaternary, in part associated with the cumulative transfer of

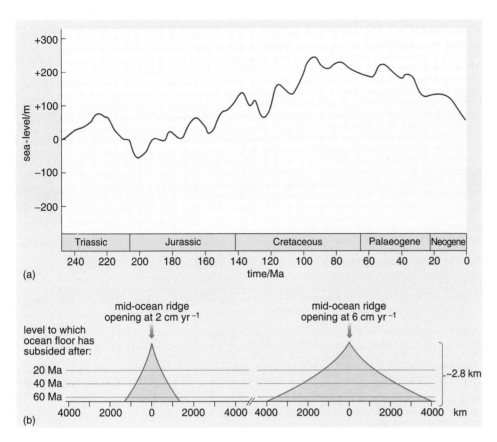

(a)

(b)

Figure 1.22 (a) Post-Palaeozoic global sea-level curve (excluding the effects of continental ice volume, such that the curve terminates above the present level at 0 m) *(Haq et al., 1988).* (b) Cartoon showing the relationship between sea-floor spreading rate (at each side of a mid-ocean ridge) and mid-ocean ridge volume.

water to the, by then, growing continental glaciers (though the latter effect has not been included on Figure 1.22a). The Late Cretaceous eustatic maximum was associated with the expansion of seas across large areas of the continents (as seen in Section 1.1.1).

1.2.3   Climate

First-hand evidence for past climates comes from various geological clues, among which the most obvious are the distributions of certain distinctive sedimentary rock types, such as tillites (glacial deposits), coals and evaporites (reflecting humidity and aridity respectively), as well as the fossils of climate-sensitive organisms. Moreover, geochemical data also allow estimation of specific aspects, such as temperature (which can be calculated from oxygen-isotope ratios) and seasonal patterns of rainfall and weathering (e.g., from clay mineral assemblages). Taken together, such evidence shows that the Earth emerged from a Late Carboniferous to Early Permian ice age, to commence a broadly warming trend (with notable oscillations, as with global sea-level) through the Mesozoic (Figure 1.23). At no time during that era did polar glaciers ever develop on a scale comparable to those of the Permo–Carboniferous glaciation, or for that matter those of the Quaternary. As with sea-level, mean global temperatures appear to have reached a maximum in the earlier part of Late Cretaceous times, when it is probable that polar ice was entirely absent at sea-level (as seen in Section 1.1.2). There was some cooling towards the end of the Cretaceous, but climates remained basically warm through the Early Tertiary until the Oligocene, when a distinct decline of temperature towards the Quaternary glaciations set in. Hence, as with global sea-level, a maximum of climatic warmth for the Mesozoic/Cainozoic was achieved in the Late Cretaceous.

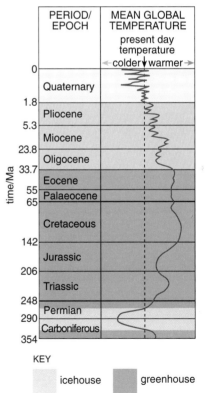

Figure 1.23 Climate through the Late Palaeozoic, Mesozoic and Cainozoic: curve of estimated relative mean global temperatures. Yellow denotes an interval of cooling, intermediate between greenhouse and icehouse states. *(Frakes, 1979.)*

1.2.4 Life

As far as organisms are concerned, the beginning of the Mesozoic marks a convenient point from which to start, because life was just recovering from the most devastating mass extinction of the entire Phanerozoic Eon. According to data assembled by the late Jack Sepkoski (1984), some 57% of families of marine animals with skeletal hard parts (the most reliable source of data for sampling past changes in biotic diversity) had disappeared from the fossil record at that time (Figure 1.24). This had possibly occurred within only 10 000–30 000 years, according to recent work by palaeontologist Richard Twitchett and colleagues (2001). Moreover, reductions in the numbers of species per family across the Permian/Triassic boundary imply a loss of about 96% of species. Collapse of the terrestrial biota, which commenced at the same time, may have taken somewhat longer, with typically Permian plant species disappearing over the next few hundred thousand years. Hence the dawn of the Mesozoic was the closest to a 'fresh start' that life on Earth ever experienced during the Phanerozoic. Renewed evolutionary diversification of the motley throngs of survivors appears to have been slow to take off again, only reaching peak rates about 10 Ma later. Although various marine animal groups that had been dominant in the Palaeozoic showed modest recovery, the new Mesozoic radiations were largely derived from stocks that had previously been less prominent (Figure 1.24). Thereafter, there was a long-term rise in the number of families towards the present, albeit interrupted by several further mass extinctions on the way, most notably at the end of the Triassic and at the close of the Cretaceous (Section 1.1.6).

A variety of reasons may account for the long-term rise in biotic diversity seen in the fossil record. To some extent, it can be interpreted as an artefact of the more comprehensive sampling that is generally possible from younger strata, and of combining the fossil record with that of extant organisms. In the latter case, only one fossil from an extant family will suffice to extend the stratigraphical range of that family from the present to at least the level of the fossil in question. In contrast, with extinct families, the top as well as the bottom of their stratigraphical range have to be estimated from fossils, which are likely to fall short of the true range at both ends. The ranges of extant families are thus more completely recorded than are those of extinct groups, especially for rarer or less easily preserved forms. The increasing proportion of extant families represented in successively younger strata therefore yields a cumulatively more complete record towards the present, giving the appearance of increasing diversity — a bias referred to as 'the pull of the Recent'. However, such biases certainly do not account for all of the observed long-term growth in biotic diversity noted above. An important geographical factor that promoted diversity was the progressive break-up of Pangaea. This created new oceanic barriers between populations of both continental and shallow marine organisms. Divergent evolution of those occupying different regions to yield new taxa then contributed to increasing global diversity. Moreover, there were increases in the number of species within local communities for ecological reasons, which will be explored in Chapter 5. Unlike the other trends discussed above, the Cretaceous did not prove to be the upper limit, as biotic diversification markedly picked up again after the terminal Cretaceous mass extinction, to reach the unprecedented levels of today (or at least those achieved before the destructive pandemic of humans). Nevertheless, levels of Cretaceous biotic diversity easily outstripped those of earlier periods.

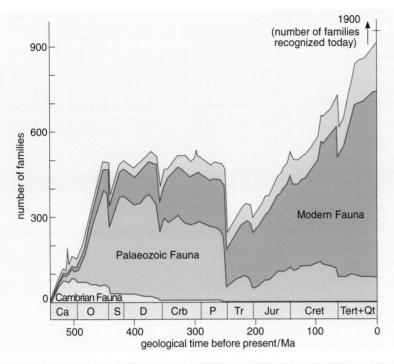

Figure 1.24 Changes in the number of families of marine animals through the Phanerozoic, as shown by the fossil record, compiled by the late Jack Sepkoski in 1984. Three overlapping phases of diversification are shown in the blue area with some representative groups illustrated beneath, while the pink area shows those families only known from instances of exceptional preservation and for which the record is patchier. *(Sepkoski, 1990a.)*

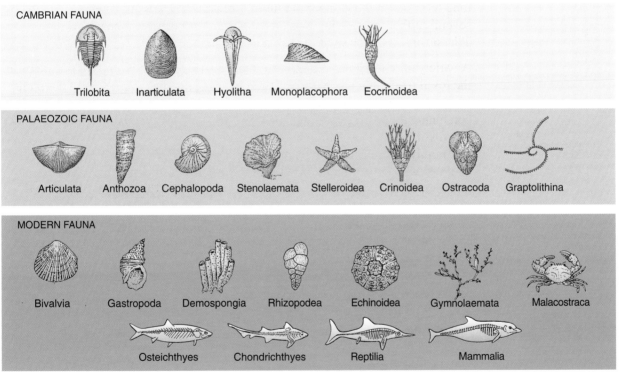

Notable among the new Mesozoic animals on land were the dinosaurs. They arose in the Triassic and reached an acme in the Cretaceous prior to their extinction at the close of that period — unless allowance is made for the birds, which were probably derived from them during the Jurassic. Among land plants, the most important event was the appearance of flowering plants, or angiosperms, which probably originated in the Early Cretaceous and then diversified rapidly in the Late Cretaceous to become the dominant land plants of the Cainozoic. Meanwhile, insects remained highly diverse from the Palaeozoic, although they too evolved many new forms, especially in tandem with the angiosperms from Cretaceous times onwards.

In the Mesozoic seas, modern types of ray-finned bony fish (Osteichthyes) arose and rapidly diversified, going on to form the most diverse group of fish known today, ranging from herring to seahorses to barracuda. They were joined by new derivatives of older stocks such as sharks and rays (Chondrichthyes). Numerous new invertebrate taxa likewise proliferated, especially among the gastropods, bivalves, echinoids, crustaceans, corals and bryozoans, all of which still feature prominently in today's seas. Alongside these flourished other new groups, such as the ammonites, belemnites and rudists, which did not survive (with the possible exception of one short-lived rudist species) beyond the end of the Cretaceous.

One major ecological innovation in the Mesozoic was the rise of the calcareous plankton, including coccolithophores and planktonic foraminifers, which became especially prolific around the start of the Late Cretaceous. As a result, carbonate skeletal material was transported in significantly greater quantities to offshore deep-water settings (as seen in Section 1.1.1), constituting an important new pathway in the global carbon cycle. Overall, the Cretaceous biota was an intriguing mixture of familiar, modern-looking (i.e. extant) groups, together with exotic-looking (i.e. extinct) groups, most of which were to be lost in the terminal Cretaceous extinction.

The Cretaceous (especially the latter half of the period) can thus be characterized as an extreme 'greenhouse' interval (Figure 1.23), with diverse biota scattered across a variegated mosaic of land, shallow sea and ocean. However, conditions were not monotonously equable. Popular images of dinosaurs scampering around in a sort of lush 'Club Méditerrané' setting (as in Figure 1.1) can give a rather misleading picture of permanently stable, benign conditions (rather like holiday snaps that omit the rainy days). In fact, the Cretaceous world was episodically affected by fairly drastic changes, including disastrous perturbations (e.g. Section 1.1.4) that were associated with mass extinctions — before the final catastrophe that brought the curtain down on the period, and indeed the Mesozoic Era (Section 1.1.6).

1.3 The Cretaceous time-scale

Although we have been speaking about Cretaceous time only in rather broad terms, it should already be clear that the Earth experienced a lot of change through this long period of nearly 80 Ma. In order to analyse its history in greater detail, we will need to refer to a standard time-scale.

1.3.1 Relative time

At the risk of stating the obvious, all that we know about the Cretaceous comes from the rocks that are around us today. In other words, it is necessary to reconstruct the chronology of the period entirely from the content of the rocks and their geometrical relationships.

○ What is the guiding principle for reconstructing the relative ages of strata when confronted with a sedimentary succession?

● Superposition — the principle that, under normal circumstances, younger strata are deposited on top of older strata.

In accordance with this principle, the pile of strata that comprise the Cretaceous System — constituting the rock record of the Cretaceous Period — is subdivided into a sequence of named stages which correspond to successive ages of time (Figure 1.25). It is important to remember that periods and ages represent relative, not absolute time. Geologists can, and do, argue about exactly when the Cretaceous

Period began and ended, in terms of Ma, but all agree that, by definition, it came after the Jurassic and preceded the Tertiary. The distinction between relative and absolute time becomes particularly crucial at the scale of ages (and at even finer subdivisions), because in some instances disparities in absolute dating may even exceed the estimated durations of the ages themselves. For example, two attempts to calibrate Cretaceous stages against absolute time that were both published in 1982 gave such different estimates for the Aptian (112–107 Ma against 119–113 Ma) that they did not even overlap. Hence, in order to avoid confusion, it is advisable not to switch uncritically between absolute and relative time-scales.

Suppose, for example, that a geologist who had found a fossil at the base of the Aptian stage stated in a publication that it came from rocks aged 112 Ma, using the first of the estimates quoted above. Another geologist, on referring to the second estimate, might then infer that the fossil post-dated other known Aptian examples and so erroneously invert their true chronological sequence. That is why we often have to discuss Cretaceous history using the arcane set of age names shown in Figure 1.25, rather than in terms of absolute time. A fossil from Aptian strata remains an Aptian fossil, no matter how much the absolute dating of the Aptian stage boundaries may change.

Also shown in Figure 1.25 is the calibration of absolute ages that is used in this book, which is taken from the compilation of Gradstein and Ogg (1996). From the previous paragraph you will realize that this is unlikely to be the last word on the issue, and later calibrations may involve yet further adjustments, albeit of decreasing magnitude as sampling coverage and dating methods improve. We employ the one scale here for the sake of consistency within this book (except when reproducing some historical figures based on older scales).

The description of stages as subdivisions of the pile of all strata constituting the system, as given above, is simplistic in that it assumes that corresponding subdivisions of sedimentary successions can be recognized around the world. In practice this is far from straightforward. In a single section of strata, it is indeed no problem to define a vertical series of units by means of successive arbitrary boundaries. The relative positions, hence ages, of fossils (or other objects) in the section could then be unambiguously recorded by reference to these fixed boundaries. However, recognition of corresponding subdivisions in other sections requires stratigraphical correlation — the matching of strata that contain equivalent objects or attributes.

Methods of correlation

For the objective of time correlation, or chronostratigraphy, the objects used for correlation must effectively serve as proxies for discrete intervals of time.

○ Would the appearance of transgressive marine deposits overlying continental beds in sections that are hundreds of kilometres apart provide a reliable basis for chronostratigraphical correlation?

● Not necessarily. If the former landscape had had relatively little relief and the rise in global sea-level had been rapid, then the transgressive deposits could have been virtually synchronous — at least on geological time-scales. However, with greater topographical relief and/or a slower change in sea-level, the marine transgression would have arrived in different areas at rather different times, so the deposit marking it would then be diachronous, hence a poor means of time correlation.

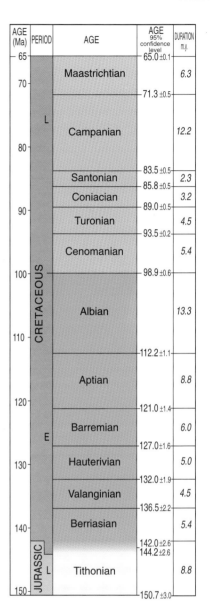

Figure 1.25 Cretaceous stages/ages, with absolute dates for the boundaries used herein. (Gradstein and Ogg, 1996, with amendments from J. G. Ogg, personal communication, 2002.)

In the case described above, the correspondence between sections relates to a common depositional process that left similar sedimentary facies, but not necessarily at the same time. The correlation of equivalent depositional formations (lithostratigraphy) is useful for reconstructing palaeogeography, but is unreliable for time correlation.

Traditionally, the most important proxy for time correlation has been fossils, the idea being that each species has occupied only a single, discrete interval of time — from its origin to its extinction, or to the present if it is extant. Hence, so long as fossils of a given species are correctly identified, their host rocks can be placed within the stratigraphical range of the species concerned (whatever that may be). Exceptions occur only where fossils have been displaced, either into younger strata as a result of exhumation and redeposition of the original specimens, or through being carried downwards into older strata in, for example, drilling operations (a particular problem for microfossils). Such problems can be avoided by careful attention to the state of preservation of specimens and by ensuring that they are recovered only from within solid rock (e.g. intact cores). Correlation by means of fossils is termed biostratigraphy. Some shortcomings of the method include the restriction of organisms to certain environments, the vagaries of fossilization and changes in the geographical distributions of species through time. In the first case, appropriate fossils may simply be lacking from a given sedimentary succession and in the second and third cases, the first, or last, appearance of a species in different successions may not mark exactly the same point in time.

○   Which biological attributes would tend to make a species most useful for biostratigraphical correlation?

●   Durable skeletal components with good preservation potential and sufficient complexity to allow specific identification are desirable, together with abundance, widespread geographical distribution and free-floating (planktonic or airborne) habits. Such a combination of features would promote preservation throughout the full time range of the species in as many different sedimentary successions as possible. Moreover, rapid evolution, or at least a short duration of the species, would offer better time-resolution for correlation purposes.

Examples of groups that fulfil these requirements and are important in Cretaceous biostratigraphy are the ammonites, belemnites, planktonic foraminifers and nanoplankton (such as coccoliths) in marine sediments and spores and pollen in terrestrial sediments (Figure 1.26). In addition, certain other benthic fossil groups are also of value for regional correlation in particular sedimentary formations where other kinds of biostratigraphically useful fossils are scarce or absent. These include various bivalves (including the rudists, mentioned in Section 1.1.3, as well as another diverse group of rapidly evolving, strangely shaped epifaunal forms called inoceramids), benthic foraminifers, calcareous algae and echinoids (sea urchins). The fossil taxa are used to establish biostratigraphical zones, which are defined on the basis of the appearance, disappearance, or sometimes acme of abundance, of one or more species, for the purposes of correlation. There are usually several zones in a stage, although exactly how many depends upon the amount of resolution offered by the available fossil taxa, and this may be quite uneven.

In recent decades, stratigraphers have developed a number of other correlation methods to supplement biostratigraphy. Globally consistent changes through time in certain isotope ratios in the Cretaceous oceans have now been documented in sufficient detail from sections around the world to provide standard reference curves for correlation purposes. Foremost among these is that for the strontium isotopes $^{87}Sr/^{86}Sr$, although the curve for the variation in the ratio of the carbon isotopes, $^{13}C/^{12}C$, is also useful for certain parts of the Cretaceous, and other curves are rapidly being developed. It should be stressed that these are still relative dating methods, based on changes in the natural balance of the isotopes concerned, and should not be confused with radiometric methods of absolute dating that rely on the rates of decay of unstable (radiogenic) isotopes in rocks. Variations in the seawater ratio of strontium isotopes occur because continental rocks contain more ^{87}Sr (derived from radioactive decay of the rubidium isotope, ^{87}Rb) relative to ^{86}Sr than rubidium-poor rocks derived from the mantle. Consequently, during periods of increased weathering of continental rocks and/or decreased oceanic volcanism, the $^{87}Sr/^{86}Sr$ ratio increases, and in the converse circumstances, it decreases.

Variation in the carbon-isotope ratio is mainly driven by fractionations involved in various living processes, which are generally biased in favour of incorporating the lighter isotope of carbon, ^{12}C, at the expense of ^{13}C in the resulting organic compounds (see Box 1.1).

(a)

(b)

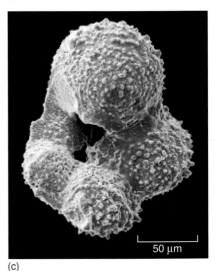

(c)

Box 1.1   Carbon isotopes and photosynthesis

Carbon consists principally of a mixture of two stable isotopes, ^{12}C and ^{13}C and, in photosynthesis, fixation of the lighter $^{12}CO_2$ is favoured over that of the heavier $^{13}CO_2$. The reasons are that $^{12}CO_2$ is taken up and diffuses into cells more rapidly and also reacts more readily in the first reaction of photosynthesis.

As a result of this isotope discrimination, organic matter produced by photosynthesis is enriched in ^{12}C and depleted in ^{13}C relative to the inorganic pool of carbon (mainly CO_2, carbonate and bicarbonate). Enrichment or depletion in ^{13}C is expressed in terms of a $\delta^{13}C$ value (δ is the small Greek letter delta).

The ratio of $^{13}C/^{12}C$ for the sample being investigated is compared with that of a carbonate standard, whose ratio is 1/87.99. The whole is multiplied by 1000 to give a $\delta^{13}C$ value in terms of parts per thousand (‰) relative to the standard:

$$\delta^{13}C \; (\text{‰}) = \left(\frac{^{13}C/^{12}C_{\text{sample}} - \, ^{13}C/^{12}C_{\text{standard}}}{^{13}C/^{12}C_{\text{standard}}} \right) \times 1000$$

○   With this formula, will the value of $\delta^{13}C$ be negative, positive or zero when: (a) the ratio $^{13}C/^{12}C$ is equal for standard and sample; (b) the ratio $^{13}C/^{12}C$ is greater in the sample; (c) the ratio $^{13}C/^{12}C$ is lower in the sample?

●   From the relative values on the top line in the brackets, for (a) the value will be zero; for (b) it will be positive; and for (c) it will be negative. So when the sample is, for example, organic matter enriched in ^{12}C and depleted in ^{13}C relative to the standard, the value of $\delta^{13}C$ will be negative.

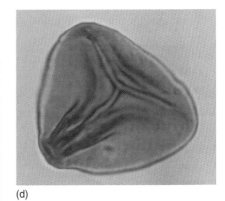

(d)

Figure 1.26   Examples of fossils employed for Cretaceous biostratigraphy: (a) ammonites *(Peter Skelton, Open University)*; (b) belemnite *(Peter Skelton, Open University, specimen courtesy of Natural History Museum)*; (c) planktonic foraminifer *(Kate Harcourt-Brown, University of Bristol)*; (d) spore *(Bob Spicer, Open University)*. Note: (c) is a scanning electron micrograph. A coccolith was shown in Figure 1.4a.

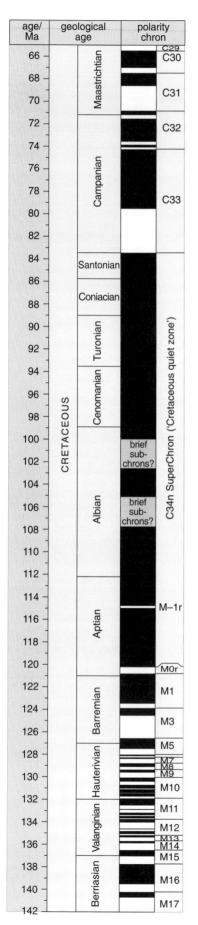

age/ Ma	geological age	polarity chron

One of the most important of these processes is photosynthesis. Thus, at times when relatively large amounts of photosynthetically derived organic material are being buried in sediments, the residual ratio of $^{13}C/^{12}C$ in seawater increases and this is reflected in the carbon in carbonate sediments produced at the time.

Both strontium and carbon ratios equilibrate globally in open seawater over geologically short time-scales (thousands of years) due to ocean circulation, so longer-term variations in the ratios can be taken as effectively global signals for correlation purposes. Besides isotopic changes, the variations in the relative abundance of certain rare elements have also proved useful at particular levels, the best known example being the famous 'iridium spike' at the Cretaceous/ Tertiary boundary (Section 1.1.6; this topic will be discussed in more detail in Part 3). Such chemical methods are grouped under the general heading of chemostratigraphy.

Another widely used approach, magnetostratigraphy, relies on the detection of remanent magnetism in suitable minerals preserved in rocks. Changes in the direction of their magnetization through time records the sequence of intervals of normal and reversed polarity of the Earth's magnetic field relative to that of today (Figure 1.27). Like the stages, these intervals, or 'chrons', have formal names or codes, which include the descriptors of 'normal' (n) or 'reverse' (r).

○ Study Figure 1.27 and identify a particular problem with this method of correlation for a large part of the Cretaceous.

● Note that for a long period, from the Aptian to the Santonian, the field remained normal with only a few minor reversals, thus providing little means for magnetostratigraphic correlation within that interval.

Although this fact rather blights the value of magnetostratigraphy for much of the Cretaceous, it does correspond closely with an exceptional episode of volcanism in the Pacific, which we will be studying in detail in Part 2, and so forms an intriguing part of the broader Cretaceous story.

A final method of correlation that deserves mention here is that of cyclostratigraphy, which relies upon the identification of regularly periodic compositional cycles within sedimentary successions. These small-scale sedimentary cycles are believed to reflect climatic oscillations which are caused by regular changes in the geometry of the Earth's rotation and of its orbit around the Sun (the Milankovich cycles of change in orbital parameters). Three kinds of periodicity interact: first, the Earth's orbit around the Sun varies between being elliptical and nearly circular; secondly, the tilt of the Earth's axis of rotation oscillates with respect to the orbital plane; and thirdly, the axis itself shows a circular rotation, like that of a spinning top. Together, these produce a complex repeating pattern of climatic fluctuation showing frequencies of tens to hundreds of thousands of years.

Figure 1.27 Magnetostratigraphic polarity zones for the Cretaceous. Normal polarity in black; reversed polarity in white. *(Ogg, 1995, with amendments from J. G. Ogg (personal communication, 2002).)*

Cyclostratigraphy offers enormous potential for high-resolution correlation in parts of the Cretaceous, because certain deposits, such as the Chalk, seem to have been highly sensitive to climatic oscillations, which are expressed as regular variations in the ratio of clay to carbonate mud (i.e. coccolith ooze). In the relatively more clay-rich Lower Chalk (now formally known as the Grey Chalk), in particular, this periodicity can sometimes be seen as a subtle banding running along quarry faces (Figure 1.28). British stratigrapher Andy Gale (1995) has shown that individual cycles can be followed for hundreds of kilometres across entire sedimentary basins, such as the Anglo–Paris Basin, at least where deposits are thick enough to contain a relatively complete record (i.e. not to have missed too many cycles). Given the characteristic frequencies of Milankovich cyclicity, this method promises correlation to within hundreds or even tens of thousands of years — a staggering resolution for rocks almost 100 Ma old.

Figure 1.28 Milankovich banding in the Grey Chalk at Arlesey Pit, Bedfordshire. *(Peter Skelton, Open University.)*

None of these methods should be thought of as, in principle, 'better' than the other methods; pragmatism has to be the watchword for correlation, because all the methods have their drawbacks as well as strengths in different contexts. Although magnetostratigraphy and strontium-isotope correlation, for example, have the advantage of being based on effectively synchronous global 'signals', not every succession will contain suitable lithologies for analysis, and besides, the technical requirements (hence costs) often make cheaper alternatives more desirable. In most instances, traditional biostratigraphical methods are still the most widely employed, because sampling fossils from measured sections is cheap and easy to do. Given the common ultimate objective of chronostratigraphical correlation, the best working approach is to use a variety of different methods as tests of each other, so as to arrive at the most robust possible conclusions.

Defining the stages

Who decides on the definition of the stages? How are the stages defined? Originally (in the 19th century), each definition was established by an authority of the day simply as the first to publish a detailed account of a discrete section within the Cretaceous System (itself already defined by Omalius d'Halloy from the sedimentary succession in the Paris Basin; see Section 1.1.1). The sections did not have to be in the Paris Basin and indeed many were not — a few even straying outside of France(!). Many of the Lower Cretaceous stages were named after areas in the south-east of France, including, for example, the Barremian from Barrême in the French Alps, and the Aptian from the Apt valley in Provence, although the Hauterivian was named from Hauterive, near Neuchâtel in Switzerland. Although definitive 'type sections' for the stages were often redesignated elsewhere (where thicker and more fossil-rich sections could be identified), the old stage names, as shown in Figure 1.25, became formally entrenched.

The main problem with the original approach of defining each stage from discrete type sections, was that of overlap. For example, the top of one stage might subsequently be found (through refined correlation) to be equivalent to the bottom of another, the original definitions of the Campanian and Maastrichtian stages being a case in point. The approach used today avoids this problem by simply defining the basal boundary of each stage, within a given boundary stratotype section. Each boundary is then automatically taken to set the upper limit to the underlying stage. Thus a seamless sequence of stages can be guaranteed. Criteria for defining the boundaries in the stratotypes themselves vary (and indeed many boundary stratotypes still await formal international approval). Most are still fossil-based, referring to the first or last appearance of a given fossil species. The base of the Aptian Stage, however, is the first to have been defined by a palaeomagnetic criterion — the base of the short-lived magnetic reversal (called M0r) that precedes the long normal interval (Figure 1.27). Note, however, that no boundary is defined according to absolute age, as discussed earlier; absolute dates for stage boundaries can only be estimated after they have been defined by the above means.

So far, so good. Each boundary stratotype fixes, by definition, the exact base of each chronostratigraphic stage. Elsewhere, however, we can only approximate to that notional time plane through correlation by means of the various methods discussed above. An important corollary of this is that the recognition of stages in geological sections (other than in stratotypes) is in a strict sense a theoretical statement. It always remains open to testing, and may occasionally be wrong (although constant improvement and integration of correlation methods should progressively minimize such errors).

For parts of the Cretaceous, there remain some correlation problems between low latitude, Tethyan marine deposits and those of higher latitudes, because of faunal differentiation between these regions. One particular area of awkwardness is the base of the Cretaceous. In the Tethyan realm, the lowest stage of the Cretaceous and uppermost stage of the Jurassic are known as the Berriasian and Tithonian, respectively. In northern high latitude ('Boreal') areas, however, a different set of biostratigraphical criteria had long been used to recognize a basal Cretaceous 'Ryazanian Stage', overlying a terminal Jurassic 'Volgian Stage'. As a step towards resolving this conflict, there is now an international consensus that the boundary stratotype should be located in the Mediterranean region (i.e. in the Tethyan realm) and that the Berriasian should thus formally constitute the basal stage of the Cretaceous. However, at the time of writing it has still not been agreed exactly how and where its base should be defined. This explains the ambiguity of the absolute dating of the start of the Cretaceous mentioned at the beginning of this Chapter. The reluctance of stratigraphers to make up their minds concerning the base of the Cretaceous, even after nearly two centuries of stratigraphical effort, may seem to some like the modern version of Medieval scholars arguing about how many angels can dance on the head of a pin. In this case, however, the delay reflects the continuing need to gather significant amounts of data from many different sites, so that ultimately the best possible boundary stratotype, offering the greatest potential for precise international correlation, can be selected for this major stratigraphical boundary.

1.3.2 Absolute time

As we noted earlier, absolute dating is distinct from relative dating. Although it is desirable to attempt to calibrate chronostratigraphical boundaries with the best available absolute age determinations, results from the two approaches should not be confused. If some part of a succession has been dated by correlation to the top of the Maastrichtian, then it should thus be described, and not as being 65 Ma old. Furthermore, something that has been radiometrically dated to 65 Ma ago should, again, be described in those terms.

Several radiometric methods are now available, all of which are based upon the decay of radiogenic 'parent' isotopes, to 'daughter' isotopes. In all cases, the first step in the procedure is to measure the amount of the daughter isotope present in a sample relative to the remaining amount of the parent isotope. From the known proportional rate of decay of the parent isotope, the time required to yield the relative amount of the daughter isotope can then be calculated. The decay rate (parent/daughter) for each radiogenic isotope is a constant, which is determined in the laboratory. Decay schemes commonly used for Cretaceous rocks include: potassium–argon (^{40}K/^{40}Ar) and a refined derivative, argon–argon (^{40}Ar/^{39}Ar); rubidium–strontium (^{87}Rb/^{87}Sr); and uranium–lead (^{238}U/^{206}Pb and ^{235}U/^{207}Pb). Fresh volcanic rock is generally the best material for dating, although isolated crystals in volcanic ash layers (bentonites), as well as certain kinds of clay minerals, are also often widely used.

Note that a calculated 'age' is always an estimate, based on assumptions concerning decay rates, initial isotopic ratios and the reliability of the retention of daughter products in samples. Therefore, wherever accuracy and precision are an issue, estimates of errors need to be quoted. Thus, for example, the absolute age for the end of the Cretaceous (as shown in Figure 1.25) should strictly be quoted as 65.0 Ma $\pm$ 0.1 Ma (at a confidence level of 95%). This means that 95% of estimates are expected to fall between 65.1 Ma and 64.9 Ma. It also means that any sequence of events that occurred within a time-frame of 100 000 years around that critical time cannot be reliably ordered by radiometric dating methods alone. How problems of time-resolution such as these are dealt with will be explored in Part 3 of this book.

Finally, there remains the question of calibration — linking absolute dates to stage (or other) boundaries. Obviously, it would be most desirable to obtain absolute age determinations on the boundaries themselves in the boundary stratotypes. However, this is often not possible simply because the minerals suitable for radiometric dating are not available at the boundary. In such cases, ages either have to be interpolated from determinations above and below in the succession (involving assumptions about intervening relative rates of sedimentation), or to be imported from dateable horizons that have been correlated with the boundary in question. Either way, there is scope for further error. The absolute time-calibration of the chronostratigraphical time-scale is thus effectively determined between a succession of tie-points scattered across the world, rather like a washing line pegged at several points. Consequently, parts of it are better secured than are others. Indeed, in some cases, the subdivision of chronostratigraphical intervals through refined correlation methods can, in principle, offer a finer degree of temporal resolution than absolute dating methods can. The cyclostratigraphic correlation of parts of the Chalk based on the Milankovich cyclicity is a case in point.

1.4   Summary

- The contrasts between the Cretaceous world and that of today provide instructive insights on changes in the workings of the Earth system through time. Features of the Cretaceous world that are of particular interest include:

   (i) attainment of considerably higher sea-levels than we see today;

   (ii) development of widespread climatic warmth, extending to high latitudes;

   (iii) repeated growth and demise of vast, shallow carbonate platforms in low latitudes;

   (iv) episodic development of widespread marine anoxia;

   (v) extensive oceanic volcanic activity, especially in the Pacific;

   (vi) occurrence of a catastrophic event that marked the end of the period.

- At the start of the Mesozoic, the continents were assembled into a single supercontinent, Pangaea. By the beginning of the Cretaceous, this had split into two supercontinents, Laurasia in the north and Gondwana in the south, separated by an equatorial seaway consisting of the narrow Central Atlantic in the west and the wedge-shaped Tethys Ocean in the east. During the Cretaceous, the Central Atlantic was joined by ocean basins to the north and south, to create the narrow, north–south oriented Atlantic Ocean. Meanwhile, the Indian Ocean opened to the south. The growth of these new oceans was accompanied by the progressive closure of the Tethys Ocean, as oceanic crust was consumed along its northern margin. This led to continental collisions in the Cainozoic that gave rise to mountain chains running from the Pyrenees, via the Alps to the Himalayas and beyond.

- The global sea-level rose (with fluctuations) through the Mesozoic, to reach a maximum in the Late Cretaceous. This resulted in part from the growth of new mid-ocean ridge systems, but also from increased rates of production of oceanic crust in the Cretaceous, both of which gave rise to large areas of young, hot, hence relatively uplifted, ocean floor, which displaced water from the ocean basins. After further fluctuations, a distinct trend of decline set in from the Oligocene.

- Following a Late Carboniferous to Early Permian ice age, the global climate warmed, likewise reaching a maximum in the Late Cretaceous, when the polar ice appears to have been entirely absent at sea-level. Once again, a decline set in from the Oligocene, culminating in the Quaternary ice age.

- Throughout the Mesozoic and Cainozoic, life showed a trend of diversification from the survivors of the devastating mass extinction that had marked the close of the Permian. However, this pattern was punctuated by several further mass extinctions, including some within the Cretaceous, but most notably by that at the close of the period (the K/T boundary extinction). Cretaceous life was thus comprised of a mixture of groups that are still important today, e.g. angiosperms and insects on land, and ray-finned fish together with numerous invertebrate groups in the sea, and extinct groups, e.g. the dinosaurs on land and the ammonites, belemnites and rudist bivalves in the sea. The Cretaceous also saw the first major flourishing of the calcareous plankton; the sinking of their skeletons to the ocean floor added a new, significant element to the global carbon cycle.

- The succession of rocks that comprise the Cretaceous System is subdivided into stages, corresponding to successive ages of relative time. The names of the Cretaceous stages are derived from the areas (mostly in France) where they were originally described, although their bases are now (or will be) defined at designated boundary stratotype sections, each of which also serves to define the top of the previous stage. The stages are recognized elsewhere by means of chronostratigraphical correlation, i.e. shared possession of recognized proxies for time-equivalence. Correspondence of similar successions of depositional units (lithostratigraphy) is considered unreliable for this purpose, because of the possibility of diachronism, although it is useful for palaeogeographical reconstruction. The preferred methods employ the succession of fossil taxa (biostratigraphy), changes in isotope ratios or abundance of distinctive rare elements (chemostratigraphy), the record of repeated reversals in polarity of the Earth's magnetic field (magnetostratigraphy) and that of orbitally forced climatic oscillations (cyclostratigraphy).

- Absolute age is estimated from the ratios of daughter to parent isotopes produced in rock samples by the decay of various radiogenic isotopes. It is distinct from relative age and the two should not be confused. The relative time-scale can be calibrated to, though never defined by, absolute time estimates.

1.5   References

ALVAREZ, L., ALVAREZ, W., ASARO, F. AND MICHEL, H. V. (1980) 'Extraterrestrial cause for the Cretaceous–Tertiary extinction', *Science*, **208**, 1095–1108.

GALE, A. S. (1995) 'Cyclostratigraphy and correlation of the Cenomanian Stage in Western Europe', pp. 177–197 in HOUSE, M. R. AND GALE, A. S. (eds), *Orbital Forcing Timescales and Cyclostratigraphy, Geological Society, London, Special Publications*, **85**.

GRADSTEIN, F. M. AND OGG, J. G. (1996) 'A Phanerozoic timescale', *Episodes*, **19**, 1–2.

OMALIUS D'HALLOY, J. J. (1822) 'Observations sur un essai de carte géologique des Pays Bas, de la France et de quelques contrées voisines', *Ann. Mines*, **7**, p.373.

SEPKOSKI, J. J. JR. (1984) 'A kinetic model of Phanerozoic taxonomic diversity. III. Post-Paleozoic families and mass extinctions', *Paleobiology*, **10**, 246–267.

TIŠLJAR, J., VLAHOVIĆ, I., VELIĆ, I., MATIČEC, D. AND ROBSON, J. (1998) 'Carbonate facies evolution from the Late Albian to Middle Cenomanian in southern Istria (Croatia): influence of synsedimentary tectonics and extensive organic carbonate production', *Facies*, **38**, 137–152.

TWITCHETT, R. J., LOOY, C. V., MORANTE, R., VISSCHER, H. AND WIGNALL, P. B. (2001) 'Rapid and synchronous collapse of marine and terrestrial ecosystems during the end-Permian biotic crisis', *Geology*, **29**, 351–354.

WINTERER, E. L. AND SAGER, W. W. (1995) 'Synthesis of drilling results from the Mid-Pacific Mountains: regional context and implications', pp. 497–535 in WINTERER, E. L., SAGER, W. W., FIRTH, J. V. AND SINTON, J. M. (eds), *Proceedings of the Ocean Drilling Program, Scientific Results*, **143**, Texas A & M University in cooperation with National Science Foundation and Joint Oceanographic Institutions, Inc. USA.

2   The mobile palaeogeographical framework

Peter W. Skelton

The Cretaceous Period saw a change from the general arrangement of continents and oceans of the Mesozoic to that of our own era (Section 1.2). Contrast the palaeogeographical maps for just before the beginning, and the end of the period, shown in Figure 2.1a and 2.1b respectively.

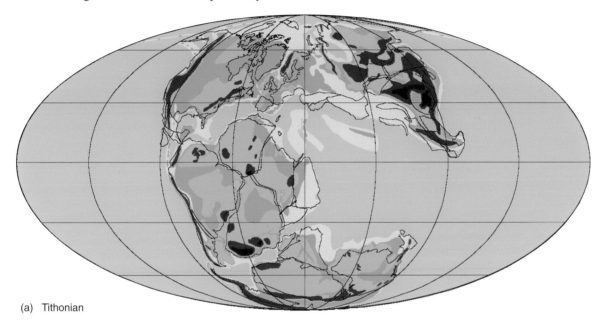

(a) Tithonian

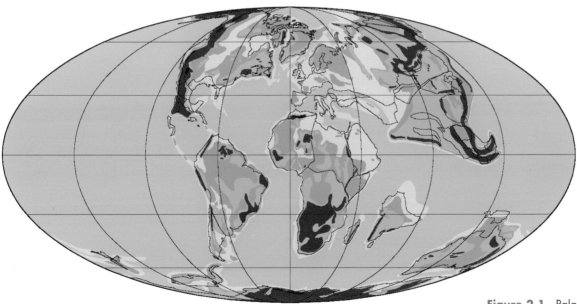

mountains (>2000 m)

mountains (1000–2000 m)

uplands (200–1000 m)

lowlands (0–200 m)

shelf (–200–0 m)

deep ocean (<–200 m)

(b) Maastrichtian

Figure 2.1 Palaeogeographical maps for: (a) the end of the Jurassic, and (b) the end of the Cretaceous Period. *(University of Chicago 'Paleogeographic Atlas Project'.)*

At the dawn of the Cretaceous, the supercontinents Laurasia (in the north) and Gondwana (in the south) were still widely separated in the east by the Tethys Ocean (Figure 1.21b). In the west, the Tethys Ocean was a constricted sluice, in what is now the Mediterranean region, leading to the almost enclosed Central Atlantic (Figure 1.21b). By the end of the period, the southern and northern parts of the Atlantic had joined the Central Atlantic to create a continuous north–south oceanic split between the Americas and the Old World continents (Figures 1.21d, 2.1b). Meanwhile, the anticlockwise rotation and northward drift of the African/Arabian continent had all but closed the great 'jaws' of the Tethys Ocean, with India advancing northwards as the Indian Ocean opened behind it. Thus, the continuous equatorial seaway between the northern and southern continents, which was a characteristic motif of the Cretaceous world, gradually yielded to the expansion of new oceans in higher latitudes. The palaeogeography at the close of the period now began to resemble that of today.

In this Chapter, we chart the plate tectonic events that brought about these changes. However, before launching into details, we should stop to consider how this history might be relevant to our investigation of the Cretaceous Earth system.

2.1 The role of plate tectonics in the Earth system

Recall the major issues encountered in the 'reconnaissance tours' of Chapter 1, Sections 1.1.1–1.1.6 and consider, in principle at least, how plate tectonics might have impinged on them.

2.1.1 Mountains

One of many revelations of the theory of plate tectonics has been the explanation of how mountain chains form within and around the continents (Figure 2.2).

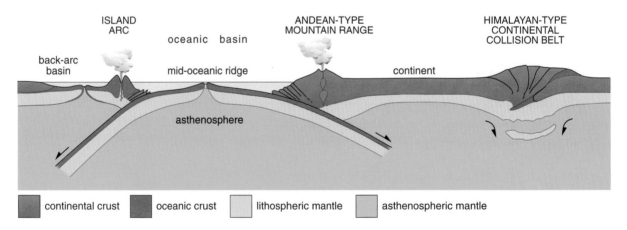

Figure 2.2 Plate tectonics and mountain formation.

○ Study Figure 2.2 and summarize the processes involved in the different modes of mountain formation that are shown there.

● On the left of the Figure, subduction of the oceanic lithosphere is associated with island arc formation where the subducting slab sinks below the opposing oceanic lithosphere, while in the centre, Andean-type mountains are formed where the subducting slab plunges directly beneath a continental margin. In both cases, crustal thickening is due to a combination of compressive stacking of ocean floor rocks in an accretionary prism and the rise of buoyant magmas derived from partial melting above the descending slabs. On the right of the Figure, earlier subduction has led to the closure of a former ocean, such that opposing continental margins have collided. The resulting compression has again caused crustal thickening, and hence uplift, with extensive stacking of thrust sheets at higher (brittle) levels in the crust.

In the case of collision, even greater uplift over a large area may ensue with isostatic rebound following the breaking off of a dense mantle lithospheric root, as shown in Figure 2.2. Such a process is now thought to have triggered the latest phase of widespread uplift of the Tibetan Plateau.

So much for the link between plate tectonics and mountain formation, but what about the connection between mountains and climate? The most obvious way in which mountains can influence regional climatic patterns is by affecting atmospheric circulation. In this respect, their most extreme effect is for the mountains to act as a high altitude heat source during the summer months (through solar heating), with air rising above them initiating a low-pressure cell. The opposite happens during the winter, giving rise to seasonally reversing winds, or 'monsoons'. If moisture is drawn into the summer low-pressure system, for example from a nearby ocean, heavy precipitation can ensue as the rising air expands and cools (Figure 2.3). Usually such effects are asymmetrical, with the rainfall largely confined to one side of the mountain watershed, while the other side remains dry in a 'rain shadow'. A well-known example of both effects is provided today by the Himalayas, with monsoonal rains to the south of the range affecting northern India, while to the north a vast rain shadow gives rise to the semi-arid steppe and desert of Tibet.

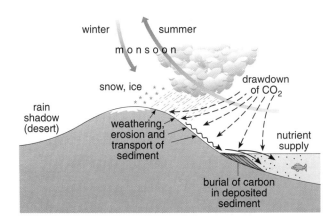

Figure 2.3 The links between mountains and climate.

To these meteorological effects of mountains must be added the consequences for the Earth's albedo, or reflectivity, i.e. the extent to which its surface directly reflects solar radiation back into space. Both the snow cover on the mountains and the generally pale scrubland and desert in the rain shadow behind them, will have a relatively high albedo, losing much solar energy to space.

Finally, the erosion of mountains can further influence climate over much longer time-scales by drawing down atmospheric CO_2 (an important greenhouse gas) in two ways. First, the weathering of silicate rocks consumes CO_2 according to the general equation: $CaSiO_3 + 2CO_2 + 3H_2O \rightarrow Ca^{2+} + 2HCO_3^- + H_4SiO_4$. Secondly, the resulting regional sedimentation may bury large amounts of organic material, itself derived from atmospheric CO_2 by photosynthesis. Hence, mountain building is certainly one aspect of the Earth system that we will need to keep track of while reviewing Cretaceous plate tectonics.

2.1.2 Ocean circulation

Continental break-up and the associated eustatic rise in sea-level (Section 1.2) would also have influenced continental climates by facilitating the transport of humidity to the continental interiors. Moreover, the changing distribution of continents and ocean basins would have affected oceanic circulation, a major driver of the Earth's climate system. Today's oceans are stirred to the deepest levels by the 'global thermohaline conveyor' (Figure 2.4). The growth of sea-ice in high latitudes leaves behind cold, more than normally salty water, which is relatively dense, and so sinks and flows away to lower levels in the ocean. The cold, dense water that forms in the present North Atlantic Ocean flows southwards along the ocean floor, eventually spreading into the Indian and Pacific Oceans. At shallower levels, there is a compensatory return flow of water that has been warmed at low latitudes. This transports heat and moisture to high latitudes, thus fuelling the precipitation of the snow that helps to maintain the polar glacial regime.

Figure 2.4 The 'global thermohaline conveyor'. Variants of this may be seen elsewhere, as not all ocean currents are as well charted as they are for the Atlantic Ocean.

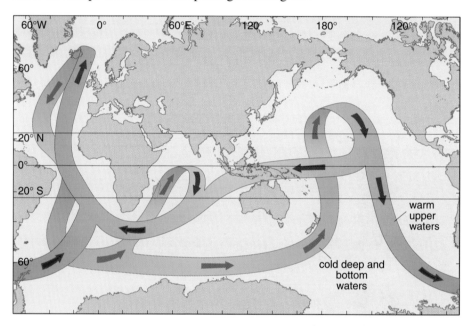

Besides the differing arrangement of ocean basins in the Cretaceous, the absence of ice at sea-level (Section 1.1.2), as well as relatively high eustatic sea-levels (Section 1.1.1) for much of the period, together imply that ocean circulation must have differed in many ways from that of today. This has implications not only for the Cretaceous climate system, but also for the mixing, and hence oxygenation, of deeper waters in the seas and oceans of the period.

○ Which of the tour stops in Chapter 1 is especially pertinent to this point?

● Tour stop 4 (Section 1.1.4), where you saw that widespread anoxia occasionally afflicted the marine realm of the Cretaceous.

2.1.3 Magmatism

Another major player in the system that needs to be considered is magmatism and its surface manifestation (volcanism). Figure 2.2 illustrates the links between magmatism and plate margins, both constructive (along mid-oceanic ridges within the ocean basins) and destructive (along the volcanic chains flanking subduction zones).

○ To which physical aspect of the oceans are the spreading rates and the total extent of mid-oceanic ridges especially relevant?

● Global sea-level (as explained in Section 1.2). Increases in spreading rate and total length of mid-oceanic ridge systems augment their total volume, thus reducing the capacity of the ocean basins, which leads to sea-level rise. Decreases in spreading rate and loss of ridge length have the opposite effect (Figure 1.22b).

In addition, mid-oceanic ridges are sites of extensive hydrothermal convection through the ocean crust, which behaves like a gigantic ion-exchange column (Figure 2.5). One of the most important processes involved in hydrothermal alteration of the ocean crust is the sequestration of magnesium from the circulated seawater and its replacement by calcium. The higher the spreading rate, the greater the rate of such exchange which, in turn, has consequences for ocean chemistry. Sub-aerial volcanism, meanwhile, impacts upon atmospheric composition, furnishing both cloud-inducing aerosols and greenhouse gases such as CO_2 (a theme that will be discussed in detail in Chapter 7).

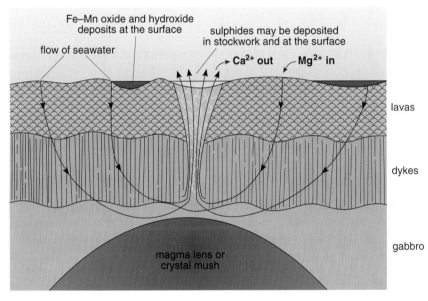

Figure 2.5 Diagrammatic cross-section to show hydrothermal circulation at the centre of a mid-oceanic ridge system and its main effects.

However, especially for the Cretaceous, we will need to consider not only the implications of these 'normal' aspects of plate tectonics for the Earth system (Figure 2.6, left), but also the exceptional effects of 'superplume' activity (Figure 2.6, right). Tour stop 5 (Section 1.1.5) illustrated the widespread growth of volcanic plateaux and seamounts that accompanied such superplume activity in the Pacific Ocean.

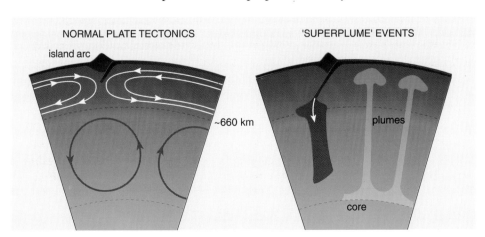

Figure 2.6 Idealized cross-sections to show contrasting modes of mantle convection. 'Normal' plate tectonics (left) is associated with upper mantle convection that is separated from activity in the lower mantle. During superplume events (right), sinking of accumulated subducted material, which is relatively cold and dense, into the lower mantle is matched by the rise of multiple plumes of hot material from the core–mantle boundary.

2.2 Major palaeogeographical changes of the Cretaceous

Palaeogeographical maps are but theories, which need to be repeatedly tested for their ability to explain observations other than those upon which they are based. Early in the evolution of the ideas that led to plate tectonic theory — or 'continental drift' as it was then known — the distribution of fossil fauna and flora, together with glacial and other climate-sensitive deposits, were used as clues for indicating past continental movements. Today, the primary sources of data for reconstructing former continental positions are often geophysical.

○ What are the two main kinds of geophysical data used for plotting Cretaceous reconstructions?

● On the continents themselves, palaeomagnetism, recorded in rocks that contain magnetic, iron-rich minerals, can provide information on palaeolatitude (as was noted with reference to Alaska in Tour stop 2, Section 1.1.2). On the ocean floor, the magnetic stripes that record the successive change in the Earth's magnetic polarity (Figure 1.27), as the ocean crust is continuously generated along constructive margins, can be back-tracked (rather like playing a film backwards) to infer the former size of ocean basins.

The latter technique is obviously limited to the ocean crust of appropriate age that remains, i.e. has not disappeared down subduction zones. Fortunately, there is still a substantial amount of Cretaceous ocean crust (as you saw in Section 1.1.5), in contrast to that of Jurassic age, which has already been largely subducted. Consequently, reconstructions of Cretaceous (and later) palaeocontinental positions are more soundly based than are those for earlier periods. One way to test them is to check for their consistency with the distribution of land-dwelling organisms through the period. Figure 2.7 shows, in schematic form, the distribution of selected Cretaceous dinosaur groups on the main continental blocks from a review by Milner *et al.* (2000).

○ Compare the dinosaur distribution data given in Figure 2.7 with the palaeogeographical maps for the ends of the Jurassic and Cretaceous shown in Figure 2.1. How consistent are the dinosaur data with what the maps show concerning: (a) the opening of the Atlantic Ocean; (b) the opening of the Indian Ocean; and (c) the maintenance of a Tethyan–Central Atlantic seaway separating the continents of Gondwana and Laurasia?

● The dinosaur data agree well with (a) and (b). The connections between South America, Africa and Madagascar/India at the end of the Jurassic (Figure 2.1a) are confirmed by the spread of abelisaurs and titanosaurs across them (Figure 2.7a). Thereafter, no dinosaur links are shown between these three continental blocks (Figure 2.7b,c). There is less agreement with (c). Towards the end of the Early Cretaceous and in the Late Cretaceous, abelisaurs and titanosaurs appear to have spread from Africa to Europe, while spinosaurs migrated the other way (Figure 2.7b). Moreover, an exchange of titanosaurs and hadrosaurs is shown between North and South America in the Late Cretaceous (Figure 2.7c).

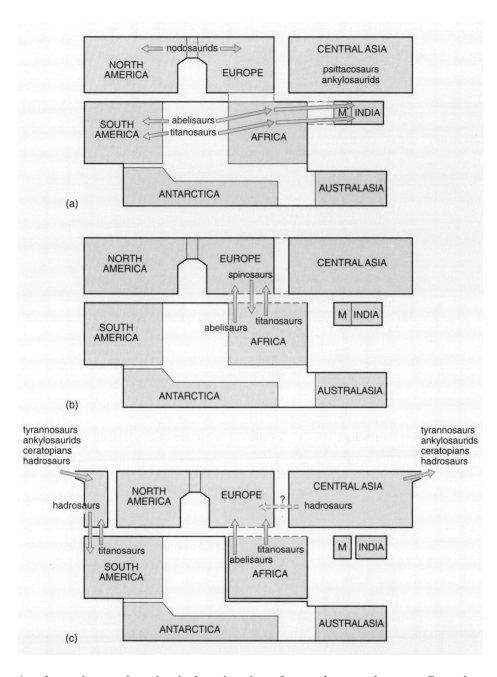

Figure 2.7 Schematic illustration of the distributions of selected dinosaur groups on the main continental blocks: (a) in the Late Jurassic to Early Cretaceous; (b) towards the end of the Early Cretaceous; and (c) in the latter part of the Late Cretaceous. *(Milner et al., 2000.)* Arrows represent extensions of geographical ranges implied by fossil evidence.

Another point worth noting is the migration of several groups between Central Asia and western North America in the Late Cretaceous. This was a consequence of the opening of the North Atlantic Ocean, which caused the two continents to approach one another on the other (Pacific Ocean) side of the globe where they became connected via a narrow corridor ('the Bering corridor').

The faunal links between the northern and southern continents noted in the answer to the above question suggest that, if not land bridges, at least island stepping-stones sometimes existed between them. Evidence that parts of the central Tethyan carbonate platforms occasionally emerged comes from the discovery of rare dinosaur remains on them (Figure 2.8), associated with plant fragments.

Figure 2.8 Vertebra of a dinosaur in a shallow underwater exposure of Lower Cretaceous (Hauterivian) limestone on the coast of Istria (Croatia). *(Peter Skelton, Open University.)*

Earlier plate tectonic models tended to treat the continents as coherent masses that kept their shape and simply drifted around like cork mats floating on water. However, more recent map making has attempted to come to terms with the complexities of plate interactions. At the outset of continental separation (Figure 2.9a), the crust becomes stretched and thinned to form extended passive margins consisting of more or less tilted fault blocks. Sedimentary building over these blocks then leads to oceanward expansion of the continental margins. The failed third arms of rift systems provide a variation on this theme. Continental rifting above a rising mantle plume tends to start with a triple junction of rifts. Subsequently, one of these fails and further rifting, leading to continental separation, is concentrated along the other two. Nevertheless, the wedge-shaped area of extension along the failed arm will have caused the new passive margin on that side to open slightly, relative to its initial shape. Such extensional distortions of continental margins need to be reversed when reconstructing pre-rift continental shapes.

Where, at a later stage of ocean development, a subduction zone has become established alongside a continent, thickened portions of crust that cannot be consumed become welded onto the continental margin. Such accreted fragments are termed terranes, and may consist of microcontinental fragments, volcanic arcs or oceanic seamounts and plateaux. Figure 2.9b shows the complex collage of accreted terranes that make up south-east Asia today. Each terrane has its own distinctive, internally consistent stratigraphy and they are separated from each other by highly deformed suture zones that include ophiolites (remnants of ocean crust) and volcanic arc suites, indicating the former presence of oceanic areas between them. Reviewing their complex geology, Metcalfe (1988) concluded that they were all fragments that had rifted away from Gondwanaland at various times in the Late Palaeozoic and then collided with each other (Figure 2.9c), before finally docking as an ensemble onto North China in the Late Triassic. However, that was not the end of the story, as south-west Borneo appears to have rifted away from Indo-China (with which it shares pre-Cretaceous faunal similarities) in the Late Cretaceous, migrating south-eastward to its present position. The subsequent (Tertiary) collision of India with Asia, moreover, caused some south-eastward extrusion and clockwise rotation of this south-east Asian collage.

Figure 2.9 Some tectonic causes of change in continental shape: (a) simplified cross-section of the Western Approaches continental margin on the northern side of the Bay of Biscay, based on seismic reflection profiles, showing the effects of rift extension *(Zeigler, 1990)*; (b) collage of accreted terranes that make up south-east Asia today (black line with arrowheads: subduction-related thrusting); and (c) their reconstructed positions in the Triassic prior to accretion *(Metcalfe, 1988)*. (d) Strike–slip movement along a curved fault, leading either to extension and subsidence ('transtension') or to compression and uplift ('transpression'), in each case shown in plan view above, and vertical section below.

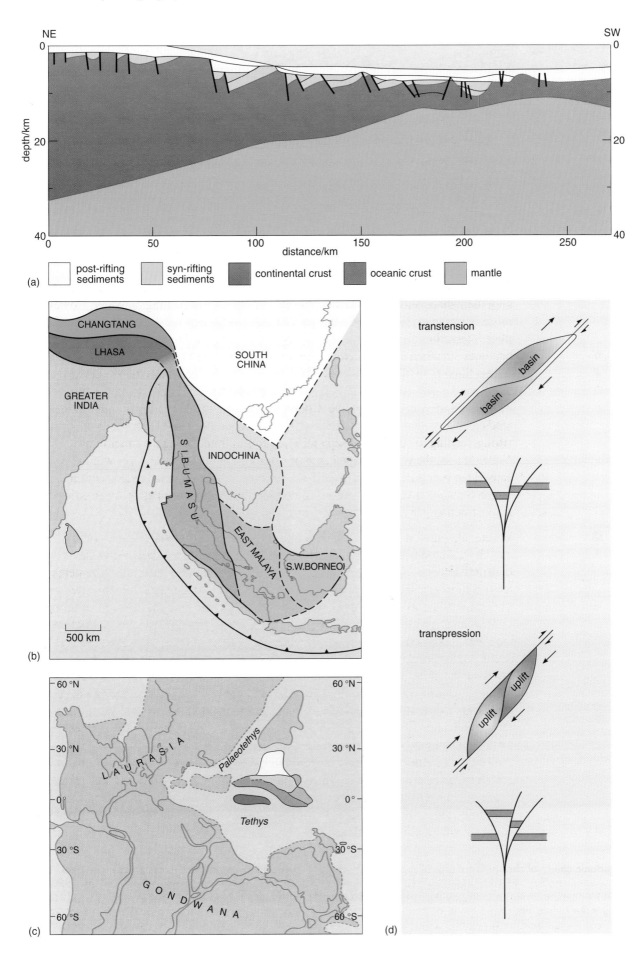

(a)

post-rifting sediments   syn-rifting sediments   continental crust   oceanic crust   mantle

(b)

(c)

(d)

A rather better analogy for such behaviour than the cork mats mentioned earlier would seem to be scum or froth floating on water, variously separating and coalescing. Further complications arise wherever there is a component of strike–slip movement between plates, either along transform faults or where subduction is oblique, as was often the case along the Tethyan margins. Besides the relative horizontal transport of terranes (which can be considerable), there may also be regional subsidence and/or uplift associated with the movement (Figure 2.9d).

So, although the 'big picture' of continental movements may seem relatively clear, the problems arise in the details, as far as important factors for the Earth system, such as mountain formation, and oceanic connections and currents, are concerned.

2.2.1 Growth of the Atlantic Ocean

At a given site along a passive margin, the final rupture of the continental crust — heralding the onset of ocean floor formation — can be dated from the earliest evidence of exposure on the sea-floor of mantle rocks. On this basis, cores drilled from off the north-western Iberian coast by the Ocean Drilling Programme (ODP) suggest that opening of the North Atlantic Ocean between that margin and the Grand Banks of Newfoundland began in the Early Cretaceous (Figure 2.10). It then 'unzipped' northwards along successive dog-legs, continuing into the Cainozoic. The separation of Greenland from Scandinavia, for example, took place in the late Palaeocene, and even then a plume-related ridge of thickened ocean crust between Greenland, Iceland and Scotland prevented the formation of a deep-water passage to the Arctic Ocean thereafter. However, there were some false starts along the way, such as a failed episode of rifting in the Labrador Sea through the Late Cretaceous.

Figure 2.10 Evidence for the Early Cretaceous start of North Atlantic Ocean opening, between Iberia and Newfoundland. Photographs show cores recovered from beneath the ocean floor by the Ocean Drilling Program (ODP), about 250 km west of Oporto, in water depths of about 5 km (scale in cm). (a) Breccia consisting of clasts of ultramafic rocks (mainly amphibolites) encased in chalk of Aptian age demonstrates that faulting had exposed lower crustal rocks on the sea-floor during Early Cretaceous rifting that heralded sea-floor spreading. (b) Serpentinite (hydrated mantle peridotite (labelled S)) and Aptian marls (arrowed) containing serpentinite pebbles show that mantle rocks were exposed at the sea-floor. The mantle was probably exposed during the final stage of rifting that heralded sea-floor spreading. (Chris Wilson, Open University.)

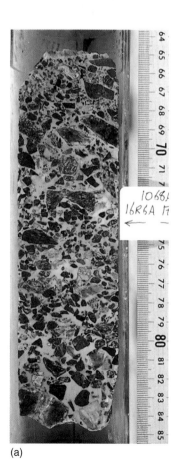

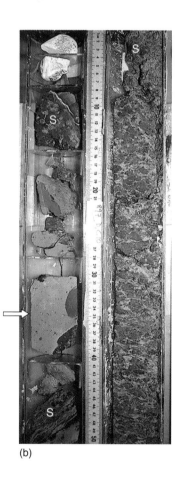

(a)   (b)

Unlike the North Atlantic, the Early Cretaceous South Atlantic was separated from the pre-existing Central Atlantic basin. Older reconstructions involved a northward 'unzipping' of a wedge-shaped ocean, like the North Atlantic. However, a recent synthesis by Hay *et al.* (1999), in which particular attention was paid to the kinds of deformations of continental shape discussed above, shows an initially parallel-sided opening, similar, although on a larger scale, to the Red Sea today. According to their study, the southern part of South America was originally wrapped around the Cape region of southern Africa, but then opened out from it due to failed triple junction rifts along its eastern margin, as Africa rifted away. It was this opening-out that gave the South Atlantic its wedge-shaped configuration. The connection of the South Atlantic with the Central Atlantic was at first blocked by the side-by-side juxtaposition of north-eastern Brazil and the southern margin of West Africa, which at that time were merely sliding past each other by dextral transform motion.

Later in the Early Cretaceous, there may have been intermittent shallow marine connections across north-eastern Brazil, via localized sedimentary basins, which contain a similar fish fauna (Figure 2.11). Evidence for marine incursion can be seen in the Araripe Basin, where lake deposits and evaporites are followed by marine deposits of the Santana Formation, of probably Albian age (Figure 2.11b). An open marine connection, involving the spread of ammonite species between north and south, eventually ensued in the latest Albian, as the Brazilian and West African margins finally separated from one another. However, up to that point, the South Atlantic had remained a relatively restricted ocean basin.

Figure 2.11 Initial shallow marine connection between the South and Central Atlantic Ocean across north-east Brazil: (a) map of connecting basins; (b) diagrammatic cross-section showing sedimentary fill of the Araripe Basin. *(Martill, 1993.)*

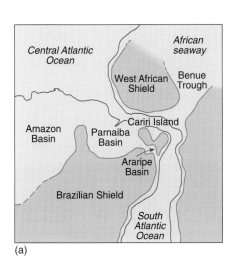

(a)

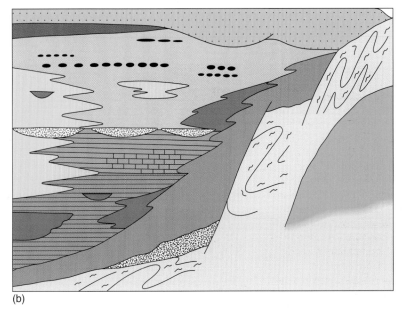

(b)

KEY for (b)

- Coarse-grained sands and grits of the Exu Formation
- Laminated silty mudstones of the Simoes Member
- Santana Formation; mainly mudstones with thin limestones and bands of concretions with fish
- Fine-grained sands, silts and silty mudstones attributable to the Batateiras Formation and tongues within the Crato, Ipubi and Santana Formations
- Evaporites of the Ipubi Formation
- Organic-rich mudstones and laminated carbonates of the Crato Formation
- Red and green mudstones of the Missao Velha Formation
- Medium- to coarse-grained sandstone bodies of fluvial units within the Missao Velha Formation
- Conglomerates, grits and sandstones of the Cariri Formation
- Igneous basement in places, may be overlain unconformably by the Exu Formation, but there may be some intrusive contacts
- Metamorphic basement; usually high grade Proterozoic gneisses: some pelitic rocks in south of basin

Understanding how the Central Atlantic and the Pacific were related during the Cretaceous is clouded by one of the more vexed issues of palaeogeographical reconstruction — the tectonic history of the Caribbean region. Today, the Caribbean Sea is floored by anomalously shallow, hence thickened, oceanic crust constituting a discrete tectonic plate (Figure 2.12). To the west, the Cocos Plate is being subducted alongside the Pacific coast of Central America, while to the east, the western Atlantic Ocean floor is being subducted on the eastern side of the Lesser Antilles. The latter process is responsible for the current island arc volcanism in places such as Montserrat. To the north, transform faults cut across the Greater Antilles chain (Cuba to Puerto Rico), separating the Caribbean Plate from the North American Plate, and to the south, complex zones of deformation likewise separate it from the South American Plate. You also need to be aware that along the Greater Antilles are igneous and volcaniclastic (volcanic-derived sedimentary) rocks of Cretaceous age. These indicate the presence of at least one volcanic arc at the time, albeit strongly dissected and deformed by subsequent tectonic activity.

Figure 2.12   The main geographical and geological features of the Caribbean region today. Black lines indicate subduction-related thrusting (with arrowheads) and strike–slip faulting. *(Draper et al., 1994.)*

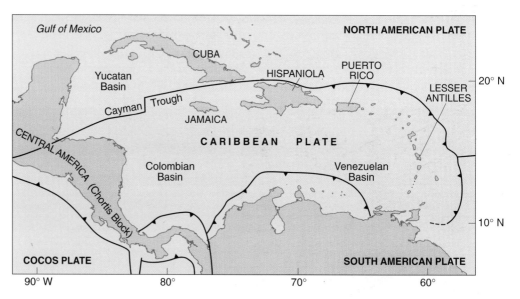

Most authorities now agree that the Caribbean Plate probably originated in the Pacific Ocean in Cretaceous times and moved eastwards like a great tongue between the Americas. There is debate, however, concerning exactly how this came about, and two of several competing models are illustrated here (Figures 2.13, 2.14). In the first model (Pindell, 1994; Figure 2.13), the Greater Antillean volcanic arc was related to south-westward-directed subduction of ocean floor (Figure 2.13a). As the American plates on either side moved towards the Pacific Ocean, the segment of ocean floor that was to become the Caribbean Plate slid in between them, with the Greater Antillean volcanic arc forming along its leading edge. This arc is known to have become extinct in the Late Cretaceous (Campanian; Figure 2.13b), during which time a new subduction zone and associated volcanic arc formed in the West. Thereafter, the Caribbean Plate continued to push eastwards between the Americas, with extensive transform deformation along its northern and southern margins (Figure 2.13c). The new western arc eventually formed the Central American isthmus.

Another model (Kerr *et al.*, 1999), shows many similarities, but differs most importantly in assuming north-eastward-directed subduction on the *Pacific* side of the Greater Antillean volcanic arc (Figure 2.14a–d), in other words, with the

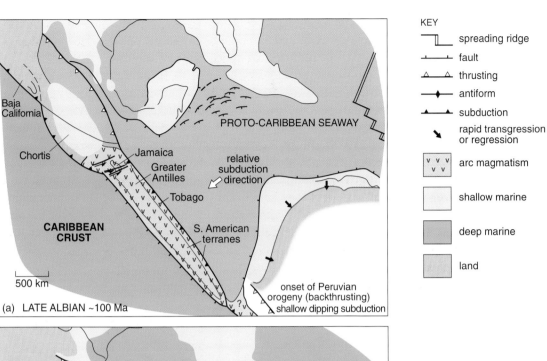

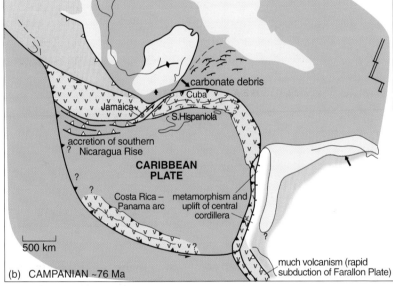

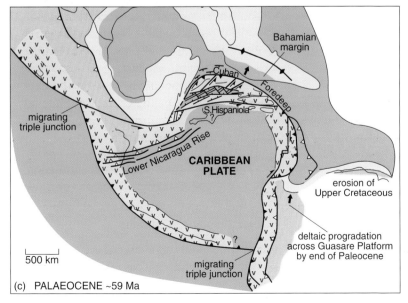

KEY

⌐‾⌐	spreading ridge
⊥‒⊥	fault
△‒△	thrusting
◆‒◆	antiform
▲‒▲	subduction
➤	rapid transgression or regression
v v v v v	arc magmatism
	shallow marine
	deep marine
	land

Figure 2.13 Pindell's (1994) model of Caribbean tectonic evolution during: (a) the Albian; (b) the Campanian; and (c) the Palaeocene. An explanation of the evolution is given in the text.

opposite polarity to that postulated by Pindell. These authors cite geochemical evidence for back-arc basalts situated on the north-eastern side of the arc, which are flanked in turn by the remnants of a second, short-lived arc system where older, proto-Caribbean ocean floor was consumed. This entire complex then became compressed, with much stacking of thrust slices, against the unyielding Bahamian promontory of North America, as the latter moved westwards. Kerr *et al.* also documented an important mid–Late Cretaceous episode of plume-related magmatism in the Caribbean Plate, which accounts for its anomalous thickness. Perhaps it was the latter that eventually blocked the Greater Antillean subduction, in Campanian times, and instigated that to the west (in Central America). The detailed geological evidence for the differing interpretations need not concern us here, but the outline given above illustrates the difficulties of determining even such major features as the polarity of subduction in tectonically mobile regions such as this.

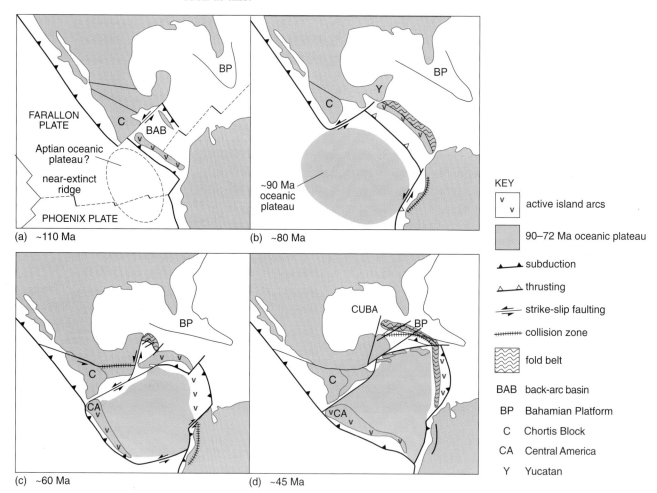

Figure 2.14 An alternative model of Caribbean tectonic history by Kerr *et al.* (1999): (a) *c.* 110 Ma; (b) *c.* 80 Ma; (c) *c.* 60 Ma; and (d) *c.* 45 Ma.

With regard to oceanic circulation, the insertion of the anomalously thickened Caribbean Plate between the Americas, not to mention the associated volcanic arcs, must have limited the extent of deep water flow between the Atlantic and the Pacific for much of the Cretaceous. However, similarities in the shallow marine fauna, such as rudists, on either side confirm the maintenance of shallow connections between the two oceans.

2.2.2 Development of the Southern Ocean system

According to the reconstruction of Hay *et al.* (1999), South America and Antarctica remained in contact throughout the Cretaceous, with the Antarctic Peninsula connected to the southern Andes. Likewise, Australia remained attached to Antarctica. The connection between these three continents, maintained well into the Tertiary, explains the present-day residual distribution of marsupial mammals in South America and Australia. Hence, the main expression of the break-up of Gondwana in the Early Cretaceous was the 'unwrapping' and retreat of the other conjoined southern continents from around the southern margins of Africa, creating a peri-African seaway extending from the Atlantic Ocean to the incipient Indian Ocean. Moreover, the Early Cretaceous drift of India (together with Madagascar) from Antarctica extended the Indian Ocean eastwards, although in this case complications were caused by the growth of a plume-related cluster of oceanic promontories, consisting of the Kerguelen Plateau, Broken Ridge and Ninety-east Ridge (Figure 2.15). This complex was emergent at times, and may have provided stepping-stones, if not an occasional land bridge, between India/Madagascar and Antarctica into the Late Cretaceous, before eventually subsiding in 'Atlantis'-style to considerable depths as the oceanic crust cooled. For example, drilling at the ODP site 750, on the margin of the Kerguelen Plateau — now at a depth of 2030 m — found Albian non-marine sediments, including coals, unconformably overlain by Upper Turonian marine chalk and marl. Moreover, some kind of land connection is implied by the presence in India of some Upper Cretaceous dinosaur genera that are shared with South America, and which presumably spread to that area via Antarctica.

80 Ma

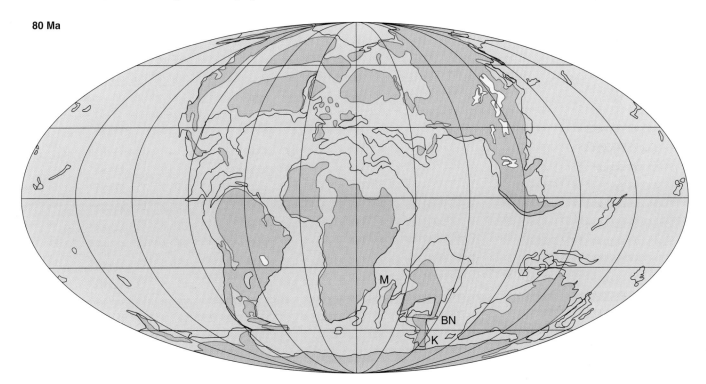

Figure 2.15 World palaeogeography reconstructed for 80 Ma ago, by Hay *et al.* (1999). Land is shown in brown. Black outlines denote continents, with adjustments to their shapes as described previously. Key to blocks around southern India: BN, Broken Ridge and Ninety-east Ridge complex; ,K, Kerguelen Plateau; M, Madagascar.

2.2.3 Tethyan history

In general terms, broad passive margins flanked the southern side of the Tethys Ocean, from North Africa to Arabia, while microcontinental fragments that had started to rift away from that margin before the Cretaceous supported extensive shallow carbonate platforms in the central Tethyan zone (Figure 2.16).

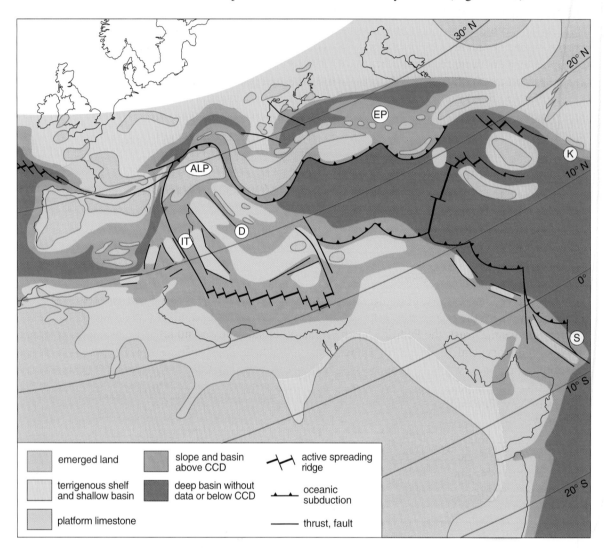

Figure 2.16 Cenomanian palaeogeography of the Tethys Ocean. ALP, regions eventually caught up in the future Alpine mountain chain; D, blocks later incorporated in the Dinaric mountain chain (Balkan Peninsula), consisting of the External Dinarides in the west and the Internal Dinarides in the east; EP, Eastern Pontide volcanic arc (northern Turkey); IT, platforms subsequently forming the Italian mainland; K, Kohistan volcanic arc (already extinct), now in northern Pakistan; S, ocean crust emplaced onto the Arabian foreland later in the Cretaceous, as the Semail Ophiolite. (Steuber and Löser, 2000.)

Legend:
- emerged land
- terrigenous shelf and shallow basin
- platform limestone
- slope and basin above CCD
- deep basin without data or below CCD
- active spreading ridge
- oceanic subduction
- thrust, fault

○ In which of the Tour stops in Section 1.1 have you already encountered some of the distinctive deposits of one of these central Tethyan carbonate platforms?

● In Tour stop 3 (Section 1.1.3), the Cenomanian rudist limestones of Istria (Figures 1.11, 1.12) were described. These formed on part of the Adriatic carbonate platform, which now comprises the 'External Dinarides' (the western block of the pair labelled D in Figure 2.16).

On the northern Tethyan margin, by contrast, subduction consumed oceanic crust alongside the Laurasian continent, with its patchwork of accreted terranes. Most of the subduction was directed towards the north. However, in the Alpine zone the subduction was directed southwards, beneath a tongue-shaped microplate that carried a cluster of platforms northwards (Figure 2.16). In Tertiary times, the latter were either thrust onto one another to form the Italian mainland (IT in Figure 2.16) or they became caught up in the Alpine and Dinaric mountain chains (ALP and D, respectively, in Figure 2.16). Remnants of Cretaceous volcanic arcs are likewise included within other mountain chains that ensued from the closure of the Tethys Ocean, such as the Eastern Pontides of northern Turkey, and Kohistan in northern Pakistan (EP and K, respectively, in Figure 2.16). In places, segments of oceanic crust also became thrust onto the opposing continental margin, as highly altered and deformed masses, termed ophiolites. Several small examples formed along the northern Tethyan margin, but perhaps the largest one was emplaced late in the Cretaceous on the southern Tethyan margin, in eastern Arabia (the Semail Ophiolite of Oman, S in Figure 2.16).

The anticlockwise rotation and then northward movement of Africa brought about the gradual closure of the Tethyan Ocean (Figure 2.1a,b) that was to lead to the Cainozoic mountain-forming episodes in which the various terranes mentioned above were crushed together. Iberia, which had originally been situated adjacent to western France, served as the pivot for the rotation of Africa. As it rotated, the Bay of Biscay opened behind it (Figure 2.16), and the Pyrenees then started to form towards the end of the Cretaceous as the northern margin of Iberia collided with southern France.

2.2.4 Pacific Ocean events

So far, we have concentrated upon the complex geographical changes on the side of the Earth where the continents were (and still are) clustered. Although entirely floored by oceanic crust, the Pacific Ocean on the other side showed no less dynamic a history. Unlike the much younger Atlantic Ocean, the Pacific is ringed by subduction zones, with associated earthquake belts and volcanic chains constituting what is popularly referred to as the 'Pacific ring of fire'. The hot springs of New Zealand, the emblematic Mount Fuji of Japan, the devastating earthquakes of Alaska and California in the last century, and the towering Altiplano of the South American Andes are but a few of its many manifestations. Subduction around the Pacific rim dates from well before the Cretaceous, but since Late Jurassic times, at least, the consumption of oceanic crust has been notably asymmetrical.

○ From what you have already read, on which side of the Pacific Ocean was Cretaceous ocean crust largely consumed, and what small segment appears to have escaped destruction in this manner?

● From Tour stop 5 in Chapter 1 (Figure 1.15), it was demonstrated that a large tract of Pacific Ocean floor of Cretaceous age remains today only in the western part of the ocean. Hence, its counterpart to the east must already have been largely subducted on the eastern side of the ocean, beneath the western margins of the Americas. The exception to the latter process is the Caribbean Plate, which, as described above, is believed to have slotted in between the two American plates, from the Pacific Ocean, starting from Cretaceous times.

As a consequence of this asymmetrical subduction, the network of mid-oceanic ridges responsible for the production of Pacific Ocean crust is today situated towards the eastern side of the ocean, with only small residual eastern oceanic plates flanking the Americas. Apart from the Caribbean Plate, all that remains of the previous Cretaceous ocean floor in the east — termed the Farallon Plate —

are various terranes derived from oceanic plateaux and guyots that became accreted onto the American margins. We can assume that this subduction must have proceeded quite rapidly in the Cretaceous, both because of the western advance of the American plates as the Atlantic opened and because magnetic stripes on the ocean floor provide evidence for greatly increased spreading rates there between 125 Ma and 70 Ma ago. One consequence of this ravenous consumption of ocean floor was the ascent of copious amounts of magma, due to partial melting over the descending slabs along the western American margins, giving rise to large Cretaceous batholiths including, for example, the Sierra Nevada of California and the Peruvian Andes.

In addition to the increased rate of sea-floor spreading in the Cretaceous, superplume activity (Figure 2.6) beneath the Pacific lithosphere contributed to the development of numerous plateaux and seamounts, which frequently reached to sea-level (Section 1.1.5). The overall effect of this activity was a huge relative enhancement in the volumetric rate of oceanic crust production throughout much of the Cretaceous (Figure 2.17), a phenomenon that you will explore in detail in Chapter 7.

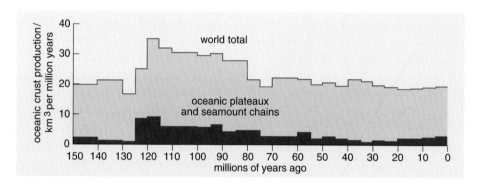

Figure 2.17 Variation in the production of oceanic crust through time, as estimated by Larson (1991). (Larson, 1995.)

2.3 Implications for the Cretaceous Earth system

We can now briefly review the implications of the palaeogeographical changes outlined above for the issues raised in Section 2.1. As before, let us first consider mountain formation. Recall from your study of Figure 2.2 that mountain chains in and around the continents arise either primarily as a result of continental collision or alongside subduction zones.

○ From a consideration of plate tectonics in the Cretaceous, where were the main continental mountain chains of the period likely to have been?

● In the outline given in Section 1.2 and the above, it is clear that Mesozoic palaeogeographical history was dominated by the progressive break-up of Pangaea. There were no major continental collisions either leading up to or within the Cretaceous; the great collision belts that were to be generated by the closure of the Tethyan Ocean — from the Alps to the Himalayas and beyond — were products of the succeeding Cainozoic Era. And those that had been produced during the Palaeozoic Era, although still in evidence (as are, for example, the Caledonian Mountains of Scotland, even today), could be expected to have acquired more modest relief by Cretaceous times. Hence, the main (actively uplifting) mountain chains of the Cretaceous should all have been associated with subduction zones, including sites of terrane accretion. In that case, the largest and oldest-established system of subduction zones was that rimming the Pacific Ocean, followed by that along the northern margin of the Tethys Ocean.

The absence of any active continental collision belts, and hence of the possibility of enhanced uplift due to the breaking off of a dense lithospheric root, means that there is no evidence for mountain-building on the size and areal extent of today's Himalayan/Tibetan system. In the subduction-related mountain chains of the Cretaceous, the most rapidly uplifted areas are likely to have been those where strike–slip transport of terranes led to transpression (Figure 2.9d), although this was probably quite localized, as perhaps in the case of south-west Borneo (Figure 2.9b). However, the massive intrusion of low-density plutons along the American margins, resulting from the copious subduction in that area, initiated rapid uplift later in the Cretaceous (and thereafter), which led to large amounts of sediment being shed.

Figure 2.18 shows a reconstruction of global palaeotopography by Hay *et al.* (1999) for their 80 Ma map that was previously shown in Figure 2.15. Note the relatively low relief shown for most of the Tethyan hinterland, especially alongside the southern (passive) margin. Some uplift is indicated along the northern margin, associated with the earlier accretion of terranes and formation of volcanic arcs, especially in the east. As expected, the main mountain chains are shown around the Pacific Ocean rim. Lesser zones of uplift (with up to 1 km of relief) are shown alongside the newly opening Atlantic and Indian Oceans, reflecting the residual thermal effects of rifting. It should also be remembered in this context that the anomalously high sea-levels of the Late Cretaceous, in particular, would have further contributed to the relatively modest nature of mountain elevation at the time.

Figure 2.18 Global palaeotopography 80 Ma ago, as reconstructed by Hay *et al.* (1999). Generalized contours of relief, based on modes of tectonic deformation, are in km.

80 Ma

KEY land under 1 km
above 1.0 km
above 2.0 km
above 3.0 km

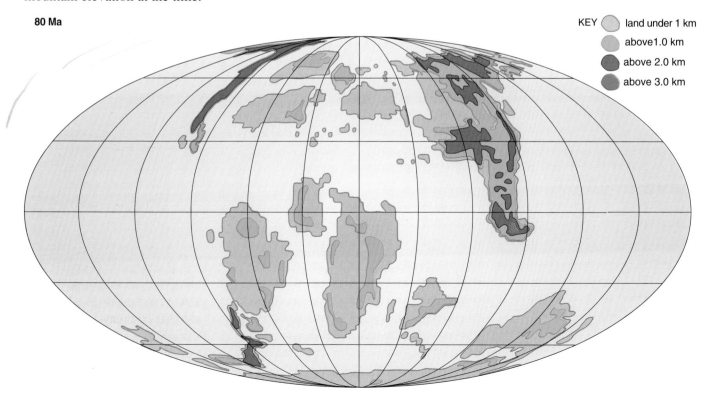

Turning now to marine circulation and climate, the combination of continental break-up and (again) high sea-levels that developed through the Cretaceous would have brought increased maritime influence into the continental interiors. Earlier Mesozoic geography was dominated by the two supercontinents of Laurasia and Gondwana straddling the Equator, and their correspondingly extreme temperature fluctuations, especially in mid-latitudes, would have created intense seasonal changes in pressure systems and precipitation. However, in the Cretaceous, this intense seasonality would have yielded to more widely equable climates. Nevertheless, localized monsoonal cells were doubtless associated with some of the more mountainous areas.

Also important was the augmentation of the total length of mid-ocean ridges and their spreading rates. The cumulative displacement of water from the ocean basins that resulted, further exacerbated by the growth of superplume-related oceanic prominences, accounted for the exceptional eustatic transgression of the period, which we will explore in the next Chapter. However, notwithstanding the spread of shallow seaways across the continents, deep-water connections between the Atlantic basins and neighbouring ocean basins remained limited for much of the period.

○ What restricted deep-water connections between: (a) the South and Central Atlantic; and between the Atlantic basins, and (b) the Central Pacific; (c) the South Pacific; (d) the Tethys Ocean; and (e) the Arctic Ocean, for much or all of the Cretaceous?

● The relevant features were mentioned in the earlier discussion of plate tectonic history. (a) The juxtaposition of west Africa and north-east Brazil along a transform fault prevented a deep connection between the South and Central Atlantic until Late Albian times. (b) Atlantic connection with the Central Pacific was limited by the anomalously thick (hence shallow) Caribbean Plate interposed between the Americas. (c) South America remained closely connected to Antarctica throughout the period, restricting connection with the South Pacific. (d) The narrow, sluice-like connection between the western Tethys Ocean and the Atlantic Ocean was partially obstructed by the constellation of shallow platforms, especially in the Early Cretaceous. (e) The North Atlantic Ocean opening had not yet extended as far as the Arctic Ocean.

The partial restriction of the Atlantic Ocean basins for much of the Cretaceous, coupled with intense evaporation in arid regions, may have provided favourable conditions for the formation and accumulation at depth of hypersaline water, which would occasionally have spilled into neighbouring basins. Hay *et al.* (1999) suggest that this circumstance could indeed have been a major driving force for a link between ocean circulation and climate very different from that of today (Figure 2.4):

> 'At this time [Late Cretaceous] the Atlantic was probably the global source of ocean deep water. We suspect that it supplied warm saline water that could be upwelled in the polar regions of the Pacific as well as in the South Atlantic, and that this resulted in the warm equable climate of the Late Cretaceous.' (Hay *et al.*, 1999, p. 26.)

Similar conditions arose from time to time in the semi-restricted ocean basins of the Tethys Ocean, as well as the surrounding shallow seas spread over the continents. Stagnation through 'ponding' of the relatively dense hypersaline water was probably an important contributor to the sporadic outbreaks of marine anoxia, evidence for which you saw in Tour stop 4 (Section 1.1.4), and which we will explore further in later Chapters.

Finally, we should briefly return to the issue of high ocean-floor spreading rates and subduction-related volcanism in the Cretaceous, to consider chemical consequences.

○ What should have been the main consequence for ocean chemistry of the enhanced hydrothermal circulation implied by the increased spreading rates?

● The ion-exchange mechanism discussed in Section 2.1 (Figure 2.5) should have decreased the Mg/Ca ratio in seawater.

In later Chapters, we will consider the evidence for, and the possible effects of, this change together with those of the associated volcanic outgassing.

2.4 Summary

- During the Cretaceous, the configuration of the continents changed from the prevailing Mesozoic pattern of two supercontinents (Laurasia and Gondwana) straddling an equatorial ocean (Tethys), to one with several continents separated by oceans, which extend into high latitudes, more like that of today.

- Palaeogeographical influences on the Earth system include the following:

  (i) Mountains (produced either by subduction or by continental collision) can affect atmospheric circulation, in turn causing changes in precipitation and albedo. In the longer term, weathering and deposition of sediment eroded from them can also influence climate by causing drawdown of atmospheric CO_2 (a major greenhouse gas).

  (ii) Oceanic circulation is an important driver of climate because of its effects on the redistribution of heat and moisture in the atmosphere, as illustrated by the 'global thermohaline conveyor' today.

  (iii) Magmatism affects global sea-level over long time-scales because of variation in the relative displacement of ocean water due to the changing total length of mid-ocean ridges and of their spreading rates. Moreover, hydrothermal circulation can modify the chemistry of ocean water, while the release of gases into the atmosphere affects climate.

- Today, past arrangements of continents and oceans are largely inferred from palaeomagnetic data. The record from Cretaceous times onwards is better than that of older configurations because of the lack of older ocean floor. The resulting maps need to be tested for their agreement with other kinds of data, such as the past distributions of organisms.

- Earlier attempts at palaeogeographical reconstruction tended to treat continents as solid slabs moving around the surface of the globe. More recent efforts attempt to correct for ways in which their shapes have been altered by, for example, the extension of passive margins and their distortion by failed third arms of rift systems, and accretion of terranes.

- The Central Atlantic already existed at the beginning of the Cretaceous. The North Atlantic started to extend northwards from it in the Early Cretaceous, though did not reach high latitudes until the Palaeocene. The South Atlantic initially opened as a discrete basin between South America and southern Africa in the Early Cretaceous, eventually achieving a deep-water connection with the Central Atlantic in the latest Albian. Connection between the Central Atlantic and Pacific was restricted by the anomalously thickened oceanic crust of the Caribbean Plate, which was probably generated in the Pacific but then inserted eastwards between the two Americas (although the details of this remain controversial).

- Connections between South America, Antarctica and Australia continued into the Cainozoic. Following the separation of Africa, the Indian Ocean began to open between India and Antarctica, though for much of the Cretaceous a plume-generated cluster of oceanic promontories between them provided occasional links for land-dwelling organisms.

- Broad passive margins flanked the southern side of Tethys, while separate microcontinental blocks supported carbonate platforms in its central part. Subduction zones formed along the northern margins, alongside previously accreted terranes. The anticlockwise rotation (pivoting around Iberia) and northward drift of Africa commenced the closure of Tethys.

- In the Pacific Ocean, loss of ocean floor by subduction was asymmetrical, with the Cretaceous Farallon Plate being consumed along the American margins. The accretion of terranes and injection of large batholiths accompanied its subduction. On the remaining western Pacific Plate, numerous plateaux and guyots testify to superplume activity, associated with high spreading rates, from mid- to Late Cretaceous times.

- The implications of the events described above are as follows:

  (i) The lack of continental collisions during the Mesozoic means that the main mountain chains of the Cretaceous were subduction-related, situated mainly around the Pacific margins, but also to a lesser extent along parts of the northern Tethyan margin. Relief was probably modest, by and large, compared with the great mountain belts of today.

  (ii) Deep-water connections between the Atlantic and neighbouring ocean basins remained limited for most of the Cretaceous. The deep Atlantic is thus thought to have served as a reservoir of hypersaline water, which could have driven a thermohaline circulation system different from that of today.

  (iii) Stagnation of 'ponded' hypersaline water, also forming in the Tethyan basins, was a likely factor in the development of oceanic anoxic events (OAEs), as well.

  (iv) Enhanced rates of production of oceanic crust, especially in the Pacific, would have had implications both for the chemistry of ocean water and for atmospheric composition.

2.5   References

HAY, W. W., DECONTO, R. M., WOLD, C. N., WILSON, K. M., VOIGT, S., SCHULZ, M., WOLD, A. R., DULLO, W.-C., RONOV, A. B., BALUKHOVSKY, A. N. AND SÖDING, E. (1999) 'Alternative global Cretaceous paleogeography', pp. 1–47 in BARRERA, E. AND JOHNSON, C. C. (eds) *Evolution of the Cretaceous Ocean-Climate System*, Geological Society of America Special Paper **332**.

KERR, A. C., ITURRALDE-VINENT, M. A., SAUNDERS, A. D., BABBS, T. L. AND TARNEY, J. (1999) 'A new plate tectonic model of the Caribbean: Implications from a geochemical reconnaissance of Cuban Mesozoic volcanic rocks', *Geological Society of America Bulletin* **111** (11), 1581–1599.

LARSON, R. L. (1991) 'Latest pulse of Earth: evidence for a mid-Cretaceous superplume', *Geology*, **19**, 547–550.

METCALFE, I. (1988) 'Origin and assembly of South-East Asian continental terranes', pp. 101–118, in AUDLEY-CHARLES, M. G. AND HALLAM, A. (eds), *Gondwana and Tethys*, Geological Society Special Publication **37**, Oxford University Press.

MILNER, A. C., MILNER, A. R. AND EVANS, S. E. (2000) 'Ch. 21, Amphibians, reptiles and birds: a biogeographical review', pp. 316–332 in CULVER, S. J. AND RAWSON, P. F. (eds), *Biotic Response to Global Change — The Last 145 Million Years*, Cambridge University Press.

PINDELL, J. L. (1994) 'Evolution of the Gulf of Mexico and the Caribbean', pp. 13–39 in DONOVAN, S. K. AND JACKSON, T. A. (eds), *Caribbean Geology — An Introduction*, University of the West Indies Publishers' Association, Kingston, Jamaica.

3   Fluctuating sea-level

Peter W. Skelton

Most successions of marine sedimentary rocks can readily be interpreted in terms of relative changes in the depth of deposition, based on the lithologies, sedimentary structures and fossils found within them. However, such changes in local bathymetry can result from tectonics or sedimentary accumulation as well as fluctuations in sea-level itself, or any combination of these. So, how can we pick out the signal of past changes in global sea-level? The ideal way to answer this question would be to find something that could serve as a permanently fixed geological 'dipstick', but unfortunately no such thing exists; therefore, we have to make do with what can be gleaned from the evidence of the Earth's restlessly moving crust.

3.1   Getting to grips with past changes in sea-level

A common early approach to the problem of estimating past eustatic rise and fall was to look at relative distributions of marine deposits of different ages over the surfaces of continents. The principle is straightforward enough; as sea-level rises, so an increasing proportion of continental crust should become flooded, whereas an overall retreat of seas is associated with a sea-level fall. We do, however, need to take account of continental relief. If, for argument's sake, the continents were virtually flat and topographically only 5 m above present-day sea-level, a global rise of just 6 m would achieve 100% flooding, while uniformly higher continents surrounded by vertical cliffs would experience no loss of area in plan view. Therefore, in order to calibrate the vertical amplitude of eustatic fluctuations against the percentage inundation of total continental area, we need some idea of how the topographical relief of continental crust is distributed, and this is where the complications begin. As an example, Figure 3.1 shows the relationship between the percentage of total area of continental crust and present elevation — termed a hypsometric curve — for the North American continental crust.

Note for Figure 3.1 that the total area of continental crust to be considered has to be arbitrarily defined, where the continental margin begins to slope towards the surrounding ocean basin, at about 60 m depth from the present-day sea-level.

○ What percentage of the North American continental crust, as defined in Figure 3.1, is covered by sea today?

● Approximately 9%, which is the value on the *x*-axis that corresponds to a present elevation of 0 m, i.e. sea-level.

○ According to one palaeogeographical analysis, about 25% of North America was covered by seas during the maximum extent of Jurassic flooding. Using Figure 3.1, estimate how much higher global sea-level was at that time compared with today.

● On the hypsometric curve of Figure 3.1, 25% flooding corresponds to about 100 m above present-day sea-level.

Figure 3.2 shows that today's continents have hypsometric profiles of broadly similar shape, although with some variation between them. However, be aware that the data in this Figure have been normalized by dividing observed heights by the mean height for each continent. This facilitates comparison of curves for

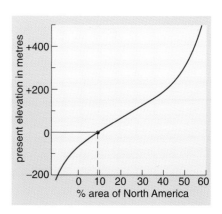

Figure 3.1 Relation of total area to present elevation for the North American continental crust (hypsometric curve). Note that 0% on the *x*-axis is arbitrarily defined to correspond to −60 m from sea-level.

different continents, but conceals the fact that their average heights differ. As the average heights here all plot at unity on the *y*-axis, absolute height scales thus differ between the continents (which is why the *y*-axis shows only relative units).

○ From Figure 3.2, describe how the profiles of (a) Asia and (b) Australia compare with that of North America. Casting your mind back to the palaeogeographical history outlined in Chapter 2, how do these differences relate to the character of the mountain ranges in the three continents?

● (a) Asia shows a slightly greater proportion of area with exceptionally high relative elevation (above *y* = 3), compensated for by a slightly lower proportion of area at moderate and low elevations. (b) Australia shows almost the opposite — a lower proportion of area at exceptionally high and moderate relative elevations, but a greater proportion at lower elevations (especially below *y* = 1). In Sections 2.2.3 and 2.2.4, you saw that both Asia and North America have relatively young mountain ranges of large extent. The former includes the Himalayan chain, produced by post-Cretaceous continental collision and showing exceptional uplift, while the latter includes the Rocky Mountain system, with considerable uplift from Cretaceous times associated with rapid subduction of the Farallon Plate. In contrast, Australia has not experienced these kinds of plate tectonic effects during this time, and so its mountain chains are of more modest relief (although, as you will see later, earlier events did affect its history).

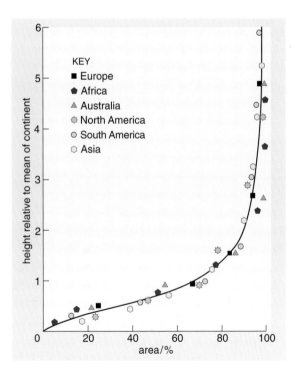

Figure 3.2 Normalized hypsometric curve for the present continents *(Kendall and Lerche, 1988)*. Normalization is achieved by dividing observed heights by the mean height for each continent (hence values on the *y*-axis are relative, not absolute). The mean height of each continent therefore plots as unity on the *y*-axis.

Thus, continental relief, as expressed in hypsometric curves, varies according to the legacy of plate tectonics (among other factors). The Cretaceous world, for example, experienced a somewhat different balance of effects from that seen today (Section 2.3). Consequently, we cannot generalize too freely in using modern hypsometric curves to calibrate the amplitude of past eustatic changes. A further problem arises from erosional loss of marine deposits on land, which leads to an underestimation of the proportional extent of former continental drowning. In short, we can expect a fair margin of error from the exercise. Nevertheless, some idea of the *relative* amount of eustatic change can be obtained, and Figure 3.3 shows some examples of estimates, which have been calibrated against a generalized hypsometric curve for today's continents. Despite considerable disagreement over amplitude, all but one of the curves concur in detecting a eustatic maximum between 90 Ma and 80 Ma ago.

A variant on this approach is to select areas thought to have been relatively stable since the time considered, and to correlate synchronous phases of marine transgression and regression over these areas (Figure 3.4a). Next, eustatic heights above present-day sea-level are estimated by adding the elevations of today's undisturbed Cretaceous deposits overlying stable massifs to their inferred depths at the time of deposition (Figure 3.4b).

○ Considering the overall trends only, what are the main points of agreement and disagreement between the curves shown in Figures 3.3 and 3.4?

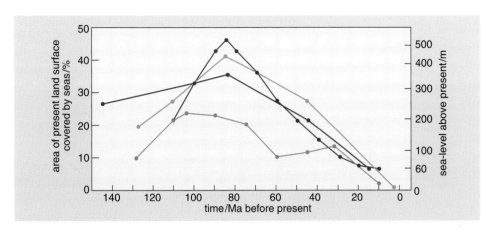

Figure 3.3 Four estimates of changes in continental area (scale on left) covered by seas over the last 150 Ma derived from different authors, with inferred height above present-day sea-level (scale on right) calibrated from a generalized hypsometric curve for today's continents *(Hays and Pitman, 1973)*. Scale on right is irregular because elevation does not vary linearly with percentage area inundated. Note that the curve of Hays and Pitman (red), was not based on direct estimation of the areas covered by marine deposits, as in the other examples, but was modelled from calculated changes in the volume of water displaced from the ocean basins (explained later).

●   With the exception of one of the curves in Figure 3.3, both Figures initially show an overall eustatic rise to a peak sometime after 90 Ma ago (Turonian). Thereafter, all the curves in Figure 3.3 and the Western Interior of the USA curve of Figure 3.4a show an overall decline (although with a brief regressive–transgressive cycle during the late Turonian–Coniacian in the latter). The major difference arises in the Northern European data of Figure 3.4a (also reflected in Figure 3.4b), which shows a resumed rise to a peak in the latest Cretaceous.

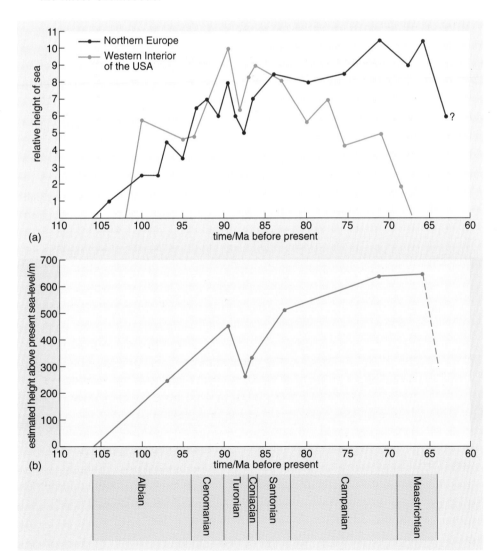

Figure 3.4 (a) Relative rise and fall of sea-level from the Albian to the Maastrichtian, interpreted from transgressive–regressive cycles in Northern Europe (predominantly Chalk) and the Western Interior of the USA (predominantly siliciclastic sediments). The scale of the *y*-axis is in arbitrary units of relative change. (b) Estimates of absolute eustatic height above the present-day sea-level, calibrated from elevations of Cretaceous deposits on stable massif areas today. *(Hancock and Kauffman, 1979.)* Note that they adopted an older time-scale than that used in this book.

Hancock and Kauffman (1979) proposed that the Western Interior data had been distorted for the latter part of the Late Cretaceous by the effects of uplift linked with mountain formation to the west. Hence, they considered the eustatic peak to have been in the latest Cretaceous, as indicated in Figure 3.4b, in agreement with their Northern European curve.

A fundamental problem for both approaches is that, in addition to the obvious mountain-building effects of plate interactions, it seems that large regions of the lithosphere slowly heave up and down on a continental scale. Hence, curves of changes in sea-level relative to a basement datum (i.e. relative sea-level) over long time-scales may differ even between regions lacking obvious evidence of local tectonic activity, such as folding and faulting. As early as 1980, the Japanese stratigrapher T. Matsumoto noted substantial divergence in long-term patterns of marine transgression and regression between different continental regions (Figure 3.5). For example, although several regions do show a transgressive peak for the Turonian, as discussed above, there are others that do not, such as Australia (in both the regions considered in the Figure). Indeed, eastern Central Australia experienced marine inundation only in the Aptian–Albian, before the main phases of transgression seen elsewhere.

Subsequent geophysical research has shown that that some of these movements can be related to the ghosts of plate tectonics past. As vibrations from earthquakes travel through the Earth, they speed up or slow down according to the composition, temperature and pressure of the material they are travelling through. From the different times of arrival of the vibrations at recording stations around the Earth's surface, it is possible to detect masses of cold, dense material that are sinking through the mantle, as well as hot, low-density plumes that are rising through it — a technique called seismic tomography. Using this method, the sinking remains of the old Farallon Plate of the Cretaceous eastern Pacific (Section 2.2.4) have been located some 1600 km below the surface of the Earth, beneath what is now the eastern seaboard of North America. Its presence helps to explain the apparent paradox of undeformed Upper Cretaceous marine deposits found today more than 1000 m above sea-level in a large area around Denver, Colorado in the Western Interior of the USA. Nobody's estimates of eustatic change had suggested a rise in sea-level on that scale, so the region must have subsided in Late Cretaceous times, when the sediments accumulated, and thereafter rose to its present altitude. The regional subsidence can now be explained as a consequence of the surface having been 'pulled down' as the shallow remains of the newly subducted Farallon Plate began to sink beneath it during the Late Cretaceous. Later, as North America drifted further westwards, the mid-continental region bobbed back up again, while the Farallon Plate continued to sink through the mantle to its present position.

○ How are these effects reflected in Figure 3.5?

● Late Cretaceous transgression is most strongly expressed in the North American regions that are considered in the Figure, especially during the Coniacian–Santonian, although with some relative shallowing already beginning in the Campanian–Maastrichtian (as also shown in Figure 3.4a).

Figure 3.5 Synoptic records of Cretaceous transgression and regression from selected regions of continental crust, based on the work of Matsumoto (1980). Note that the stages are shown in equally spaced columns, so that the chronostratigraphical scale along the top is not calibrated to absolute time.

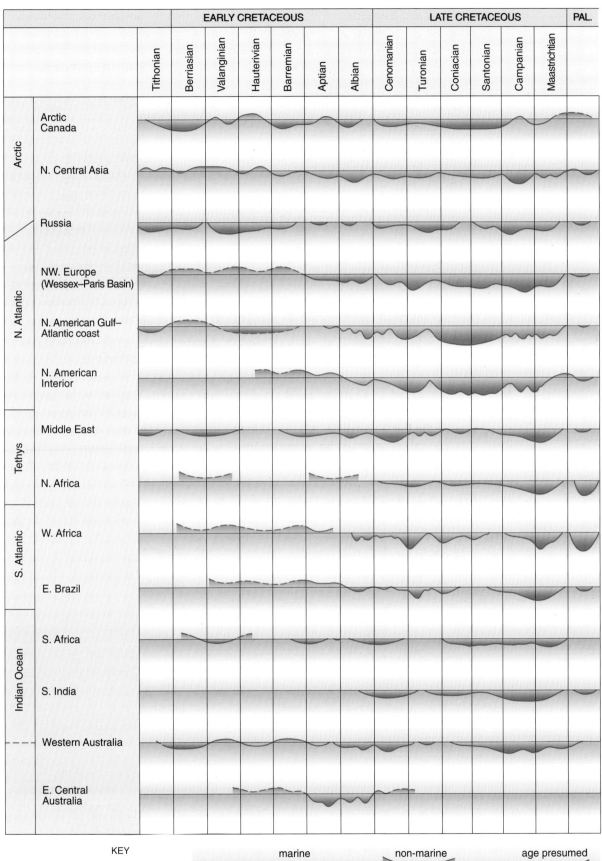

		EARLY CRETACEOUS							LATE CRETACEOUS						PAL.
		Tithonian	Berriasian	Valanginian	Hauterivian	Barremian	Aptian	Albian	Cenomanian	Turonian	Coniacian	Santonian	Campanian	Maastrichtian	
Arctic	Arctic Canada														
	N. Central Asia														
	Russia														
N. Atlantic	NW. Europe (Wessex–Paris Basin)														
	N. American Gulf–Atlantic coast														
	N. American Interior														
Tethys	Middle East														
	N. Africa														
S. Atlantic	W. Africa														
	E. Brazil														
Indian Ocean	S. Africa														
	S. India														
	Western Australia														
	E. Central Australia														

KEY

cycle of sedimentation

marine   non-marine   age presumed

transgression   regression

With the understanding of mantle dynamics provided by seismic tomography, such effects can now be simulated by computer models based on reconstructions of former plate movements. For example, American geophysicist Michael Gurnis and colleagues have modelled the subsidence of Eastern Australia as it drifted eastwards over a previously subducted slab in Early Cretaceous times, and later rose again (Figure 3.6), as with the aforementioned response of North America to the Farallon Plate.

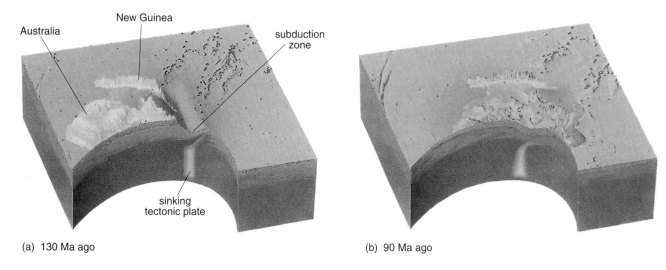

(a) 130 Ma ago   (b) 90 Ma ago

Figure 3.6 Computer model showing the effect of the eastward drift of Australia over the sinking remains of a subducted tectonic plate during the Early Cretaceous: (a) 130 Ma ago; and (b) 90 Ma ago *(D. Fierstein (illustrator) in Gurnis, 2001).*

○ As before, how might these effects be reflected in Figure 3.5?

● The initial subsidence could explain the Aptian–Albian transgression in eastern Central Australia, noted earlier, while the rebound would account for the subsequent regression that was contrary to the trend seen in most other regions.

The opposite kind of effect to that discussed above — anomalous uplift — is seen in regions overlying plumes of relatively hot, low-density material that are rising through the mantle. One of the more notable examples is southern Africa, which experienced only relatively modest marine transgression during the Cretaceous (Figure 3.5), and still has broadly elevated areas today.

Do shallow oceanic 'dipsticks' (such as guyots) offer any better prospects? After all, given the simple general relationship between the present age and depth of the oceanic crust, due to thermal relaxation from the time of formation (Figure 1.22b), we should be able to back-track its subsidence through time, and thus pick up any sedimentary signal of the changing sea-level above it. However, the enterprise may again be spoiled by the behaviour of mantle plumes, which can obscure the record by resetting the cooling clock, thus causing renewed uplift. Such processes were relatively common in the Cretaceous, especially in the Pacific (Section 2.2.4).

○ Where have you already encountered an example of an ambiguous record of sea-level change from sediments deposited on shallow oceanic crust?

● In Tour stop 5 (Section 1.1.5), you read about the problem of deciding whether the marked sea-level fall registered in 98 Ma-old sedimentary rocks on Allison Guyot was eustatic in origin or due to renewed (plume-related) regional uplift.

Such widespread vertical movements of the lithosphere do not undermine the general validity of eustatic sea-level change, but they certainly complicate its expression. Thus, the global signal can be expected to vary in amplitude from one region to another and, in some cases, the regional record may completely buck the trend seen elsewhere, as you saw above in the Cretaceous succession of eastern Central Australia.

A different approach to estimating long-term eustatic changes involves looking at one of the main causes and modelling the changing volume of mid-oceanic ridges; and hence the corresponding displacement of water from the ocean basins, over time. Hays and Pitman (1973) attempted this for the time from the Early Cretaceous to the present day. They used the palaeomagnetic record from the ocean floor to monitor past spreading rates of mid-ocean ridges, and hence their changing profiles (see Figure 1.22b), and thereby calculated their volumetric changes. Necessarily absent from their calculations was the Tethyan ridge system, which was destroyed as the ocean closed. They also omitted the Indian Ocean ridge system from their calculations, arguing that the demise of the Tethyan system was probably more or less matched by the rise of the latter, i.e. that the effects of the two cancelled each other out. From their calculations for the volume of water displaced, with some allowance for depression of the remaining oceanic and continental crust due to additional loading from the displaced water, i.e. isostatic subsidence, they calculated the associated eustatic changes. The resulting curve, which again showed a maximum eustatic elevation between 90 and 80 Ma ago, is shown in Figure 3.3.

The most widely discussed (but also contentious) approach to the problem was brought into the public domain in 1977 by Peter Vail and colleagues at Exxon Production Research, USA. Using seismic reflection profiles (Box 3.1) across continental shelves, they subdivided the sedimentary successions into partially unconformity-bounded packages of strata known as depositional sequences, or simply sequences (see Figure 3.8).

Box 3.1   Seismic reflection surveying

Seismic reflection surveying is a geophysical technique used by the oil industry and research groups to derive subsurface geological information. It has been used, in particular, in the study of sedimentary basins, and can be done on land or at sea. Seismic waves are sound waves generated by a seismic source. They travel through the subsurface and are reflected back to the surface at geological boundaries within the subsurface, where they are recorded (Figure 3.7a). Besides simple reflection, e.g. wave 1 in Figure 3.7a, waves may pass through a boundary and be refracted (bent) before being reflected off a deeper boundary (wave 2 in Figure 3.7a). From the various reflected signals picked up by the detectors, a picture of the subsurface geology can be compiled. When plotted on paper, the record of ground motion (or pressure variation at sea) against time is called a seismic trace (Figure 3.7b). The amplitude of a reflection is a measure of the strength of the ground motion, shown by the deviation of the seismic trace. In practice, complex stacking of many seismic traces (the details of which need not concern us here) is used to reduce background noise, e.g. from nearby roads in land surveys or ships in marine surveys. An array of seismic traces processed in this way and displayed side by side form a seismic section (Figure 3.7c,d). This effectively represents an acoustic profile of structures within the Earth and is produced by moving the source and detectors along the line of survey. Shots are fired at a regular horizontal distance apart (usually around 25 m). In order to improve the visibility of the reflections, the right-hand half of the seismic wave trace is usually shown in black (Figure 3.7d).

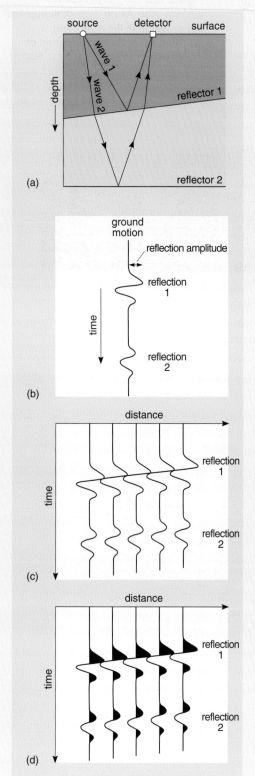

(a)

(b)

(c)

(d)

The horizontal scale of a seismic section is a measure of horizontal distance along the line of survey. The vertical scale is the two-way time of the seismic wave, i.e. the time taken for the wave to travel down to the reflector and back up again.

As a first step to interpreting its geological significance, a seismic section is examined for reflection continuity. Continuity is shown where a reflection on a trace can also be recognized on neighbouring traces, with only small changes in the arrival time. Because half of the seismic waves are shaded black, the continuities appear as black or white bands running across the section (Figure 3.7d) and the geological surfaces (more usually, closely spaced clusters of surfaces) that give rise to them are termed reflectors. They are formed by interfaces where there is a change in density and/or acoustic velocity (speed of sound in the material) of the rock and/or its contained fluid. The interfaces may be bedding planes but may also be fault planes or any other extensive boundary between rock types. In some areas, only a few or no seismic reflectors may be imaged either because there are no interfaces shallow enough to reflect the available seismic energy (seismic energy is attenuated as it travels through the Earth), or because a very complex structure confuses the signal.

The fact that the vertical scale is measured in two-way time and not in depth is important because it means that a seismic section is *not* a true geological cross-section through the Earth. To convert from one to the other, we would need to know the velocity of sound through each of the rock layers. As this velocity varies for each different lithology, this is no easy task. Oil companies need such information because they have to calculate how deep to drill in order to intersect an oil reservoir within a geological structure. The only precise way to determine the velocities is to drill a borehole somewhere along the line of the seismic section and then to measure directly the time taken for sound waves to pass through each of the rock units within the borehole. The time it takes a seismic wave to reach any specified depth in the borehole can then be calculated. Where this depth corresponds to a major change in rock type capable of forming a seismic reflector, we can use this time to locate the corresponding reflection on the seismic section. This is called 'tying the well to the seismic section' and allows information obtained on the various rocks in the borehole to be extrapolated along the seismic section.

This is a very simplified explanation of how seismic sections can be used to derive subsurface geological information. In reality, there are other subtle differences that distinguish seismic sections from a geological cross-section and much computer time is required to remove 'artefacts' which are inherent in the acquisition and processing of seismic data. However, there is insufficient space to detail these here and they will not be considered further.

Figure 3.7 Seismic reflection surveying: (a) seismic reflection and refraction; (b) a seismic trace — reflection 1 has a higher amplitude than reflection 2; (c) a diagrammatic seismic section, composed of seismic traces from successive shot points; (d) shading of the right-hand half of the wave to make the reflections more easily visible. See text for explanation.

Each depositional sequence is marked by a package of sediments representing a depositional cycle of coastal transgression followed by regression. Thus, reflectors within each wedge can be seen to overlap one another landwards (towards the right in Figure 3.8a) at the base of each wedge, while those at the top successively step back towards the basin. Towards the outer shelf, where deposition was more or less continuous, the landward unconformities between the wedges fade as they pass into conformable successions (Figure 3.8a). When

Figure 3.8 (a) Seismic section from offshore West Africa, with stratigraphical interpretation of the reflections; (b) chronostratigraphical diagram of the succession in (a); (c) chart of relative changes in sea-level inferred from (b). *(Vail et al., 1977a).* Note that Vail *et al.* used an older time-scale than that used here.

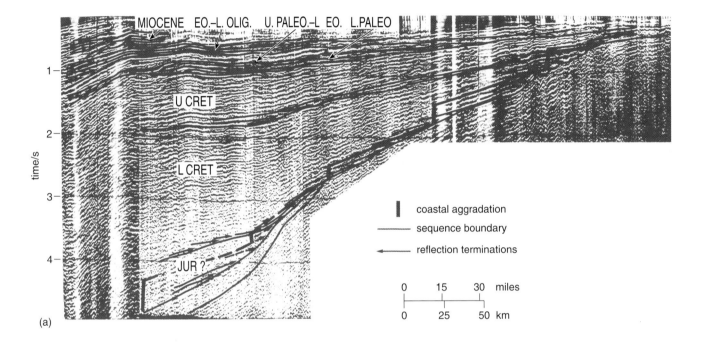

(a)

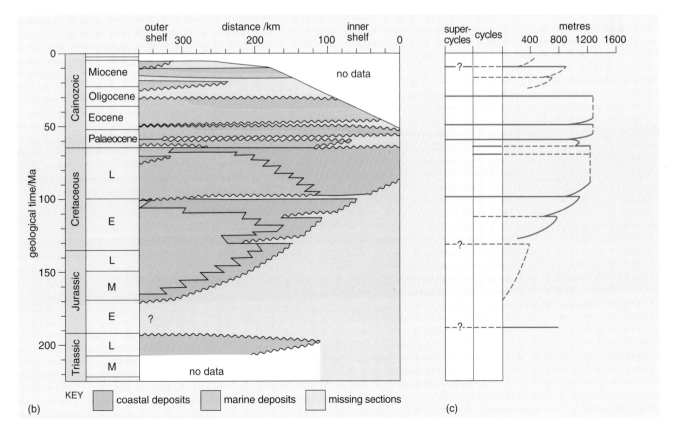

(b)

(c)

KEY: coastal deposits | marine deposits | missing sections

the deposits are arranged in a chronostratigraphical diagram (Figure 3.8b), the unconformities, which represent gaps in the sedimentary record, are expanded to fill their corresponding time intervals, and so appear as landward-expanding wedges interdigitating with the depositional sequences. It is then but a small step to infer the *relative* sea-level changes (which include a component of subsidence), signalled by the cycles of coastal onlap and offlap (Figure 3.8c). Moreover, within each major sequence, a hierarchy of constituent cycles can be detected, all the way down to the individual packets of sediment bounded by the reflectors, each recording a minor cycle of relative change in sea-level.

Few people would have problems with the argument so far. However, Vail and colleagues then made the controversial proposal that cycles of relative sea-level change correlated between different regions could be considered eustatic. The absolute amplitudes of these changes had to be inferred by calibration from other independent estimates of global sea-level (Figure 3.9a), in order to remove the bias due to cumulative subsidence on the continental shelves that were investigated. Because these were of special interest to the petroleum companies, most showed relatively thick sequences. In effect, what Vail *et al.* were proposing was a curve of relative deviations from the first-order curve (Figure 3.9b). On this basis, they proposed a chart of 'eustatic changes of sea-level' (Figure 3.9c). Later versions of this curve have appeared, most notably by Haq *et al.* (1988), based upon the same method of analysis although with a more even (as opposed to 'sawtooth') interpretation of the relative rates of eustatic rise and fall (Figure 3.10).

Given the alternative possible controls on depositional geometry, arising from local tectonic influences and factors affecting the rate of sediment supply (such as climate), that it was a *eustatic* curve was a bold assertion which inevitably caused considerable controversy. An early criticism was that many of the apparently global episodes of transgression might alternatively reflect a preponderance in their database of passive margins that were experiencing the local tectonic consequences of rifting (Figure 2.9a) at similar times. The argument for a eustatic cause relies crucially on the precision of correlation. A significant problem for the shorter-term cycles is that the available means of correlation often lack sufficient resolution to determine whether spatially separated sequences can be attributed to a common eustatic cause, or whether they merely formed independently within some finite time interval. The latter case would reveal no more than a chance clustering of genetically unrelated sequences, perhaps even belonging to different orders in the hierarchy of cycles. The optimist may thus fall into the trap of seeing an apparent, if slightly fuzzy, eustatic signal where none is warranted.

Indeed, in the polarized debate that followed, rather too many optimists ignored the absolute necessity of independently testing every purported correlation of sequences and proceeded to use the sequences themselves for correlation, thus guaranteeing a circular argument. For these and other reasons, today, the jury remains out on the extent to which sequence stratigraphy can reveal eustatic change, although it has been widely adopted as an immensely powerful tool for analysing *relative* sea-level changes and their effects on patterns of sedimentation in given areas. This is not to imply any doubt concerning the existence of global changes in sea-level, as variations in potential causal factors, ranging from ocean floor spreading rates to the spread of continental ice sheets, surely mean that there must have been. Rather, it is distinguishing eustatic causes from other, regional causes of relative sea-level change that remains problematical. Certainly, the curves should never be used uncritically, without independent checks, as a global correlation tool.

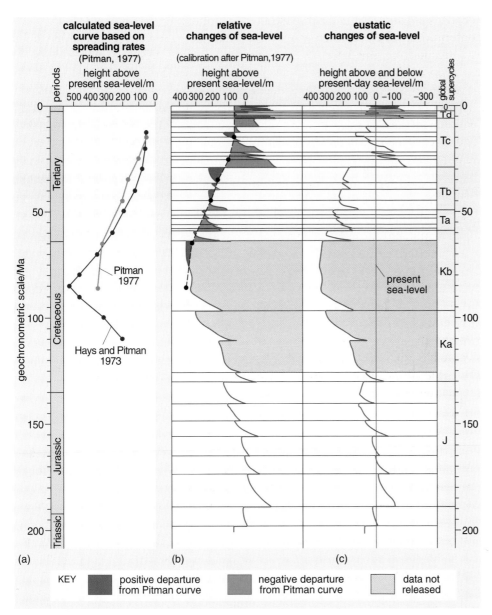

Figure 3.9 The original 'global cycles of sea-level changes' proposed by Vail *et al.* (1977b). (a) The sea-level curve estimated by Hays and Pitman (1973; see earlier discussion), together with a later version published by Pitman (1978; in press 1977). (b) Plot of correlated relative changes of sea-level, calibrated against other first-order estimates of global sea-level, principally the Pitman curve for the late Cretaceous–Tertiary. (c) Proposed eustatic curve based on (b). Details of subsidiary cycles within the grey area (comprising most of the Cretaceous) were not revealed in the original publication as the results were then confidential. Note that Vail *et al.* used an older time-scale than that used here. For a later version of this eustatic curve, published by Haq *et al.* (1988), see Figure 3.10.

3.2 Results for the Cretaceous

It would be overly simplistic to conclude from the foregoing discussion that the quest to identify past eustatic changes is a lost cause. Instead, what we have is a spectrum ranging from well-established examples, over which there is a wide consensus, to little more than contentious speculation. The full range is present in the literature on the Cretaceous.

The first-order (longest-term) pattern for the period is now generally agreed. This is often referred to as 'The Great Cretaceous Transgression', with shallow seas spreading widely across the continental shelves and continents, and this transgression has long been recognized as a distinctive hallmark of the period (Figure 1.2). Indeed it was first mooted in 1875 by the Viennese geologist Eduard Suess, who later invented the term 'eustatic' for worldwide changes of sea-level of this sort. Notwithstanding the regional variability portrayed in Figure 3.5, it is evident that marine inundation was more widespread and pronounced throughout most of the Late Cretaceous than during the early part, when global levels seem to have been closer to those of today. Exactly when it reached its maximum height is less clear, although most evidence currently points towards the Turonian, as discussed earlier. Opinions have also varied over the maximum amplitude of the rise in global sea-level, depending upon the assumptions and methods of analysis used. Watts (1982), for example, suggested that the tectonic evolution of passive margins of similar age may have played a large role in the Vail curves. Later estimates have generally been less than the several hundreds of metres proposed in some earlier estimates, but all agree with a maximum of over 200 m higher than today (Figure 3.10) — comfortably more than the estimated 60 m that would be gained simply by melting all of the present continental ice. The role of broadly uplifted areas of young, hot oceanic crust (Section 2.2) in displacing ocean water was evidently paramount.

Figure 3.10 Eustatic curves proposed by Pitman (1978) and Haq *et al.* (1988). Note: curves do not terminate at 0 m because continental ice volume has not been allowed for.

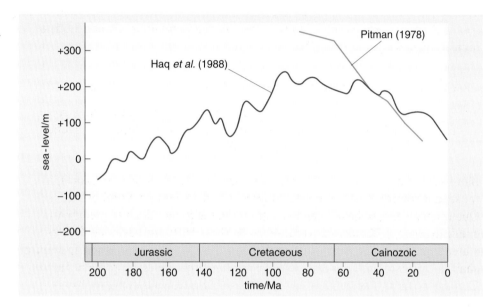

However, the shorter-term oscillations in eustatic sea-level proposed by Vail *et al.* are more problematical. Because of their smaller amplitude, they are susceptible to a greater variety of alternative local explanations, such as variations in local rates of subsidence and/or sediment supply. Moreover, the problems of precision in correlation that were mentioned earlier can become overwhelming when the limit of resolution of the available methods exceeds the duration of the cycles themselves, as with the highest-frequency cycles.

○ (a) Eight ammonite zones are recognized in the Tethyan region for the Aptian Stage. By reference to Figure 1.25, how long, on average, was each zone? (b) The combined duration of the last major glaciation and the preceding interglacial of the Pleistocene was *c.* 120 000 years. Approximately, how many combined glacial/interglacial intervals of this time-span could be fitted within the mean duration of one of the ammonite zones in (a)?

● (a) In Figure 1.25, the estimated duration of the Aptian Age is shown to be 8.8 Ma. On average, each ammonite zone thus represents 1.1 Ma. (b) Approximately nine of the combined glacial/interglacial intervals could be accommodated within the time represented by each ammonite zone.

As it happens, some geologists have suggested that the late Aptian, in particular, was a relatively cool episode, possibly even associated with minor glaciations (as you will see in Chapter 5). In this case, it would not be possible, using the currently available ammonite zones for correlation, to test whether cycles of this order were synchronous between regions.

Since the advent of sequence stratigraphy, the widely exposed (and often relatively undisturbed) marine successions of the Cretaceous have become popular targets for application of this method, although a spectrum of current opinion exists over the causes of the sedimentary cyclicity observed. Often the debate has become polarized between eustatic versus local tectonic explanations, because of the initial emphasis on the former in the seminal work of Vail *et al.* and the readily visible evidence for the latter that can be discovered through field studies. By contrast, the possible role of climate, through its effects on rates of sediment production and transport, has perhaps too often been overlooked in the past, although its importance has been recognized in relation to some kinds of high-frequency cyclicity.

○ Can you recall an example of high-frequency cyclicity from Chapter 1?

● The discussion of cyclostratigraphy (Section 1.3.1) noted the likely role of climatic oscillations in producing widespread minor sedimentary cyclicity, such as that seen in the Lower Chalk (Figure 1.28).

Short-term glacio-eustatic oscillations (with amplitudes of 50 m or more), like those that controlled sequence development in the icehouse world of the Quaternary, seem to have been exceptional in the predominantly greenhouse conditions of the Cretaceous (Figure 3.11). By default, the basic increments of water depths that could host accumulating sediment (accommodation space) for shallow marine sediments tended to be of relatively modest amplitude. Hence, Cretaceous marine sequences formed on shallow shelves and platforms are frequently found to have been built up over short time intervals through the stacking of numerous shallowing-upward sedimentary cycles (parasequences) of low amplitude (up to around 10 m) (Figure 3.12).

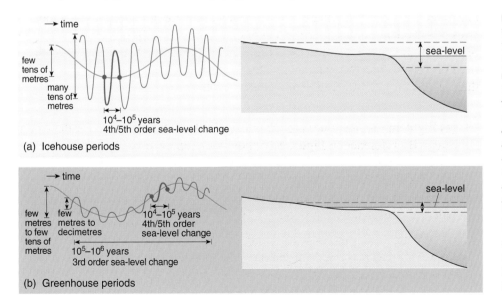

Figure 3.11 Characteristic differences in the amplitude of short-term oscillations in sea-level between: (a) icehouse, and (b) greenhouse conditions. (a) Lower mean sea-level under icehouse conditions is accompanied by high amplitude, short-term glacio-eustatic oscillations. The upper parts of continental slopes, continental shelves and coastal plains are thus subjected to frequent flooding and subaerial exposure, resulting in relatively thick sedimentary cycles. (b) In contrast, high mean sea-level during greenhouse periods is accompanied by low amplitude, short-term changes in sea-level. This circumstance results in extensive epeiric seas covering platforms and low-angle ramps, on which relatively thin sedimentary cycles are deposited. *(Wilson, 1998.)*

Figure 3.12 (a) A carbonate platform succession of Santonian age in the southern Central Pyrenees, showing the relatively low-amplitude minor sedimentary cyclicity typical of such Cretaceous successions. (b) Detail of one of the small cycles from the middle part of (a): fine-grained deposits with abundant rudists in the recessed lower part pass upwards to coarser-grained bioclastic limestones in the more resistant upper part, indicating shallowing deposition. *(Peter Skelton, Open University.)*

(a)

(b)

It is unclear how these small greenhouse cycles were generated, although orbitally forced climatic cyclicity is one possibility. Besides its effect on sediment supply, periodic climatic warming can also be expected to have had a small effect on global sea-level, simply through the thermal expansion of surface waters (in some instances, possibly augmented by the melting of polar continental ice).

○ Approximately 20 cycles make up the full platform succession shown in Figure 3.12a, which is entirely contained within the Santonian. By reference to the absolute time-scale shown in Figure 1.25, what is the maximum value for the mean formation time for each cycle?

● According to Figure 1.25, the duration of the Santonian Age was 2.3 Ma. Hence, each cycle, on average, could have taken up to 115 000 years to form.

Periodicity of this order is consistent with, but of course does not prove, orbital forcing (Milankovich cyclicity). At least for the highest-frequency cycles of this sort, the relatively much slower effects of changes in sea-floor spreading rates and vertical lithospheric movements in response to processes within the mantle can be discounted as causes. However, the earliest phases of Pyrenean tectonics had already commenced by this time (Section 2.2.3), so some local tectonic input to the pattern cannot be excluded in this instance.

Where sedimentation on tectonically stable foundations was relatively undisturbed, especially in inner platform settings, it is possible to identify likely cases of orbital forcing by analysing the periodic variations in the amplitude of the cycles. For example, Italian geologists have undertaken a detailed study of 46 m-scale cycles in the Lower Cretaceous (Valanginian–Hauterivian) of the southern Apennines. Each cycle commences with sub-tidal facies and finishes with features indicative of subaerial emergence, testifying to repeated oscillations of water depth (d'Argenio *et al.*, 1997) or variation in sediment supply. They found that these fundamental cycles were hierarchically organized into 'bundles' within 'superbundles', with regular frequencies. Given the duration of the whole succession estimated from biostratigraphic data, d'Argenio and his colleagues were able to recognize several orders of periodicity ranging from tens to hundreds of thousands of years and corresponding with various of the expected Milankovich frequencies (Section 1.3.1). Such a correspondence is unlikely to be due to chance alone, and hints strongly at orbital forcing.

Hence, at both the longest (first-order) and shortest time-scales (at least for some parasequences), there seems to be reasonable evidence in favour of eustatic changes in sea-level. It is over the intervening sequences in the hierarchy that most room for debate persists, although in the end the answer for many successions is likely to be a complex, perhaps often unresolvable, amalgam of all the various causes of relative sea-level change discussed above.

3.3 Summary

- Local bathymetric changes recorded in sedimentary successions may reflect any combination of eustatic change or regional effects (local tectonics or sediment supply).

- One way of estimating eustatic change is to use the proportional area of continental crust covered by marine deposits at different times as a proxy for global sea-level, calibrated by reference to a hypsometric curve of elevation versus percentage of area concerned. However, hypsometric curves vary between continents, and between different times, because of the varying effects of mountain formation (among other influences). Therefore, large errors may arise in estimates of past eustatic levels. Nevertheless, there is broad agreement that maximum levels (post-Palaeozoic) were reached between 90 Ma and 80 Ma ago.

- A related approach involves correlating cycles of transgression and regression in areas judged to have been stable, and calibrating the eustatic levels achieved by summing the present elevations of the deposits concerned and the inferred depths at the time of deposition. This approach likewise shows an overall eustatic rise through to the Turonian, although the pattern thereafter remains unclear.

- One problem for the two above approaches is the differential vertical movement of the lithosphere due to processes within the mantle, such as the sinking of old subducted slabs and the rise of hot plumes, both of which can now be imaged by seismic tomography. Examples include the subsidence, followed by rebound, of North America and eastern Central Australia as they passed over old sinking slabs at different times in the Cretaceous, and the uplift of much of southern Africa above a plume. Because of such processes, which can affect both continental and oceanic lithosphere alike, we cannot expect to find a wholly consistent long-term eustatic signal in all regions.

- An alternative approach involves modelling global changes in mid-ocean ridge volume and hence the water displaced from the ocean basins. This confirms the pattern noted earlier of a eustatic maximum between 90 Ma and 80 Ma ago.

- Sequence stratigraphy, based originally on seismic reflection profiles, allows subdivision of the sedimentary record into a series of partially unconformity-bounded sequences, each showing a landward-thinning wedge. Reflectors within each wedge show onlap at the base and offlap at the top, thus recording a cycle of *relative* sea-level change, involving transgression followed by regression. A hierarchy of such sequences within sequences can be observed, down to the level of individual cycles of shallowing-upward deposition (parasequences).

- Early claims that sequences of relative sea-level change can be widely correlated to reveal a global sea-level curve have proved controversial, with tectonic and climatic influences postulated as alternative explanations for the observed sequences. In many cases, especially for the shorter-term cycles, the available means of correlation lack the precision to resolve the issue.

- Not in dispute for the Cretaceous is the long-term, first-order eustatic curve peaking at over 200 m above present sea-level during the Late Cretaceous (most probably in the Turonian).

- At the other end of the hierarchy, relatively thin (up to 10 m) parasequences are typical of Cretaceous shallow-marine successions. These contrast with the thicker successions typical of icehouse times, due to large amplitude glacio-eustatic oscillations, and reflect the limited increments of accommodation space usually provided by the minor sea-level oscillations associated with the predominantly greenhouse conditions of the period. How these minor cycles were caused remains unclear, although some at least appear to reflect orbitally forced cycles of climate change operating in the range of tens to hundreds of thousands of years.

3.4 References

D'ARGENIO, B., FERRERI, V., AMODIO, S. AND PELOSI, N. (1997) 'Hierarchy of high-frequency orbital cycles in Cretaceous carbonate platform strata', *Sedimentary Geology*, **113**, 169–193.

HANCOCK, J. M. AND KAUFFMAN, E. G. (1979) 'The great transgressions of the late Cretaceous', *J. Geol. Soc. Lond.*, **136**, 175–186.

HAQ, B. U., HARDENBOL, J. AND VAIL, P. R. (1988) 'Mesozoic and Cenozoic chronostratigraphy and cycles of sea-level change', pp. 71–108 in WILGUS, C. K., HASTINGS, B. S., KENDALL, C. G. ST. C., POSAMENTIER, H. W., ROSS, C. A. AND VAN WAGONER, J. C. (eds), *Sea-level changes: an integrated approach*, SEPM Special Publ. No. 42, pp. 71–108.

HAYS, J. D. AND PITMAN, W. C. III (1973) 'Lithospheric plate motion, sea level changes and climatic and ecological consequences', *Nature*, **246**, 16–22.

MATSUMOTO, T. (1980) 'Inter-regional correlation of transgressions and regressions in the Cretaceous Period', *Cretaceous Research*, **1**, 259–373.

PITMAN, W. C. III (1978) 'Relationship between eustacy and stratigraphic sequences of passive margins', *Geol. Soc. Am. Bull*, **89**, 1389–1403.

VAIL, P. R., MITCHUM, R. M. JR AND THOMPSON, S. III (1977a) 'Seismic stratigraphy and global changes of sea level, Part 3: relative changes of sea level from coastal onlap', pp. 63–81 in PAYTON, C. E. (ed.) *Seismic stratigraphy – applications to hydrocarbon exploration, AAPG Memoir,* **26**.

VAIL, P. R., MITCHUM, R. M. JR. AND THOMPSON, S. III (1977b) 'Seismic stratigraphy and global changes of sea level, Part 4: global cycles of relative changes of sea level', pp. 83–97 in PAYTON, C. E. (ed.) *Seismic stratigraphy — applications to hydrocarbon exploration, AAPG Memoir,* **26**.

WATTS, A. B. (1982) 'Tectonic subsidence, flexure and global changes of sea level', *Nature*, **297**, 469–474.

4 Changing climate and biota

Bob Spicer

The overwhelmingly predominant source of energy that drives processes, including life, at the Earth's surface is the Sun. At any given time, the total amount of energy arriving from the Sun is virtually balanced by that returned to space, either by direct reflection (about 30% on average, today) or by re-radiation of energy absorbed by the Earth and its atmosphere. Relatively minute contributions to the budget also come from the Earth's internal heat as well as chemical energy either stored in or released from sedimentary deposits. An approximate equilibrium is maintained because the amount of energy radiated by the Earth–atmosphere system depends upon its mean temperature. Thus, when more solar energy is absorbed by the Earth–atmosphere system, the latter warms up until it is re-radiating an equivalent amount, and likewise when less is absorbed, cooling reduces the amount radiated.

The low latitudes, between the Tropics, receive the largest doses of solar radiation per unit area, because they most directly face the Sun, while the increasingly oblique incidence of the Sun's rays towards the poles reduces the intensity of solar energy received there to a minimum. Hence, today we get palm trees and pina colada in the former and penguins and polar bears in the latter. Yet the energy is re-radiated somewhat less unevenly from the top of the atmosphere (Figure 4.1), as temperatures at the top of the cloud cover (the source for most of the energy re-radiated at longer wavelengths) vary little with latitude. Given the overall energy balance, that means low latitudes actually re-radiate less energy than they receive from the Sun, while high latitudes do the opposite. Hence to balance the books, so to speak, energy is conveyed from low to high latitudes via the atmosphere and oceans, in combination. This polar heat transport is a fundamental determinant of climate.

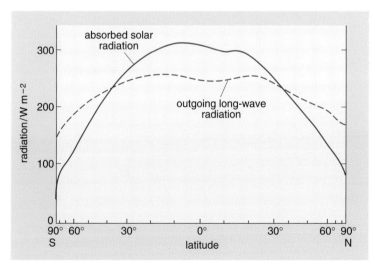

Figure 4.1 Variation with latitude of the solar radiation absorbed by the Earth–atmosphere system (solid curve) and outgoing long-wave radiation lost to space (dashed curve). Values are averaged over the year and scaled according to the area of the Earth's surface in different latitude bands. *(Van der Haar and Suomi, 1971.)*

Variation in the amount of solar radiation received by the Earth, due to periodic orbital changes as well as sunspot cycles, plays an important role in forcing climate change over time-scales ranging from decades to hundreds of thousands of years (Section 1.3.1). We have no compelling reason, however, to attribute longer-term changes such as those shown in Figure 1.23 — including the Cretaceous greenhouse episode — to such external forcing. Analogy with other stars observed today at different stages of development suggests that the Sun would have been about 25–30% less luminous when the Earth formed, and that its luminosity would gradually have increased to its present level. Hence in the

Cretaceous it should have been 1–2% less luminous than today. To understand the Earth's longer-term climatic fluctuations, then, we need to look at changes within the Earth–atmosphere–ocean system that could have modulated its equilibrium temperature and polar heat transport over time. We will lay the groundwork for this investigation — the evidence and its interpretation — in this and the next Chapter, and return to consider what lay behind these changes in Part 2.

4.1 The polar forests

One feature of the Cretaceous world that was markedly different from today was the almost total lack of polar ice caps. Instead of icy wastes, high latitudes supported luxuriant forests teeming with life. Understanding why this was the case and being able to understand the composition and dynamics of these polar environments is key to understanding possible global warming in the future. This is because any change in global mean surface temperature is always expressed most markedly at the poles. Moreover, because polar regions are the ultimate heat sink for the planet, conditions at the poles determine, to a large extent, the rate and process of global climate change. From a biological perspective, the polar regions mark the cold end of the environmental gradient running from the equator. In addition, polar warming leads to extinction of organisms adapted to cold and the polar light regime, as they have nowhere else to go. On the other hand, plants and animals that evolve in polar regions during times of global warmth are able to move toward the equator during global cooling episodes and so have a profound influence on the global biota.

The importance of polar ecosystems during the Cretaceous cannot be underestimated. Our understanding of them, and the broader polar climate, is derived from the rock record and modelling experiments; in other words, indirect evidence. We cannot use the uniformitarian principle because in today's 'icehouse' world there are no modern analogues of such ecosystems. We will now examine just how we are able to reconstruct that ancient polar environment through exemplar case studies. We start by looking in detail at the Arctic Slope of Alaska before looking more broadly at both the Arctic and Antarctic regions.

4.1.1 Arctic Alaska today

In our introductory 'tour' of the Cretaceous in Chapter 1, our second stop (Section 1.1.2) was the northernmost part of Alaska, often referred to as the 'North Slope' or 'Arctic Slope'. The term 'slope' is derived from the fact that northwards of the east–west-running Brooks (Mountain) Range the land surface slopes away northwards towards the Arctic Ocean. Today, this terrain is characterized by gently rolling foothills just to the north of the mountains, and flat boggy ground near the coast.

Rivers predominantly drain to the north and, in places, cut through and expose the sedimentary deposits that make up the region. There are almost no trees on the Arctic Slope and those that do occur are restricted in stature and location. Occasionally, small bushes survive in sheltered areas, but the predominant vegetation is tundra (Figure 4.2). Typically tundra consists of sphagnum moss, tussocks of grasses, a variety of flowering annuals, and very low-growing forms of alder (*Alnus*) and willow (*Salix*). Although long-lived and woody, these 'trees' rarely exceed a metre in height, and often are no more than 30 cm high.

The reason this vegetation is so restricted is largely a function of climate. Although climate stations are few in this region and measurements do not span long periods of time, field measurements show that the mean annual temperature is well below freezing.

Figure 4.2 Aerial photograph of tundra showing patterned ground in the left foreground. The polygonal pattern is caused by repeating freeze–thaw cycles that form convection cells in the soil. (Bob Spicer, Open University.)

Summer daily high temperatures are often in the mid to high 20s °C and winter temperatures are as low as –50 °C. Despite its more northerly position, the annual range of temperature on the Arctic Slope is moderated by the proximity of the Arctic Ocean. In fact, the lowest recorded temperature in the United States was measured not on the Arctic Slope but south of the Brooks Range along the Dalton Highway. This gravel road runs between Fairbanks and the oilfield at Prudhoe Bay and was built in the 1970s specifically to supply the oilfield development. At Prospect Creek Camp, 135 miles north of Fairbanks, the temperature fell to –62 °C on 24 January 1971. However, the primary factor that seems to limit the vegetation is not the winter minimum but the mean annual temperature. Figure 4.3 shows the mean annual temperature (MAT) along a transect from Fairbanks in the south to Prudhoe Bay in the north.

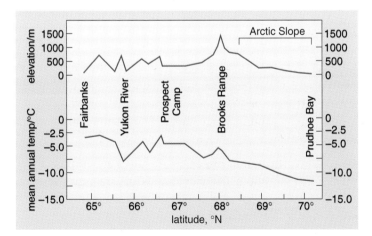

Figure 4.3 Temperature transect from Fairbanks to Prudhoe Bay. (Haugen, 1982.)

From Figure 4.3, note that the trend for declining temperature with increasing latitude is interrupted by topography. In the Yukon River region, MAT is locally depressed because the basin acts as a trap for cold air. However, in the Brooks Range, we get an unexpected rise in temperature with increasing elevation. Here mountain faces are warmed up by the strong absorption of low-angle sunlight in the summer, while adiabatic winds (descending air undergoing compression, and hence heating at lower altitudes) introduce heat into the mountain passes where the temperatures are measured. Overall, the MAT of the Arctic Slope is lower than any area to the south of the Brooks Range.

In addition to the cold, the influence of the strong polar high-pressure cell also means that the Arctic Slope region is relatively arid. However, that is not to say that the region is a desert. Again the observations are limited but at Sagwon, 350 miles north of Fairbanks on the Dalton Highway, the mean annual precipitation (MAP) has been measured at 140 mm, whereas at Bettles, south of the Brooks Range, it is 360 mm. In the mountains, the values are likely to be higher but very variable. Despite the relative drought on the Arctic Slope, the soil is mostly water-saturated in the summer because precipitation that has fallen during the winter as snow eventually melts in the summer sunlight but is unable to penetrate the surface. This is because the ground below the tundra (at a depth of a few tens of centimetres) remains frozen all year round. The permanently frozen zone, or *permafrost*, may extend several kilometres below the surface.

The depth of the soil that experiences an annual thaw controls to a large extent the type of vegetation it can support. On the Arctic Slope, the depth of the thaw is very shallow and is insufficient to support tree growth. To the south of the Brooks Range, and particularly along watercourses, the annual thaw is deeper and ribbon forests grow, consisting of the conifer white spruce *(Picea alba)*, willow *(Salix)* and balsam poplar *(Populus balsamifera)* (Figure 4.4a).

Figure 4.4 (a) Aerial photograph of ribbon forests. Such forests form along the banks of streams and rivers where the upper surface of the permafrost is at its greatest depth below the land surface. (b) Aerial photo of the Yukon River valley showing Taiga vegetation (see text). *(Bob Spicer, Open University.)*

(a)

(b)

Where the permafrost is only at a shallow depth, the black spruce *(Picea nigra)* predominates. In other areas, such as in the Yukon River valley, the permafrost is sufficiently deep that more contiguous forests are able to grow (Figure 4.4b).

Here, the forest is dominated by conifers (mostly white spruce with black spruce occupying poorer soils and areas where the permafrost is closer to the surface), but often the river margin *(riparian)* communities consist of bushes and small trees of the angiosperm species alder *(Alnus)* and willow *(Salix)*. Even further south birch trees *(Betula)* and aspen *(Populus tremuloides)* become common both along rivers, as well as in the forests themselves.

The conifer-dominated forests make up a distinctive form of Arctic vegetation known as Taiga (Figures 4.4b, 4.5). Taiga stretches across Canada, northern Scandinavia and Russia. In fact, 'Taiga' is the Russian word for these predominantly evergreen forests which typically have only small stunted trees rarely more than 10 m high.

Figure 4.5   Ground shot of Taiga showing scruffy black spruce trees. *(Gil Mull, State of Alaska Geological Survey.)*

The trunks of these trees are often not more than 15 cm in diameter at breast height (the position at which, by convention, tree girth is measured). If you were to cut across the trunk you would find that many are of great age with numerous very thin rings, reflecting the fact that annual growth is restricted to only a few months each summer. Although during much of the growing period the Sun never sets, this short growing season (typically June–September) only allows the tree the opportunity to lay down a small thickness of wood. Moreover, although temperatures may sometimes rise above 30 °C they may also plunge to near freezing for several days at a time when cold air masses spill out from over the Arctic Ocean.

The evergreen *Picea nigra* looks a rather sorry sight because its branches are short and tend to droop downwards. This architecture is effective in preventing winter snow load building up to the point where it can break the branches. In contrast, the deciduous larch *Larix*, which is a major component of Russian Taiga, sheds its leaves during the winter and in so doing reduces the area of the tree upon which snow can settle.

Not only are these trees small in stature but they may also be widely spaced.

○ Why do you think wide spacing might be advantageous to trees growing at high latitudes?

● The wide spacing allows low angle sunlight to be more effectively intercepted. This is because when the Sun is low in the sky (as it is for much of the summer) only those trees at the edge of a closely packed forest would receive direct sunlight over most of their surface; the rest would only receive sunlight at their very tips. On the other hand, widely spaced trees receive more direct sunlight, even when diffuse, over a greater proportion of their surface.

The wide spacing also allows a diverse community of annuals to grow. These herbaceous plants are often opportunists and thrive in disturbed sites. Disturbance can arise from shifting river courses, movement of the soil caused by the freeze–thaw cycle (*solifluction*; the effects of which can be seen in the patterned ground in Figure 4.2) or by fire. As surprising as it may seem, both tundra and Taiga are prone to burning during the summer months. This is because long periods of direct sunlight and low rainfall cause the vegetation to dry out sufficiently to be able to burn, and any source of ignition such as a lightning strike can start fires that may burn unchecked for several months. These fires can have the effect of killing tree seedlings or even mature trees and opening up the vegetation for colonization by weedy herbaceous (non-woody) species.

4.2 The polar light regime

We have already alluded to the long summer days and low angle sunlight of the Arctic summer. The winters are not only cold but the daily periods of direct sunlight are extremely short or even non-existent.

Figure 4.6 shows the distribution of light throughout the year at high northern latitudes. The distribution is the same in the Antarctic, but with the time of maximum daylight occurring in December instead of June. The distribution of light throughout the year at different latitudes is determined by the angle of inclination of the Earth's rotational axis from the perpendicular to the plane of its orbit around the Sun (the angle of *obliquity*). A lower (more upright) angle implies a more constant distribution, whereas a higher (more inclined) angle implies a larger summer/winter variation. Figure 4.6 uses the present-day obliquity of 23.5°, but this value is known to vary between 21.8° and 24.4° over a 41 000-year cycle and is part of the Milankovich cyclical variation in the amount and distribution of solar insolation over time (Section 1.3.1). This angle defines the position of the polar circles at a latitude of 66° and it provides a convenient definition of what we mean by 'polar' or high latitudes. For convenience, we will consider anything at a latitude of 66° or above as polar or high latitude as it defines regions of the Earth that are subject to the polar light regime where, on at least one day of the year, the Sun fails to rise above the horizon. This is not to say that there are 24 hours of darkness, but just that there is no direct sunlight and midday is marked by twilight. The higher the latitude, the darker the sky is on that day. The distinction between twilight and darkness is arbitrary.

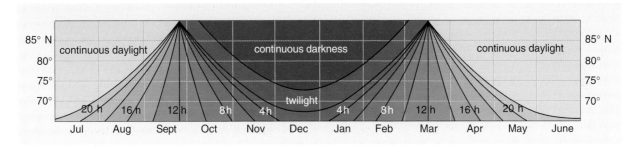

Figure 4.6   The distribution of light and darkness at high latitudes with present-day obliquity of 23.5° (see text). Numbers indicate the duration of daylight (in hours) in any 24-hour period. Beyond 22 hours, daylight is 'continuous' because, although the Sun goes below the horizon, light scatter in the atmosphere provides illumination. *(Anon., 1978.)*

○ By referring to Figure 4.6, determine what the duration of continuous twilight is at 71° latitude.

● About 1.75 months.

At northernmost Alaska (e.g., Barrow, latitude 71° 16′ N) the Sun stays below the horizon for a continuous period of 1.75 months each winter. The brightest it gets is for a period around noon each day when there is twilight. For the rest of the 24-hour period, it is dark. In the summer, however, there are about 2.5 months during which the Sun never sets but just circles the sky, dropping lower to the horizon each 'night' but rising high in the sky around noon. These long periods of summer insolation are sufficient for rapid plant growth, particularly when coupled with warm temperatures.

The modern vegetation of the Arctic thus is characterized by conifer-dominated Taiga and, where conditions are not suitable for trees to grow, the landscape supports tundra. However, in the highest latitudes, tundra is replaced by polar desert where most, or even all, of the ground surface may be devoid of plant life. What flora there is consists mostly of lichens and mosses with a few of the hardiest angiosperms sheltering as low-growing cushion plants between boulders. In polar deserts, the insulative value of the snow is replaced by its hindrance value as a reducer of light and air temperature. Wherever the snow lies deeply, it melts late and reduces the duration of the active growing season. At such extreme high latitudes, this excludes all vascular plants. As we shall now see, the vegetation of the Arctic was very different in the greenhouse world of the Cretaceous.

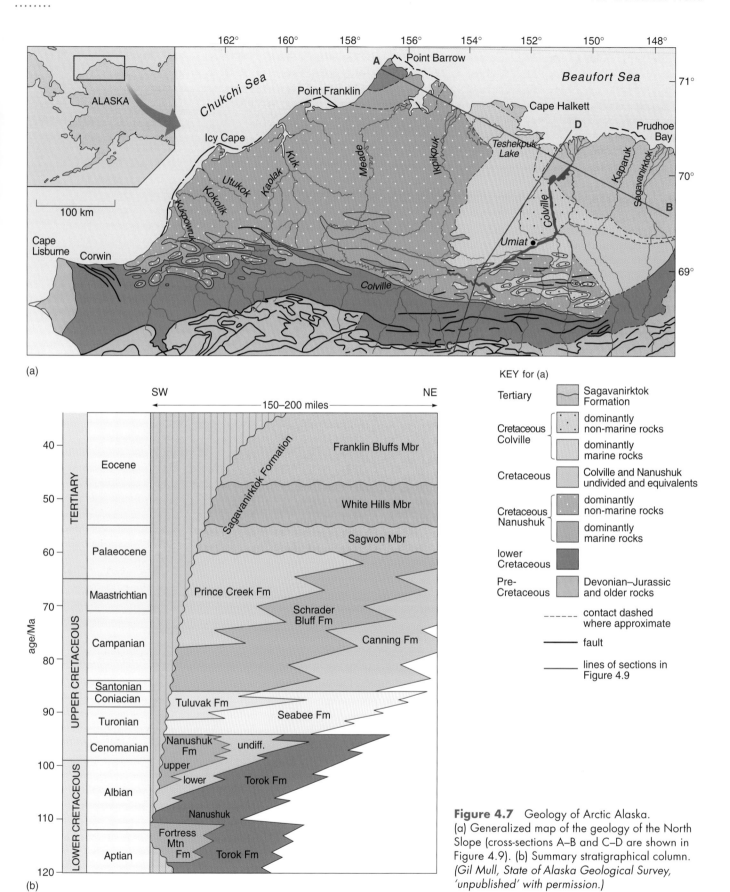

(a)

(b)

KEY for (a)

Tertiary	Sagavanirktok Formation
Cretaceous Colville	dominantly non-marine rocks
	dominantly marine rocks
Cretaceous	Colville and Nanushuk undivided and equivalents
Cretaceous Nanushuk	dominantly non-marine rocks
	dominantly marine rocks
lower Cretaceous	
Pre-Cretaceous	Devonian–Jurassic and older rocks

- - - - - contact dashed where approximate

——— fault

——— lines of sections in Figure 4.9

Figure 4.7 Geology of Arctic Alaska.
(a) Generalized map of the geology of the North Slope (cross-sections A–B and C–D are shown in Figure 4.9). (b) Summary stratigraphical column. *(Gil Mull, State of Alaska Geological Survey, 'unpublished' with permission.)*

4.3 The Cretaceous Arctic: a case study from northern Alaska

Today, the Arctic Slope of Alaska is a barren place and yet the surficial rocks contain a wealth of geological information that shows that it once supported a lush, highly productive ecosystem. That Cretaceous ecosystem has no close modern analogue because it grew in relative warmth under a polar light regime. This Section shows how information about this ecosystem was revealed, and how it may enhance our understanding of the greenhouse mode of operation of the Earth system.

The Arctic Slope of Alaska occupies an area slightly larger than the British Isles (Figure 4.7). Moreover, the surface rocks across this region are mostly of Cretaceous age and yield a huge amount of information about the Arctic environment at this critical time in the history of the Earth. This was a time that saw the close of the Mesozoic, the evolution of the flowering plants that characterize today's global vegetation, and extreme global warmth. Most importantly, because polar conditions are the most sensitive to climate change, and are a key component in defining the global heat cycle, the rock record of environmental conditions in northern Alaska throughout the Cretaceous plays a critical role in understanding the evolution of the Earth system.

4.3.1 The development of the Arctic Slope

The wealth of data on the now extinct polar forest ecosystem of northern Alaska results from the tectonic environment, as well as the climate and composition of the vegetation. The development of the Arctic Slope began with the anticlockwise rotation of a tectonic microplate known as the Arctic Alaska terrane. Much of Alaska and western North America are composed of accreted terranes (Section 2.2). These may be distinguished from each other in that each is a fault-bounded entity characterized by a distinctive stratigraphical succession or rock assemblage that differs markedly from those of adjacent or nearby terranes. The Arctic Alaska terrane rotated about a hinge near to the present position of the Mackenzie River delta (north-western Canada) and collided with a more southerly terrane, the Angayucham terrane, that today makes up part of the interior of Alaska (Figure 4.8).

Figure 4.8 Tectonic reconstruction diagrams to show the evolution of the Arctic Slope: (a) 150 Ma ago; (b) 70 Ma ago. AA indicates Arctic Alaska. *(ODSN Plate Tectonic Reconstruction Service.)*

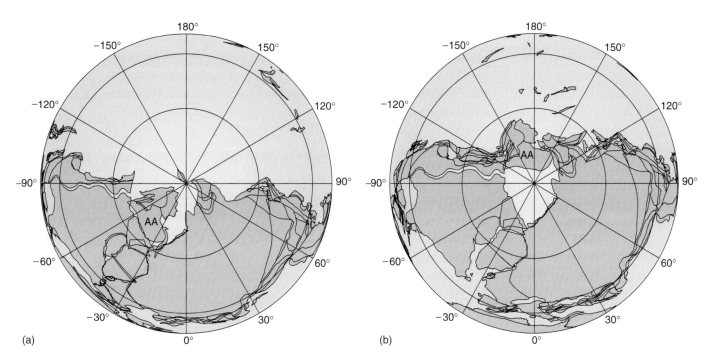

(a)

(b)

This collision gave rise to the ancestral Brooks Range. The similarity of the stratigraphical succession of the Arctic Alaska terrane to the rocks of the north-western Canada cordillera and the Arctic Sverdrup Basin in northern Canada strongly suggests that it is related to the North America Craton and is not exotic to North America. The Angayucham terrane, however, yields early Cambrian trilobites and Carboniferous (Mississippian) floras and faunas which show that it has an affinity to Siberia (Angara).

Lower Cretaceous sediments exposed within, and adjacent to, the western and central Brooks Range indicate an abrupt shift in sediment dispersal from a late Palaeozoic and early Mesozoic northern source to a Cretaceous source to the south-west. This south-western source was the young Brooks Range and it records the start of the Brooks Range orogeny. This is also the beginning of the deposition of a succession of sedimentary rocks known as the Brookian Sequence that was derived from the newly uplifted mountain belt. The composition of the Brookian Sequence sediments is therefore largely determined by the composition of the Brooks Range rocks.

The orogenic belt exposed in the western and central Brooks Range is composed of multiple sheets of mafic igneous and sedimentary rock which developed following the anticlockwise rotation of the Arctic Alaskan crustal plate. As this plate rotated, in a relative sense it moved southward, into a south-dipping subduction zone. In so doing, a stack of several thousand metres of imbricated sedimentary and igneous rocks was formed, overlain by a sheet of imbricated oceanic rock. These transported sheets or allochthons are discrete, fault-bounded stratigraphical successions consisting of several rock units that collectively have been moved and juxtaposed to adjacent allochthons by thrust faulting. Within the allochthons thrust faulting is also common, but within an allochthon formations generally have a consistency that enables the allochthon to be recognized over a wide area. Formations and stratigraphical successions in different allochthons are distinct enough to allow one allochthon to be distinguished from another. Allochthons of the Arctic Alaska terrane have stratigraphical successions that are so similar that it is likely they were all originally part of the same continuous depositional basin.

The major orogenic activity which formed the Brooks Range began during Berriasian time and continued into the Aptian, and was therefore mostly pre-Albian in age. In the south-central Brooks Range, granite plutons were intruded during Cenomanian to Turonian times. This suggests that the core of the Brooks Range was mobilized and subjected to dominantly vertical uplift beginning in the Albian and coinciding with the deposition of the sedimentary rocks that contain the oldest Cretaceous floras of the region.

4.3.2 Arctic Slope Cretaceous stratigraphical framework

Figure 4.9 diagrammatically illustrates the major sedimentary packages that make up the Arctic Slope. The Brookian Sequence sedimentary deposits unconformably overlie older deposits and fill the asymmetrical Colville Basin, prograding onto the gently dipping south flank of the Beaufort Sill (Figure 4.9b). The sequence is composed of Cretaceous and Tertiary siliciclastic sediments, which from the Albian onwards are primarily confined to the area north of the Brooks Range, but older early Cretaceous sedimentary deposits continue to depths of 10 km beneath the thrust-faulted mountain front. These older sedimentary deposits are entirely marine and exhibit features typical of turbidites

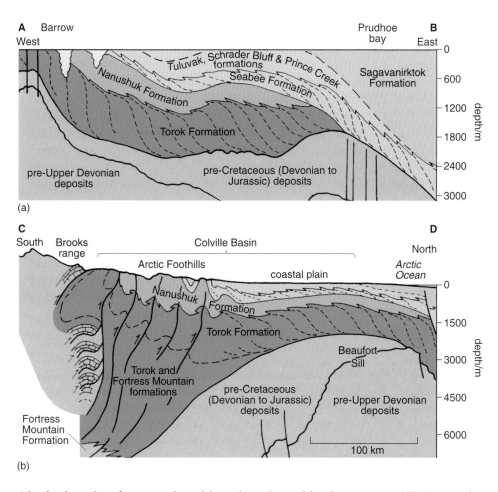

(a)

(b)

Figure 4.9 Diagrammatic vertical cross-section through the Arctic Slope geology along the transects shown in Figure 4.7: (a) east–west along transect A–B; (b) north–south along transect C–D. Note that the vertical scale is greatly exaggerated. All the sediments overlying the pre-Cretaceous rocks make up the Brookian sequence. *(Bird, 1987.)*

(rhythmic units of greywacke with tool marks and load structures, siltstone and shale) together with manganese nodules. This suggests deep-water conditions and the development of submarine fans. In Hauterivian–Barremian age rocks, the sediment was clearly derived from a source to the south.

In the deeper parts of the basin, a continuous succession shows the transition from turbidite through shallow marine to non-marine deposition. This transition took place during the Albian and is documented in the Torok Formation by the change from shales in the lower part containing a restricted pyritized radiolarian fauna, to shales in the upper part containing abundant calcareous microfossils.

○ How might this tell us the water depth was decreasing?

● The presence of pyrite suggests low oxygen conditions often associated with deep water in the Cretaceous, whereas the subsequent presence of calcareous fossils shows that the sediment surface was above the carbonate compensation depth (Section 1.1.1). This is not a foolproof interpretation as anoxia can occur at shallow depths, but in this instance such evidence ties in with other indicators of shallowing.

In the central Arctic Slope area, the bulk of the Torok Formation is dated as Albian based on marine megafossils. The Torok marine shales grade upwards into the late Albian to Cenomanian Nanushuk Formation. This is a passive margin deltaic sequence which is at least 3444 m thick in outcrops at Corwin Bluff and thins eastward to a pinchout edge close to the present Colville Delta. It crops out in the Northern Foothills and is present as subsurface rocks over much of the western and central Arctic Slope.

The lower part of the Nanushuk Formation consists of a thick succession of intertonguing shallow marine sandstone, shale and siltstone which grades seaward into predominantly prodelta shale and siltstones of the Lower Cretaceous Torok Formation. Seismic and borehole data indicate the Nanushuk and Torok Formations are laterally equivalent units and some seismic reflectors are continuous from topset to bottomset positions. The topset seismic reflectors include alluvial deltaic and shelf deposits of the Nanushuk Formation and upper part of the Torok Formation. The foreset reflectors represent the middle part of the Torok Formation while the bottomset reflectors are the basin floor deposits of the lower part of the Torok Formation.

The Nanushuk and Torok units apparently prograded east and north-east across the Colville Basin, suggesting a major sediment source to the south-west. This prograding succession is interpreted to have been formed by a large river-dominated deltaic complex known as the Corwin Delta, comparable in size to the modern Mississippi Delta. Compositional variation in the lithic components of Nanushuk Formation sandstones also suggests an additional source area to the south. This source supplied an eastern deltaic lobe, the Umiat Delta, which exerted little influence over the Corwin Delta except along the south side of the basin.

Within the Nanushuk Formation, basal marine sandstone predominates and grades upwards into a predominantly fluvial succession of fine-grained sandstones, mudstones and coals. These non-marine rocks are not present everywhere and are totally lacking in the eastern part of the Arctic Slope where correlative rocks are entirely turbidites deposited north-east of the prograding deltaic complex. We shall be examining the non-marine units in some detail later.

Figure 4.10 Chart showing the stratigraphical relationships of the major geological units comprising the surface rocks of the Alaskan Arctic Slope.

The generalized Cretaceous stratigraphy following Mull *et al.* (2003) is shown in Figure 4.10.

ERA	Period	Epoch	Formation		approx. thickness range/m	dominant lithologies and depositional environments
CAINOZOIC	Quaternary		Gubik		90–180	clay, silt, sand – marine and fluvial
	Tertiary		Sagavanirktok		0–1140	sandstone, conglomerate, siltstone, mudstone, coal – delta plain and alluvial plain
MESOZOIC	Cretaceous	Late	Prince Creek Schrader Bluff		0–345 to 1500?	sandstone, siltstone, mudstone, coal, bentonite, basal fissile shale – delta and alluvial plain, offshore transgressive lower part
			Seabee			
		Early	Nanushuk		0–3300+	lithic and quartzose sandstone, siltstone, mudstone, coal, bentonite – delta plain, delta front, alluvial plain
						siltstone, wacke, conglomerate, mudstone, minor coal – turbidite fore deep and fluvial
			Fortress Mountain Torok		1000–3000 310–5700	mudstone, siltstone, some wacke – offshore neritic shelf, slope, basin

4.3.3 Palaeogeographical setting

Today, the northernmost part of the Arctic Slope is at a latitude of 71° 16´ N and before the concept of plate tectonics and plate movement the fossil floras of the region posed a dilemma. The abundance of leaf fossils and, in particular, those representing warm conditions (thermophilic taxa), suggested that the climate of the region was much warmer than that of the present day. Rather than suggesting Alaska was further south, this was taken as evidence that the polar regions, and therefore global climate, was considerably warmer in the past. Now that we know crustal plates can move, we need to establish the palaeoposition of the Arctic Slope before we can interpret the global significance of its palaeoclimatic signal.

Early palaeomagnetic work positioned the Arctic Slope during the period of time when the Nanushuk Formation was deposited (Albian–Cenomanian) at between 80–85° N. Subsequently, however, direct measurement of the palaeomagnetism in Nanushuk Formation rocks indicated a lower latitude of 74.5 ± 7.5°. Note however that this is still some 5° poleward of its present position. As more data have been accumulated, the palaeogeographical position has become more secure and a palaeolatitude of approximately 75° N in the early Cenomanian now seems robust. However, during the rest of the Cretaceous the Arctic Slope appears to have moved closer to the palaeomagnetic North Pole, and by late Maastrichtian times northernmost Alaska was at approximately 85° N.

Palaeomagnetic studies can only provide positions relative to the magnetic pole that may, or may not, be the same as the rotational pole. For living organisms and the ocean/climate system, the position of the magnetic pole has only minor significance; it is the rotational pole that is important. Although theories about the Earth's interior and the generation of the Earth's magnetic field make a link between the rotational and magnetic poles, the difference in them can be quite considerable and constantly changes. For example, in 1994 the magnetic North Pole was measured by the Geological Survey of Canada to be offset from the rotational pole by 12.7° latitude at 78.3° N and 104.0° W. Moreover, this was 150 km north-west of the position where it had been located a decade earlier. To test the congruence, or otherwise, of the average rotational and magnetic poles during the Cretaceous, Ann Lottes of the University of Chicago examined the distribution of Cretaceous climatically sensitive sediments such as coals and evaporites. She found that the best fit for the distributions, indicative of the position of the rotational pole, was within 4° of the magnetic pole in the Maastrichtian. This separation is less than the errors of most palaeomagnetic studies, so for most purposes we can regard the two poles as being essentially congruent. There is certainly no evidence for a sustained separation and so the fossil biota of northern Alaska therefore represents a truly polar ecosystem.

4.3.4 The Cretaceous flora of the Arctic Slope

The Arctic Slope Nanushuk, Tuluvak, and Prince Creek formations yield an abundance of plant fossils ranging in age from Albian to Maastrichtian. Large-scale plant remains (the *megaflora* — leaves, wood, etc.) are restricted to non-marine lake-bed shales, fluvial sandstones, and ironstone nodules and sheets within bentonitic clays. All these sediments represent the uppermost alluvial floodplain deposits associated with the Cretaceous Corwin and Umiat Deltas. Pollen and spores are abundant in almost all sediments, the exceptions being the coarser-grained sandstones and the bentonites. Taken together, these plant remains and their association with different sedimentary environments help us to build up a picture of the composition, structure, and ecological development of the Cretaceous polar forests. Inevitably, the kinds of plants that made up these forests are less familiar to us than the plants that live today. All of the Cretaceous species are now extinct, as are most, but not all, of the genera (plural of *genus*). However, with the exception of the seed ferns and the bennettitales, all the major groups that exist today were present in the Cretaceous. Box 4.1 provides an overview of the evolution of the major plant groups.

Box 4.1 The evolution of major plant groups

In Figure 4.11, you will see a simple 'balloon' diagram that shows the major past and present plant groups. The point at which each balloon starts marks the first fossil evidence of the group and the end of the balloon indicates its apparent extinction. The width of the balloon represents how 'important' the group was at any given time, where importance is a combination of species diversity and abundance as measured by the number of individuals and their biomass. This, of course, is a very imprecise measure but in qualitative terms it provides us with an insight to past plant life.

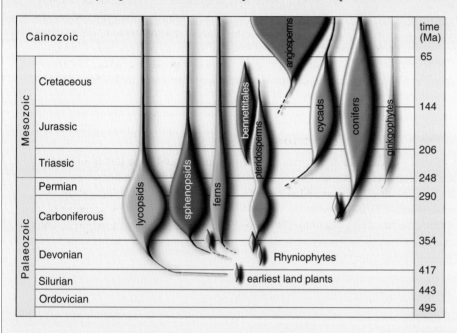

Figure 4.11 Balloon diagram of major plant group evolution.

The lycopsids

This group of plants reproduced by means of spores that were produced in kidney-shaped sporangia attached to the 'upper' side of the microphylls. The helically arranged microphylls and kidney-shaped sporangia borne on the upper side of the microphylls along the sides of the stem are features that characterize all the lycophytes or lycopsids. Today, these plants are represented by the clubmosses *Lycopodium* and *Selaginella*. Figure 4.11 shows that the lycopsids began to diversify in the Late Devonian and by the Late Carboniferous they dominated much of the world's vegetation. By the Cretaceous, however, they were smaller plants, fewer in number and less diverse. However, lycopsid spores are quite commonly preserved in Alaskan Cretaceous rocks and although their stems and leaves are hardly ever preserved they must have been present in the polar forests.

The sphenopsids

Another group that played a major role in the Carboniferous is the sphenopsids. The modern representatives of this group are the 'horsetails' or 'scouring rushes' which have the Latin generic name *Equisetum*. Today, these plants are rarely more than a few metres tall and this seems to have also been the case in the Cretaceous Arctic forests. *Equisetites* (the fossil form of *Equisetum*-like sphenopsids) is extremely abundant in most Cretaceous sediments on the Arctic Slope. Sphenopsids were apparently some of the first plants to colonize river banks and any disturbed ground and before the appearance of grasses in the mid–Late Cretaceous formed much of the ground cover together with ferns. Like ferns, all of the sphenopsids, past and present, reproduce by means of spores.

The ferns

The ferns have a long evolutionary history stretching back into the Late Devonian. Their spores are usually (but not always) borne in sporangia situated on the underside of fronds. Ferns come in a variety of shapes and sizes; some have a small stem or rhizome that sits on or under the soil surface, while some so-called tree ferns produce a trunk made up of intertwined roots and stems. Some have extremely elongated fronds that scramble over and climb up other plants, while some are aquatic and float on the surface of ponds and lakes. The tree habit dates back to the Late Devonian but has evolved many times over in different lineages of ferns.

The seed fern or pteridosperms

Sometimes it is very difficult to distinguish true ferns from the seed ferns just from the foliage. The difference, as the name suggests, is that seed ferns, otherwise known as pteridosperms, reproduced by seeds and not spores. Seed ferns were particularly abundant in the Late Carboniferous, Triassic and Jurassic, but during the Cretaceous they succumbed to competition from the flowering plants (angiosperms) and eventually became extinct in the Late Cretaceous.

The cycads

The cycads are a group that has survived at least from Permian times and probably had their origins in the Carboniferous. They reproduce by means of seeds, mostly borne on modified fronds grouped together to form cones. The only living exception is the most primitive living example, *Cycas*, where modified fronds bearing seeds are separate from one another and do not form a cone. Plants are either female or male; the male plants always producing pollen from cones. Modern cycads typically have a squat trunk bearing whorls of leathery evergreen fronds. Today, they are found only in regions where frosts are non-existent or not severe, and each genus occurs on several different continents. This fragmented distribution indicates that the modern plants are the remnants of a previously more widespread distribution; i.e. they are relictual. This is confirmed by fossil evidence that shows that the group was distributed worldwide in the Mesozoic, and even made up a significant component of polar vegetation. However, unlike their modern counterpart these polar cycads were deciduous and had vine-like stems.

The bennetitales

This group of cycad-like plants were restricted to the Mesozoic. They produced seeds in cone-like structures sometimes surrounded by protective scales or bracts. As with the cycads, male and female plants were sometimes separate but they could also be bisexual and the reproductive structures sometimes even had both male and female parts arranged in a similar way to that of flowering plants. The fossil foliage of cycads and bennetitales are often very difficult to distinguish between and distinctions are usually made on the basis of microscopic details of the epidermal cell walls.

The ginkgophytes

This group of plants is only represented by one species today, *Ginkgo biloba* or Maidenhair tree. In the Mesozoic, this tree was global in distribution but was particularly abundant in polar forests. Today, it only has a presumed natural occurrence in China. The tree is often planted in gardens or as a street tree but usually only the male is used in this way because the female tree produces fleshy fruits that have the odour of rancid butter when ripe.

The conifers

The conifers originated in the Late Carboniferous and rose to dominate many plant communities in the Mesozoic. They are, of course, still a highly successful group and thrive in the forests across North America and Eurasia. Most conifers are woody trees although at least one species is a small parasite living in the branches of other trees. Most conifers reproduce by means of seeds borne in cones, but some cones have been modified to be fleshy, berry-like structures attractive to animals as food. Juniper is one such plant. A typical conifer female cone has seeds borne on a woody scale associated with a tongue-like bract. These bracts can be seen in the picture of a Douglas Fir cone (inset). Pollen is produced in a male cone that is not as massive as the female cone.

Most modern families of conifers arose in the Mesozoic but one abundant conifer family that disappeared at the end of the Cretaceous was the Cheirolepidiaceae. These dominated the vegetation at low latitudes (<40°) and were adapted to seasonal drought. They had very thick cuticles, small leaves, and were deciduous, dropping their photosynthetic shoots when conditions became too dry. You can find abundant fossil cheirolepidiaceous conifers in the Lower Cretaceous Wealden Beds of Southern England.

The angiosperms

Angiosperms are flowering plants. It is perhaps easier to provide examples of angiosperms than it is to define them. They may be herbaceous like buttercups or tulips, or woody like alders, willows or oaks. Palm trees are also angiosperms which, like tulips and grasses, only produce one seed leaf or cotyledon when they germinate. They are therefore called 'monocots' as distinct from buttercups and willows which are called 'dicots'. Some biochemical components of angiosperms originated in the Carboniferous but morphological features typical of angiosperms began appearing in the Triassic. However, it is not until the Cretaceous that we see these features coming together and conferring significant evolutionary advantage.

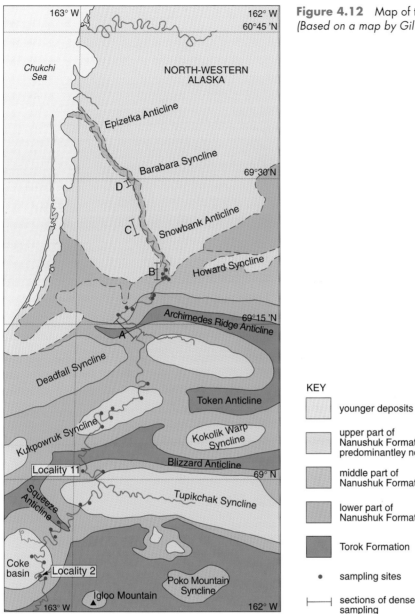

Figure 4.12 Map of the area around the Kukpowruk River.
(Based on a map by Gil Mull, State of Alaska Geological Survey.)

It is not possible here to examine all the fossil localities across the Arctic Slope.
Instead, we will look in detail at several plant-yielding sections which represent a
range of ancient depositional environments and communities. We will begin with
those exposed along the banks of the Kukpowruk River at the western end of the
Arctic Slope.

The Kukpowruk River transects a series of intertonguing marine and non-marine
sediments of the Torok (marine), and Nanushuk (both marine and non-marine)
formations (Figure 4.12). The minor folding of the region leads to repeated
exposure of the various formations present, affording the opportunity to examine
repeated associations of plant assemblages and sedimentary facies. The shallow
marine shales and sandstones of the Nanushuk Formation overlie, but also
intertongue with, the Torok Formation. Dominant rock types are shale, siltstone,
sandstone and claystone forming a package that is 1500 m thick in the south-west
to 600 m thick in the north-east.

Taken together, the Torok–Nanushuk Formations comprise a package of sediments representing an overall shallowing and transformation from shallow marine to fully non-marine environments as the Corwin Delta complex prograded.

A typical section of the upper part of the Nanushuk Formation along the Kukpowruk River is shown in Figure 4.13. For now, do not worry about what kinds of plants the different names in the units listed below represent — we will deal with that later. Figure 4.14 is a graphic log of this section; on the left-hand side of the sedimentary log are numbers that refer to beds grouped together into informal units that can be described as follows:

Figure 4.13 Typical section through the upper part of the Nanushuk Formation along the Kukpowruk River. Numbered bars refer to the units shown in Figure 4.14. The cliff is just over 20 m high. *(Bob Spicer, Open University.)*

Locality 2: 68° 49′ 11″ N, 162° 10′ 38″ W

Unit 1. The base of the documented section; this consists of an olive-grey indurated siltstone, 11m thick, within a carbonaceous mudstone.

Unit 2. Carbonaceous mudstone, rich in small *Podozamites* leaves and rare *Pityophyllum* leaves.

Unit 3. A series of yellow-weathering grey siltstones and fine-grained sandstones with a crumbly texture containing abundant *Equisetites* rhizomes, *Podozamites* fragments, wood pieces representing small branches and small compressed logs preserved parallel to bedding. Occasional small tree stumps, a few decimetres in diameter, occur normal to bedding and are rooted in the underlying mudstone.

Unit 4. A poor quality coal, containing numerous ironstone nodules and thin sandy layers, abundant logs, branchwood and *Podozamites* leaves. *Pityophyllum* is also present but less abundant.

Unit 5. An olive-grey siltstone, containing numerous nodular and sheet form ironstone concretions.

Unit 6. Another thin coal within which is rooted a small (approximately 20 cm diameter) upright tree trunk.

Unit 7. The tree trunk (shown in Figure 4.14) rooted in Unit 6 protruded approximately 1 m into this coarsening-upward, yellow/orange weathering grey siltstone rich in *Podozamites* fragments.

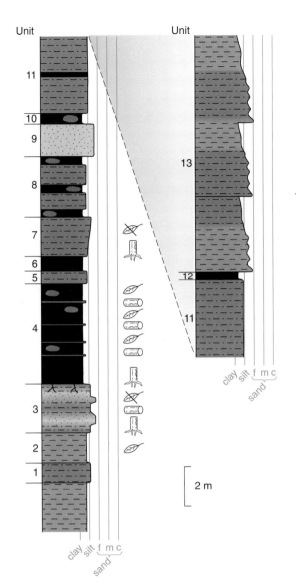

Figure 4.14 Sedimentary log of the Kukpowruk River Corwin Formation. *(Spicer and Herman, 2001.)*

KEY

sandstone

siltstone

mudstone

coal with iron nodules

λ   rootlets

upright trees

logs

plant fossils

plant fragments

Unit 8. A sequence of three thin (<0.51m) poor quality coals, containing some ironstone concretions, separated by olive-grey siltstones.

Unit 9. A grey, weathering to olive-grey/yellow, fine-grained sandstone.

Unit 10. A poor quality coal containing ironstone nodules.

Unit 11. An olive-grey to brown siltstone with a thin, poorly developed coal in the lower third, above which there are occasional (up to four) white laterally discontinuous, indurated bands.

Unit 12. A thin very poor quality siliciclastic-rich coal.

Unit 13. A yellow-weathering grey well-bedded series of siltstones and fine-grained sandstones forming a set of three fining-upward cycles (estimated to be up to 121m thick).

At the bend in the river, a high angle normal fault, of unknown throw, brings down blocky, cross-bedded, olive-grey siltstones and sandstones rich in *Podozamites*, *Ginkgo*, rare *Birisia*, *Equisetites*, *Pityophyllum*, and compressed logs. *Ginkgo* predominates in the sandier beds, while *Podozamites* is most common in the siltstones. These beds are moderately contorted and overlie, apparently with an erosional contact, yellow/grey sandstones and siltstones that dip at approximately 45° to the west.

The section is interpreted to represent a well-vegetated floodplain between, but close to, small rivers. Several environments are represented here. The most obvious feature is the abundance of coal beds that represent *mires*. The word 'mire' is a general term to describe any wet non-marine environment in which plant material accumulates to an extent that the amount of organic matter far exceeds that of siliciclastics. This can occur in a low-lying depression that fills with water and is colonized by plants (in which case we refer to it as a *swamp*), or it may be a *raised bog* that is maintained in a water-saturated state by rainfall alone. A swamp may develop into a raised mire over time if organic accumulation is facilitated by a climate conducive to growth and sufficient rainfall distributed all year round to prevent the bog drying out. As the organic matter accumulates, the surface of the mire rises above the local water table and eventually even above the raised water levels associated with occasional flooding. At this point, no siliciclastic material will enter the mire by water transport and the coal becomes relatively ash free (and therefore desirable from an economic point of view). The only sources of inorganic material are those derived from the plant material itself and airfall material associated with wind-blown dust or volcanic eruptions.

○ Based on the description of mire type and development just given, which type of mire do you think the coals in the Kukpowruk section (Figure 4.13) represent?

● In this case, the coals are of poor quality with a high inorganic content which suggests low-lying swamps rather than raised bogs.

In addition to the swamps represented by the coals, floodplain pond and crevasse splay depositional environments are also represented. The carbonaceous shales (Units 1 and 2) represent low-energy situations but with a high inorganic input relative to the accumulation of organic material. Some plant remains were being preserved, therefore wholesale decay is not an explanation for the lack of organic matter. Instead, the rate of influx of mud was high relative to that of plant material. A floodplain pond depression that was subject to continuous or frequent inundation by silt-laden water is indicated. Individual plant remains show little or no sign of mechanical fragmentation so they are unlikely to have travelled far before being deposited. That the environment was vegetated, and supported tree growth and not just aquatic plants, is shown by the *in situ* tree trunks projecting into Unit 3 (Figure 4.14). Unit 3 is made up of altogether coarser-grained sediments suggesting higher-energy deposition and one in which the energy fluctuated. This variation is indicated by the fluctuating grain size occurring in discrete layers. The sediment must have been deposited rapidly as it entombed and preserved an upright tree trunk, but also washed in fallen trees from elsewhere as evidenced by the trunk lying parallel to bedding. In the upper parts of Unit 3, fragmented plant remains were deposited as current strengths lessened. Subsequently, the surface of this newly deposited sediment was colonized by other plants whose roots penetrated the upper parts of the Unit. Such a sequence of events is typical of parts of a crevasse splay event where a nearby river broke its banks and deposited sediment on the interfluve floodplain surface.

○ In Figure 4.14 and the description above, there is very little reference to sedimentary structures. Why do you think these were not obvious in this succession of rocks?

- The abundance of preserved roots throughout the succession indicates a high degree of bioturbation due to root penetration. Preserved roots are those that were growing when the next layer of sediment was deposited, and effectively smothered and sealed the root-bearing layer. Roots of previous generations of plants occupying that layer are likely to have decayed, but in the growing process they would have destroyed all original sedimentary structures. This is common in non-marine sediments.

Figure 4.16   Shoots and leaves of *Podocarpus*. (*Bob Spicer, Open University.*)

Figure 4.15   *Podozamites* leaves. (*Bob Spicer, Open University.*)

The shoot form of *Podozamites* (Figure 4.15) is common in the coal-forming mires and associated sediments. We do not really know what the *Podozamites* plant looked like, but circumstantial evidence of frequent association of *Podozamites* leaves with upright trunks suggests that it was a tree. The trunks appear to be relatively straight, rather like that of a modern plantation-grown conifer. Given the similarity of leaf form, a good modern analogue for *Podozamites* is probably the Southern Hemisphere conifer *Podocarpus* (Figure 4.16).

Pityophyllum (Figure 4.17) was also evidently a conifer but whether it was a tree or bush is unclear. Both *Podozamites* and *Pityophyllum* are common in the coals, and because coals form in stagnant conditions, it is likely that the leaves were not washed in, but grew on plants that were components of the mire community itself. However, these plants were not restricted to the swamps, and as leaf forms they are ubiquitous in almost all the sediments. When the Nanushuk Formation was being deposited, the *Podozamites* tree and, to a lesser extent, *Pityophyllum* plant were widespread and abundant across the Corwin Delta floodplain.

Figure 4.17   *Pityophyllum* leaves forming a leaf mat in Nanushuk Formation sedimentary deposits. (*Bob Spicer, Open University.*)

One plant that had a more restricted distribution is *Ginkgo*. In the field description above, a fault brings into juxtaposition with the logged section a series of cross-bedded sands that contain abundant *Ginkgo* leaves (Figure 4.18) in addition to *Podozamites*. Such large-scale channel sandstones are absent from the logged section in Figure 4.14 as is *Ginkgo*.

Figure 4.18 A *Ginkgo* leaf from Nanushuk Formation sediments. *(Bob Spicer, Open University.)*

○ What might this suggest regarding the temporal and/or spatial distribution of *Ginkgo*?

● The throw on the fault is unknown so potentially it is possible that *Ginkgo* may have a different stratigraphical distribution to *Podozamites*, although clearly, at the time interval represented by the *Ginkgo*-bearing blocks, both *Ginkgo* and *Podozamites* co-occurred. Another possibility is that *Ginkgo* had a more restricted ecological distribution and its close association with channel sandstones suggests it might have been a stream margin plant.

In fact, this association of *Ginkgo* with channel-fill sandstones and crevasse splay sediments is repeated many times in the Arctic Slope successions which suggests that *Ginkgo* was indeed a plant of river margins. We never find *Ginkgo* in coal beds and only rarely do they occur in carbonaceous shales and siltstones.

The single living species of *Ginkgo*, *Ginkgo biloba*, is a native of a small area of China and given that it has been prized as an ornamental tree for a long time its natural ecological distribution is uncertain. The modern form is a tree and there is every reason to suppose that this is a good analogue for the ancient forms. However, the Cretaceous of the Arctic Slope saw not just one species of *Ginkgo*, but a whole diversity of related plants all belonging to the ginkgophytes. These leaf forms have a common similarity in that they look like highly dissected *Ginkgo* leaves and the most common are assigned to the genus *Sphenobaiera* (Figure 4.19).

Figure 4.19   *Sphenobaiera. (Bob Spicer, Open University.)*

If *Podozamites*, *Pityophyllum* and *Ginkgo* were the main forest trees represented in the Kukpowruk section, what was growing beneath them? In Unit 3 (Figure 4.14), there are abundant rhizomes of *Equisetites*. Rhizomes are horizontal stems and in the case of *Equisetites* these grew underground and at intervals produced upright stems, often with cone-like reproductive structures that produced spores. The rhizomes are quite easy to identify because, like all members of this plant group (the sphenophytes), they have striations running along the stems and at intervals there are breaks in these striations at nodes (Figure 4.20). Often at these nodes other rhizomes, upright stems or roots are produced and in many instances there are whorls of sub-spherical nodules. These are characteristic of *Equisetites*: such nodules are also produced by the modern relative *Equisetum* (scouring rush or horsetail, Figure 4.21) and are the site of concentrations of nitrogen-fixing bacteria. This association with nitrogen fixers is common in plants growing in nutrient-poor conditions, particularly vigorous-growing early colonizers of river banks and point bars. Not surprisingly, *Equisetites* rhizomes are one of the most commonly identifiable plant fossils in many Cretaceous floodplain sediments. Before grasses evolved and became ecologically important, *Equisetum* 'meadows' formed the dominant ground cover in disturbed environments.

Other ground cover plants were ferns. In the siltstones and sandstones of the downthrown side of the fault, remains of *Birisia* were found (Figure 4.22). This is a common Cretaceous Arctic fern and its frequent association with *Equisetites* in the absence of other plants suggests that, like *Equisetites*, it was an early colonizer. In some settings, even marginal marine siltstones and sandstones, *Birisia* leaves are the only fossils found and they occur as large, minimally broken fronds in some abundance. This suggests a community local to the environment of deposition composed solely of *Birisia* ferns. Other ferns have a more scattered distribution, are always mixed with other types of plants, and not limited to any particular sedimentary facies. Consequently, these are most likely to have been present in mixed communities and are likely to have formed ground cover in more mature communities.

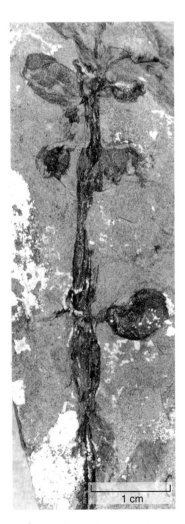

Figure 4.20   *Equisetites* rhizome with nodules. *(Bob Spicer, Open University.)*

Figure 4.21   Modern *Equisetum*. Plants are *c.* 30 cm high. *(Bob Spicer, Open University.)*

Figure 4.22   *Birisia* frond from the Nanushuk Formation. *(Bob Spicer, Open University.)*

Figure 4.23 shows a graphic log of part of another section representing an exposure of the lower part of the Nanushuk Formation along the Kukpowruk River. The units have the following characteristics:

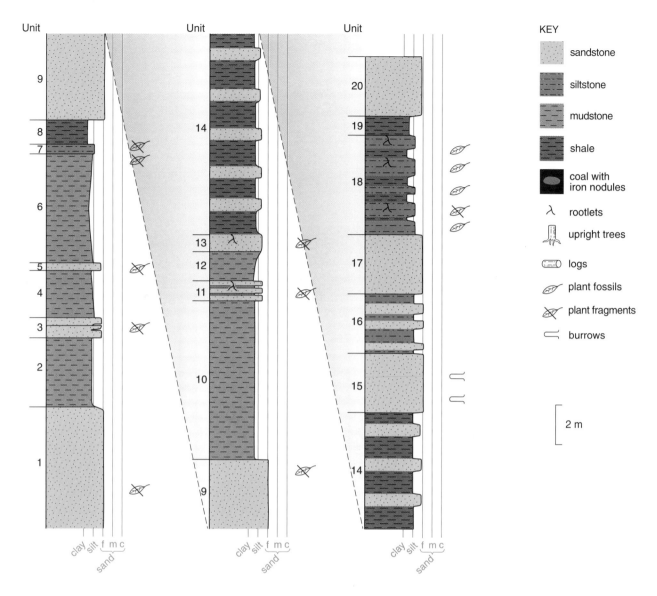

Figure 4.23 Graphic log of part of the Nanushuk Formation along the Kukpowruk River at Locality 11 (Figure 4.12). *(Spicer and Herman, 2001.)*

Locality 11 (Figure 4.12): 69°00′02″ N, 162°57′44″ W

Unit 14. Interbedded dark grey shale and paler-coloured more indurated siltstones and fine-grained sandstones with few, if any, ripple marks. They have a lumpy texture with numerous layers of ironstone nodules. There are at least eight layers of siltstones/sandstones, each approximately 0.5 m thick, dispersed more or less evenly throughout the thickness of the unit. Thickness: 16 m.

Unit 15. Lenticular sandstone (lens-shaped), fine-grained, pale grey, weathering to yellow, well-bedded. Horizontal 'U'-shaped burrows. Thickness: 3 m.

Unit 16. Poorly indurated brown siltstone interbedded with three fine-grained sandstone layers: the bottom sandstone layer is about 0.1 m thick, the middle layer is 0.21m thick, and the upper layer is 0.5 m thick. Total thickness: 2.5–3 m.

Unit 17. Flaggy, lenticular, sandstone (as all the sandstones in the exposure). Thickness: 3 m.

Unit 18. Lumpy interbedded siltstones and claystones containing abundant plant remains that are often minimally fragmented. *Birisia alata* dominates. There are abundant 'Y'-shaped roots. Thickness: about 5 m.

Unit 19. Shale. Thickness: 1 m.

Unit 20. Sandstone. Thickness: 3 m.

Interpretation: The basal portion of this section represents marginal marine facies of shallow but varying water depth and varying terrigenous siliciclastic input. While this section is in most part marine, a blocky/lumpy brown-bedded siltstone dipping to the south occurs at the southern end of the exposure. This contains abundant plant remains, predominantly *Birisia alata* as large frond fragments, *Equisetites* as rhizomes (but some branched aerial parts do occur), *Parataxodium*, *Arctopteris*, *Heilungia*, *Pityophyllum*, and in the coarser-grained units *Ginkgo*. Burial was evidently rapid and frequent as plant remains occur on numerous bedding surfaces over at least 3 m of section. Often the plants are preserved cross-cutting bedding planes, indicating rapid and repeated influxes of sediment. *Birisia* fronds are by far the most abundant element and their occurrence adjacent to a marginal marine setting must indicate that *Birisia* formed coastal 'marshes'. The succession is rooted throughout, suggesting that after each successive influx of sediment destroyed the marsh vegetation, *Equisetites* and *Birisia* rapidly recolonized the fresh sediment. Evidently, this succession clearly demonstrates marsh vegetation very close to the shoreline and it is likely that the plants were, to some degree, tolerant of mildly or periodically saline conditions.

The marine units are characterized by predominantly organic-rich shales, probably formed in deeper water, interspersed by coarser-grained siltstones and fine-grained sandstones exhibiting a variety of ripple forms, mud drapes, worm traces and *Rhizocorallium*. No megafaunal remains were seen which possibly indicates a near-distributary environment with variable salinity. These siltstones are mostly devoid of plant remains except water-worn wood fragments.

This set of observations is consistent with the discussion above but introduces some new plant genera: *Arctopteris*, *Parataxodium* and *Heilungia*. *Arctopteris* is another common fern while *Parataxodium* is a conifer whose nearest living relatives are the bald cypress, *Taxodium* and the dawn redwood *Metasequoia*. Both *Taxodium* and *Metasequoia* have a relictual distribution that today is confined to warm temperate climates. *Taxodium* is a tree typically found in swamps of south-eastern USA, while *Metasequoia* was only known as a Cretaceous and Tertiary fossil until its discovery as a living tree in China in the 1940s. Both species are deciduous (they have synchronous leaf loss so that the plant is bare of leafy shoots for part of each year); a trait that clearly could be advantageous during long periods of winter darkness experienced at high latitudes. It is likely that *Parataxodium* was also deciduous because the leafy shoots are always shed as complete discrete units and are commonly found as 'leaf mats'. Leaf mats are bedding surfaces covered in masses of shed leafy shoots implying synchronous leaf loss during the (short) period over which the individual bedding surface was being deposited.

Heilungia belongs to the cycadophytes, a group that is widespread in the Cretaceous Arctic in the form of *Nilssonia* leaves (most notably *N. serotina*, *N. alaskana* and *N. yukonensis*). The modern representatives of this group are the cycads that today have a relictual distribution across Mexico and the West Indies, East Africa, India, China, Japan, and Australia. Previously, members of the group were global in distribution and the fact that they are now all evergreen and frost sensitive (today, they all grow in warm temperate to tropical climates) was one of the first pieces of palaeobotanical evidence used as an argument for ancient global warmth. However, recent discoveries have demanded a rethink of this interpretation.

Like *Parataxodium, Nilssonia* leaves also formed leaf mats and moreover not only single leaves were shed, but sometimes whole clusters of leaves attached to a short shoot (Figure 4.24). This short shoot with leaves attached is called *Nilssoniocladus*. The genus *Nilssoniocladus* was first described in 1975 based on material from the lowermost Cretaceous Oguchi Formation, Central Honshu, Japan.

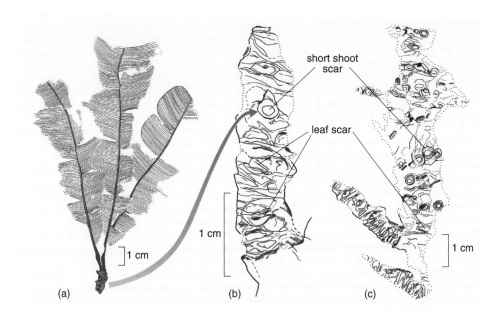

Figure 4.24   The leaves and leafy shoots of the *Nilssoniocladus* plant: (a) cluster of leaves still attached to a short shoot, but shed from the parent plant; (b) short shoot showing both leaf scars and a scar of a subsidiary short shoot; (c) main shoot showing attached short shoots.

The type material, *Nilssoniocladus nipponensis*, represents a plant with a thin vine-like stem bearing spirally arranged dwarf short shoots with persistent leaf scars. The leaves were assigned to *Nilssonia nipponensis* Yokoyama and were borne in clusters of three to seven or more at the distal ends of the dwarf shoots. Based on these characteristics, the *Nilssoniocladus* plant was interpreted to have been deciduous. In 1996, the *Nilssoniocladus* plant was also found on the Arctic Slope of Alaska and in north-eastern Russia at palaeolatitudes greater than 73° N.

Specimens of stems from the lower Nanushuk Formation exhibit scars where short shoots were attached, and attached short shoots devoid of leaves. The frequent leaf mat accumulations also suggest synchronous leaf shedding and that *Nilssoniocladus* species in the Arctic, like those from Japan, were deciduous.

Nilssoniocladus is widely regarded as representing an extinct member of the Cycadales. The deciduous behaviour was quite unlike that of the modern cycad relatives. The modern plants bear their long leathery frond-like leaves directly on squat trunks and when the leaves die after several years they remain attached to the trunk until they rot off.

The presence of cycad foliage at high palaeolatitudes in the Cretaceous of Asia and North America has been used as evidence for frost-free, even warm temperate, near-polar conditions. This idea, perhaps more than any other, contributed to the concept of the Cretaceous being a time of extreme global warmth. This, however, assumed that the Cretaceous cycadophytes had a similar biology and climatic tolerance to their living relictual relatives, in spite of the fact that no bulbous trunks, typical of many living cycads, had ever been found as Arctic fossils.

Additionally, the assumption that Cretaceous polar vegetation included elements that were evergreen raised important questions regarding the Earth's obliquity.

○ *Nilssonia* leaves typically occur at palaeolatitudes of 75° N and, if the obliquity of the Earth were the same as at present, what period of continuous winter darkness would the parent plants have experienced?

● The parent plants must have experienced continuous winter darkness of at least six weeks bounded by a continuous twilight period of three weeks in the spring and autumn.

A warm temperate to subtropical frost-free climate, combined with this light regime and evergreenness in which a living leaf load was borne throughout the year, would inevitably have led to a high respiratory drain on the plant's resources during the winter. In particular, young plants would have been vulnerable.

One way around this conundrum was to argue for a reduced obliquity. Such a reduction would have shortened the duration of winter darkness so evergreen plants might have been able to survive in near-polar environments. However, with the re-interpretation of *Nilssoniocladus* as a deciduous plant, no change in obliquity is required to explain survivorship through a long winter dark period because the plants would merely shed their leaves and enter dormancy until the light returned in the spring.

4.3.5 An insight into Cretaceous Arctic vegetation dynamics

The repeated association of leaf forms with particular sedimentary facies allows us to reconstruct patterns of plant associations. Also, because some depositional environments represent more stable conditions than others, we can reconstruct the sequence of plant community development (Figure 4.25). We can distinguish early pioneer communities from those of mature forest and persistent mires.

We have already considered the pioneering role of *Equisetites* and *Birisia* and their association with wet or marshy environments. Similarly, we have considered the association of *Ginkgo* with river margin or riparian settings and the more widespread occurrence of *Podozamites* and *Pityophyllum*. There is, however, one plant on the diagram we have not yet encountered: *Desmiophyllum*.

Desmiophyllum is a leaf form that is locally abundant in sandstones and shales of the lower Nanushuk Formation (Figure 4.26). It is a broad elongated leaf with parallel venation, a rounded to pointed tip and a tapering base. It looks like a large leaflet of *Podozamites* or fragment of a large digitate *Ginkgo* leaf. In fact, it is easy to confuse all three leaf forms when they are found as fragments. *Desmiophyllum* is regarded as a gymnosperm (a seed-bearing, non-flowering

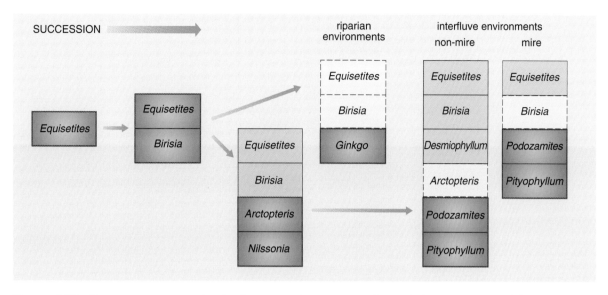

Figure 4.25 Diagram representing the development and composition of typical plant associations of the lower Nanushuk Formation. Genus names in boxes with orange shading represent the most common and ecologically significant plants in each association. Dashed boxes indicate minor components of the associations.

plant) but apart from that its taxonomic affiliations are unknown. For this reason we assign it to a category called '*incertae sedis*', a sort of holding category that carries with it no implications of relationships to known taxa. However, because we usually find it in large leaf mats where all the leaves are in a similar state of decay it is likely that, as with the other taxa considered so far, it was a woody deciduous plant. It seems to be absent from sediments representing mires, e.g. coals and coal-associated shales.

Figure 4.26 A leaf form assigned to the genus *Desmiophyllum*. (Bob Spicer, Open University.)

4.3.6 The flowering of the Arctic

The land vegetation of our present world is characterized by the flowering plants, otherwise known as 'angiosperms'. You can see from Figure 4.11 in Box 4.1 that the angiosperms are the most recent major group of plants to evolve, and yet they have become very diverse in a relatively short time. As we shall see shortly, it is very difficult to say exactly when or how the angiosperms arose, but before the mid-Cretaceous they were not ecologically important in that they made up only a very small proportion of terrestrial biomass. It was during the Cretaceous that the angiosperms transformed global vegetation. By the mid-Cretaceous, fossil pollen, leaves and wood that clearly look angiospermous, and even flowers themselves, become increasingly common. During the Late Cretaceous, feature combinations that characterize many modern families became well established and in many environments, particularly those that suffered frequent disturbance such as river margin sites, flowering plants became numerous, if not dominant. Today, angiosperms are not only numerous: they are extremely diverse. There are between 200 000 to 300 000 species (depending upon the classification used) belonging to 300–400 families falling into two apparently natural groups: the monocots (palms, grasses, lilies, etc.) and the dicots (alders, willows, buttercups, etc.), and almost all our food crops are angiosperms. Perhaps more importantly for our consideration of the Cretaceous Earth system, angiosperms have developed an intimate relationship with climate and this provides quantitative information of what the Cretaceous world was like. It is therefore important to try to understand the evolutionary history of the group of plants that, to a large extent, define our present global vegetation.

The above examples of the flora of the lower part of the Nanushuk Formation are fairly typical of Arctic vegetation before the arrival of flowering plants or angiosperms. Although individual angiosperm characteristics are seen in pre-Cretaceous plants, the combination of features which we see in modern flowering plants, and which we consider typical of angiosperms, did not appear consistently grouped in individual plant lineages until the Cretaceous. Even then, plant parts with angiosperm characteristics only started to become numerous after the Aptian. In our modern world, the angiosperms are such an important group in terms of biodiversity, biomass and as a source of food that we need to take a few moments to consider the characteristics that define an angiosperm.

When dealing with the fossil record, we have to rely on morphological characters: physical characters that make up the body architecture of the organism. The morphological characters must also be likely to be preserved because they are robust and resist decay. Flowers are regarded as a key element of an angiosperm even though they are only rarely preserved, so it is useful to begin with these.

If we think of a typical flower, say a buttercup *Ranunculus repens*, it is made up of a ring of scale-like *sepals* surrounding a ring of brightly coloured *petals* (Figure 4.27).

Within the petal ring (known as a *corolla*), whose function it is to attract pollinating insects, there is a ring of male pollen-producing organs (commonly called *stamens*). The most internal structure is the female part of the flower where we find the *ovaries* containing the potential seeds. From the ovary protrudes a stalk-like *style* and on this is the pollen-receiving surface, the *stigma*. From this you can see that such a flower is bisexual (it has both male and female

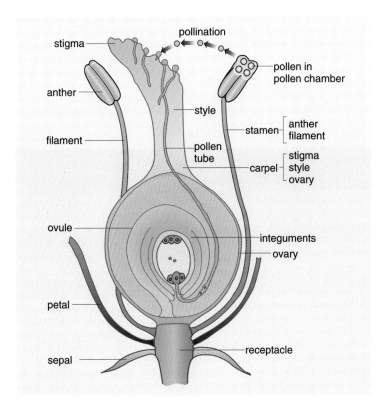

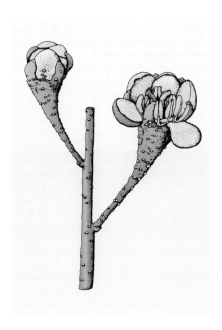

Figure 4.28 Reconstruction of *Silvianthemum suecicum*. The colours are purely indicative to aid interpretation of its parts. *(Friis, 1990.)*

Figure 4.27 Diagrammatic representation of a typical bisexual angiosperm flower.

parts), the potential seeds or *ovules* are enclosed and pollen is received on a special stigmatic surface because the grains cannot have direct access to the enclosed ovules. The reconstruction of the Late Cretaceous flower shown in Figure 4.28, *Silvianthemum suecicum*, shows that such features were present relatively early in angiosperm evolution.

However, if we go even further back in time to the mid-Jurassic, we can see that angiosperms were not alone in having bisexual reproductive organs. A good example is to be found in the 'flowers' of the bennettitales such as *Williamsoniella* (Figure 4.29).

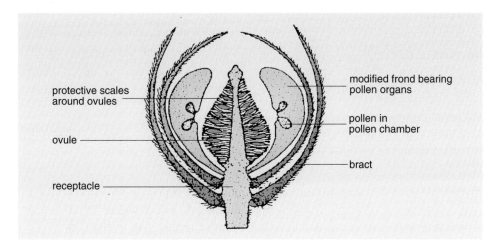

Figure 4.29 Longitudinal section of a reconstructed reproductive unit of the bisexual *Williamsoniella*. Note the similarity of organization to that of an angiosperm (Figure 4.27). *(Harris, 1944.)*

If flowers do not define an angiosperm, what does? Angiosperm leaves tend to be quite different to those of ferns, cycadophytes and conifers. They usually have a network of veins that divide and reconnect many times (Figure 4.30). However, this reticulate venation is not unique to angiosperms because some plants, such as the modern non-angiospermous *Gnetum* and the fossil leaves of the Jurassic seed ferns, are also reticulate. Moreover, some angiosperms such as the palms and grasses have parallel veins rather than a network.

Figure 4.30 The net-like, or reticulate, venation of an angiosperm; in this case, a leaf of the flowering cherry. *(Bob Spicer, Open University.)*

The wood of many modern angiosperms is characterized by having fluid-conducting cells called vessels as well as tracheids and other types of cells which are also found in conifers. In transverse section, these vessels can be quite large and may be organized in a diffuse pattern or grouped mostly in the wood that is produced early in the growing season — the so-called earlywood (Figure 4.31). While most modern angiosperms have vessels, some do not. In the Cretaceous, fossils indicate that there were several evolutionary routes to vessel formation and some groups that today possess vessels probably used not to. Moreover, some non-angiosperms such as members of the Gnetales, also have vessels. Vessels, therefore, are not uniquely diagnostic of flowering plants.

There are several other features commonly found in angiosperms which could be discussed. However, none of these is exclusive to the flowering plants, or found in all angiosperms, or fossilizes frequently enough to make it helpful when using the fossil record in the study of angiosperm origins. However, there is one feature that can be preserved. It is at the core of the concept of angiospermy, and has been recently used to rewrite the early history of the angiosperms. This feature is the enclosure of the ovules by an outer covering to form a carpel (Figure 4.27).

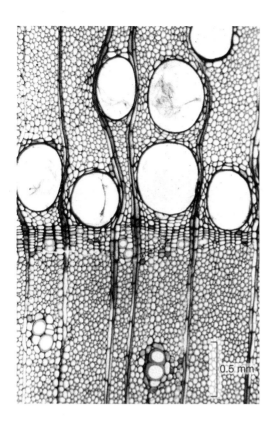

Figure 4.31 A cross-section of a piece of angiosperm wood (oak) showing the large vessels. In this species, vessels produced in the spring, when the leaves are growing and demand for water is high, are large, whereas those produced in the autumn, when the leaves are beginning to die, are small. *(Bob Spicer, Open University.)*

Ovules (unfertilized seeds) are attractive to animals because they are very nutritious, although even in angiosperms they have little in the way of food reserves for the developing embryo. Enclosing the ovules in a carpel affords extra protection against being eaten. Perhaps more important is that the fertilization process itself takes place in a controlled moist environment within the carpel wall. Moisture is essential for fertilization and more primitive plants, such as ferns, cannot complete their life cycle in dry conditions. By enclosing the ovule in a carpel, angiosperms are immune to drought in the external environment during this critical phase in their life cycle.

Although building an extra wall around the ovules requires an extra investment of resources, such an evolutionary innovation clearly has advantages, so that we see a trend towards ovule enclosure in several lines of Mesozoic plants. Partial enclosure is seen in some cycadophytes, glossopterids and, in particular, the seed fern *Caytonia*, named after Cayton Bay near Scarborough, Yorkshire, where it was first found. The partially enclosed ovules of *Caytonia* and the fact that the *Caytonia* plant also produced *Sagenopteris* leaves with reticulate venation led early researchers to suggest an angiosperm affinity for this plant. However, we now know that total enclosure of the ovules so as to require a stigmatic surface for pollen germination was absent in *Caytonia*. The first convincing evidence of the evolution of a true carpel comes not from the Yorkshire Middle Jurassic, but the Late Jurassic or Early Cretaceous (there is some dispute over the exact age) of China.

The fossil known as *Archaefructus* consists of a specialized reproductive branch with helically arranged fruits. The simplest way to envisage these fruits is to think of a leaf with two to four ovules attached along the midrib and the leaf folded along the midrib, so that the leaf margins meet and enclose the ovules completely. This structure meets the criteria required for it to be considered a carpel. No corolla or male reproductive parts are known, nor are normal leaves, but these fruit-bearing branches are associated with leaf-like structures, although these are poorly preserved in the specimens so far recovered.

Is *Archaefructus* a true angiosperm? Well, if we define angiospermy purely on the enclosure of the ovules then it does appear to be so. However, even in the modern angiosperm *Drimys*, enclosure of the ovules is not complete at fertilization. Other features that could, if taken together, indicate that *Archaefructus* was an angiosperm are lacking.

Is *Archaefructus* the ancestor of modern flowering plants? This question is even more difficult to answer. The evolutionary tendency for enclosure of ovules was widespread in Mesozoic times, as was the evolution of reticulate leaf venation and bisexual reproductive structures. Clearly, Mesozoic environmental conditions favoured features now seen as characteristic of flowering plants. At some point in time, probably the Late Jurassic or Early Cretaceous, these features came together and allowed the plant or plants that possessed them to outperform many competitors. As a consequence of this, the angiosperms underwent a rapid geographical spread and, by the middle part of the Cretaceous, they were ecologically significant in many environments from pole to pole. The evidence for this is the widespread associated occurrences of a diversity of leaves and pollen (occasionally wood and rarely flowers) that display angiospermous characteristics. While the characters taken in isolation are not proof of angiospermy, their frequency of occurrence and association argue strongly for them having been produced by flowering plants.

An important key to understanding the angiosperm evolutionary radiation is to be found in the work of Jim Doyle and Leo Hickey on the Potomac Group. The Potomac Group includes many of the rocks that are found around Washington DC on the eastern seaboard of the USA. This work generated a paradigm shift in the way that the fossil record contributed to understanding the early global radiation of the angiosperms.

Using well-defined European pollen zones, and fossils from associated marine rocks, Doyle and Hickey (1976) were able to correlate their non-marine rock units with those that were better dated in Europe, and to provide a stratigraphical framework for the interpretation of the associated leaf fossils. Another significant feature of their work was the attention paid to the association of particular fossil forms with particular rock types and their original environments of deposition. This facies analysis, like that for pre-angiosperm floras in Alaska, has proved to be of great importance when trying to reconstruct early angiosperm communities.

One of the reasons why early angiosperm pollen grains were so useful stratigraphically was that changes in pollen morphology occurred frequently throughout the section, reflecting rapid evolution. Added to this was the fact that pollen and spore assemblages changed as the pre-existing fern and gymnosperm vegetation was infiltrated, and in many areas eventually replaced, by an angiosperm-rich flora. These community changes appear to be independent of climate or other physical environmental factors and largely reflect inter-plant competition.

At the base of the stratigraphical interval studied by Doyle and Hickey (Figure 4.32), ferns, cycadophytes, ginkgophytes and conifers dominate the spore/pollen assemblages and angiosperm pollen is rare, small and with a single simple groove-like aperture (monocolpate).

At the top of zone I (lower Albian), the first pollen appears with three groove-like apertures (tricolpate). This is significant because tricolpates are highly distinctive of dicotyledonous angiosperms. Both monocolpates and tricolpates diversify in

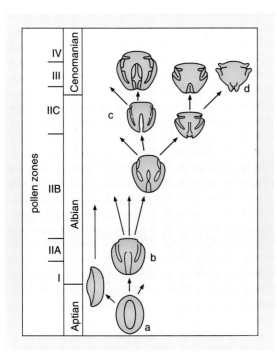

Figure 4.32 Diagram of angiospermous pollen evolution in the Potomac Group, on the eastern seaboard of the USA. Note the increasing complexity of pollen architecture from grains with simple single groove-like apertures in the Aptian ('a'), to three groove-like apertures in the Albian ('b'), to those with three grooves and pores ('c'), and three pores typical of many modern angiosperms ('d') such as birch (*Betula*) and the *Normapolles* complex of the Late Cretaceous. *(Thomas and Spicer, 1987.)*

subzones IIA and IIB, and many tricolpates develop thin areas (almost pores) in the centre of the colpi. These 'tricolporoidates', as they have been called, diversify further in size, shape and sculpture in subzone IIC (upper Albian to Cenomanian). These forms presumably gave rise to the true tricolporates of zone III. In zone IV (Cenomanian), grains with round pores rather than furrows appear. These are the first triangular triporates of the so-called *Normapolles* complex which became a significant element in assemblages in eastern North America and Europe during the Late Cretaceous and Early Tertiary.

Much of the Potomac pollen zonation was based on subsurface borehole records as these provide good vertical stratigraphical control. The surface outcrops yielding the leaf fossils were positioned stratigraphically using the pollen/spore assemblages they contained. Doyle and Hickey were able to derive a reasonably comprehensive picture of the range of leaf morphologies and ecological heterogeneity of the angiosperm flora at several points in their stratigraphical succession. Leaf architectural features (venation pattern, margin characteristics and overall leaf organization) were used to document evolutionary trends in stratigraphical succession. Conventional taxonomic partitioning was largely ignored because previously it had been misleading and it proved to be largely irrelevant to the study of evolutionary change. What was important were accurate descriptions of the plant part architecture.

In the lower Albian zone I assemblages of Virginia and Maryland, both simple and lobed leaves all have poorly organized venation (Figure 4.33 leaves a–f). In the middle of subzone IIB (middle to upper Albian, Figure 4.33 leaves g–i), palmately veined forms occur for the first time and exhibit more regularity in their vein courses than the leaves of zone I. In leaf forms such as *Menispermites virginiensis* we even find that the tertiary veins have a tendency towards a degree of regularity. By the middle of subzone IIB, we also see leaf margins with double convex glandular serrations and both pinnately and palmately lobed forms (Figure 4.33 leaves h, i and j).

Figure 4.33 Sketches of angiospermous leaf fossils from the Potomac Group, eastern seaboard of the USA. Note the increasing vein organization as time progresses. *(Thomas and Spicer, 1987.)*

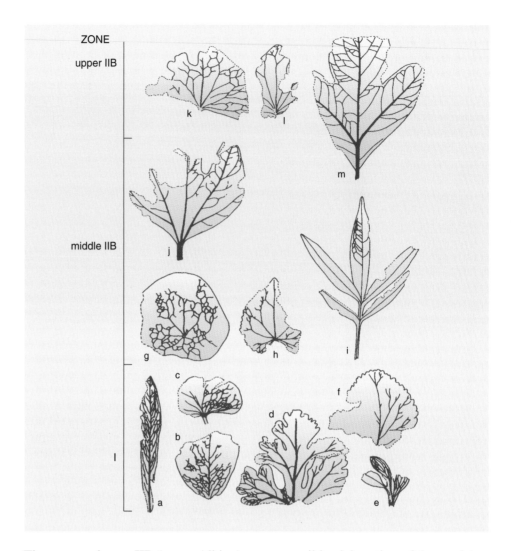

The upper subzone IIB (upper Albian) sees a possible elaboration of the cordate–reniform complex as typified by the genera *Menispermites* and *Populophyllum* (leaves k and l). Truly pinnately compound leaves first appear in upper subzone IIB and the leaflets have secondary and tertiary vein orders that are readily distinguishable. Palmately lobed leaves are also present in upper subzone IIB and later become locally dominant in subzone IIC (uppermost Albian to lower Cenomanian). Confined to the coarser-grained fluvial facies, they are typified as '*Sassafras*' *potomocensis* Berry. The genus name '*Sassafras*' is in quotes because its relationship to modern *Sassafras* is unknown.

Mid-Cretaceous angiosperm leaves (Figure 4.34) may be broadly grouped into a number of rather loosely defined categories, only some of which will be discussed here. Architecturally, the simplest are the pinnately veined forms with entire margins. Lateral branches of the midvein (secondary veins) usually loop near the margin (Figure 4.34a). In the earliest forms, the venation is poorly organized. Later, more regular venation does occur, but there are extant leaves of this general form that also have a low level of vein organization. These vein patterns are seen in modern members of the Magnolia group and for this reason the fossil forms may be termed magnoliid-like, not because they are necessarily related to the magnoliids but because they display similar vein organization.

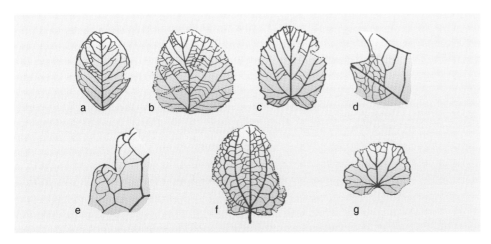

Figure 4.34 Sketches of a selection of mid-Cretaceous angiosperm leaves. *(Thomas and Spicer, 1987.)*

A second major type of leaf is typified by the so-called *Menispermites* form (Figure 4.34g). These are more or less reniform, with a series of major veins that radiate from a single point. Some may represent aquatic plants, but most were probably climbers because rare specimens show attachment to thin woody stems. A third major form, and one in which there is considerable morphological diversity, are platanoid-like leaves (Figure 4.34b–d). These are fundamentally palmately veined, often with toothed margins (although some of the early forms had entire margins), and are typified by leaves of the *Araliopsoides*, *Sassafras*, and *Platanophyllum* types. A fourth form, the 'trochodendroides' (Figure 4.34e, f), have rounded teeth (e) and major veins that leave the midvein and then curve in towards the apex.

Although the Potomac Group leaves and pollen show clear evolutionary trends, this does not mean that the eastern seaboard of the USA was the centre of origin of the group. Early Cretaceous angiosperm pollen is known from several places around the world but in an analysis of pollen occurrences Hickey and Doyle (1997) demonstrated a migration pattern from low to high latitude (Figure 4.35). This trend of poleward migration is also seen in leaves. In Alaska it is not until the end of the beginning of the Cenomanian, in the upper part of the Nanushuk Formation, that we see the first angiosperm leaves arriving on the Arctic Slope. When angiosperms first appear, they tend to occur in fluvial channel-fill sandstones or crevasse sedimentary deposits. Rarely are they seen in interfluve floodplain siltstones, and never in coals.

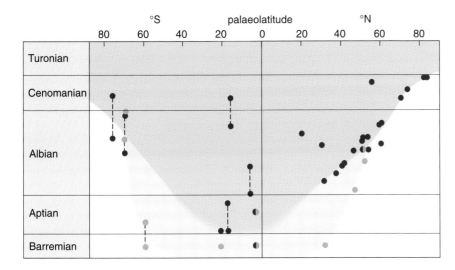

Figure 4.35 Illustration of the poleward migration of angiosperms as evidenced by pollen occurrences. Orange circles represent monocolpate grains while red circles represent tricolpates. Split circles indicate the presence of both types. *(Thomas and Spicer, 1987.)*

4.3.7   Building a polar forest

We are now at an appropriate point to try to put all this information together to reconstruct a typical Arctic forest in the Cenomanian. Figure 4.36 is a diagram showing a composite forest reconstruction based on fossils from the upper part of the Nanushuk Formation. The angiosperms, mostly platanoids, occupy the river margin sites. These riparian margins are shown here as levées, but bear in mind that levées may not have been formed along the sides of all Arctic rivers at this time. Ginkgophytes are shown as being less abundant and are in the process of being ousted by the platanoids. Angiosperms are also occupying lakeside sites but here, in addition to platanoids, trees or bushes bearing leaves with magnoliid features sometimes occur. Lake margins are also the locations where we are likely to find cycadophytes such as *Nilssoniocladus*. In more highly disturbed settings, before woody plants have time to grow, we find *Equisetites*. Ferns frequently join *Equisetites* in associations alongside rivers, in marshes and as understorey plants in forests. In low-lying areas between the rivers, these forests are dominated by taxodiaceous conifers such as *Parataxodium*, *Metasequoia* and *Cephalotaxopsis*. *Podozamites* is also locally present but together with *Pityophyllum* is mostly found dominating the swamps.

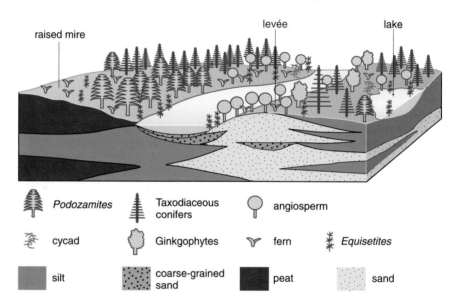

Figure 4.36 Diagrammatic reconstruction of a typical Cenomanian forest on the Arctic Slope of Alaska.

Peat accumulation in the swamps may eventually raise the peat surface above the local water table. This does not necessarily mean that the peat will stop accumulating. If rainfall is sufficiently high and distributed more or less evenly throughout the year, continued growth of the mire into a raised bog can be sustained by direct rainfall alone.

As the raised mire continues to develop, sustained by high rainfall, changes take place in the composition of the mire community. The most obvious change is a decline in species richness and this can be observed in modern raised mires in south-east Asia. It is linked to the declining nutrient status of the mire surface. As the nutrients become recycled repeatedly, inevitable losses from the system increasingly limit the stature, diversity and density of plants that can survive.

Chemical analyses of thick coals from the Nanushuk Formation show that the most labile and biologically important elements such as phosphorus and potassium are concentrated in the upper parts of coal seams. As the raised mire develops, the influx of plant nutrients from stream waters decreases and, apart from wind-blown dust or volcanic ash, the peat-forming community becomes starved of the supply of essential elements. However, biologically labile (useable) elements released during the partial decay of the peat are available and these become concentrated in the living biomass while becoming depleted from the peat. These inorganic elements are a major contributor to the ash component of these otherwise low-ash coals.

Figure 4.37 is a histogram of ash content versus depth of a Nanushuk Formation coal from the westernmost Arctic Slope. Note the increasing ash content closer to the upper surface of the coal seam.

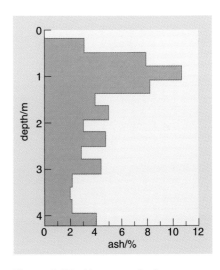

Figure 4.37 Histogram of ash content versus depth of a Nanushuk Formation coal (coal 203c). *(Youtcheff et al., 1987.)*

○ What other mechanism could account for this different inorganic chemistry in the upper parts of the seam?

● Ions of soluble elements (those important to plant growth) could be washed down from overlying sediments during burial, consolidation of the rock sequence and, in the case of rocks exposed at the surface, weathering processes.

In this case, however, there is corroborating evidence in the form of the pollen content of the coal which supports a raised mire interpretation.

Figure 4.38 shows a simple palynological diagram for the same coal as in Figure 4.37. Note the changes in the pollen content over the thickness of the coal.

○ What might this say about the plant community development on the mature raised mire?

● The community begins with a mix of conifers and ferns and minor amounts of mosses. Cycadophytes do not appear to make a significant contribution to the pollen flora. As the mire develops, the conifer component declines, while the fern contribution increases. This suggests an opening up of the community and a shift from woody trees to a more herbaceous community.

This loss of diversity and biomass slows down peat development and, in the absence of other factors, peat growth would be self-limiting. However, other factors do play a role. As in modern Arctic peats, wildfires occurred on a frequent basis. The evidence for this comes from the coal itself. Polished blocks of Nanushuk coal, when viewed in reflected light under a microscope, reveal thin layers of charcoal and naturally produced ash. In the Nanushuk coals, these burnt layers are rarely more than a few millimetres thick, suggesting the fires were of short duration and quickly swept over the surface of a predominantly herbaceous vegetation. The charcoalification of root systems is limited suggesting the fire did not burn down into the underlying peat, perhaps because it was wet or there was insufficient fuel for it to take hold. This would be the case if the vegetation was mostly herbaceous and contained little wood. Nanushuk coals have few large charcoal pieces embedded in them.

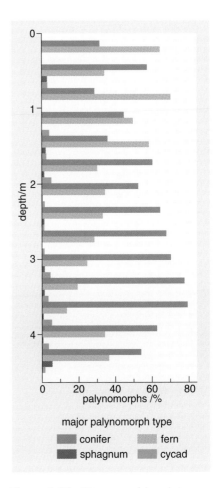

Figure 4.38 Histogram of the relative abundances of the palynomorphs representing major plant groups in the same North Alaskan coal as in Figure 4.37. *(Youtcheff et al., 1987.)*

○   What effect do you think these fires had on the vegetation of the mires?

●   Apart from the destruction of much of the living biomass above ground, the fires would have quickly returned inorganic nutrients back into the surface layers of the peat to be utilized by a new flush of growth. However, this effect would have been limited, due to the shallow depth of the burning. Had the fire penetrated deeper, some additional nutrients would have been released, but this effect would have been small in a mature raised mire, because the deeper peat layers were nutrient depleted.

Note that the Nanushuk coal 203C was sampled over a thickness of almost 3 m. Such thick coals are common in the Nanushuk and given a compaction ratio of between 12 : 1 and 20 : 1, the original peat may have been 36–60 m thick. The low inorganic content of the peat suggests that no sediment was washed into the mire from nearby floodplain rivers despite the mire sitting on a delta floodplain surface. This also strongly suggests that coal 203C represents a raised bog.

○   If we assume a compaction ratio of 20 : 1, does this mean that the peat surface was 60 m above the surrounding floodplain?

●   No, it does not. Because the mire is on a delta floodplain, that surface is likely to be subsiding under gravitational compaction and dewatering. So, as the peat is growing, its base is sinking and provided subsidence and peat growth are approximately the same then very large thicknesses of peat can accumulate.

○   What other subsidence mechanism might have been operating on the Arctic Slope in the Late Cretaceous?

●   Residual collision movement of the Arctic and Angayucham terranes would have given rise to continued tectonic deepening of the foredeep to the north of the Brooks Range. Moreover, as sediment was shed into the basin, loading would have generated syndepositional fault systems that may have contributed to the lowering of the delta plain surfaces.

The tectonic context, in combination with the prevailing climate, was crucial to the development of thick peat accumulations as well as to their eventual preservation. Figure 4.39 summarizes this relationship and is particularly relevant to Cretaceous coals.

Coals seams exposed along the Colville River and occurring within the Nanushuk Formation range from a few centimetres to over 5 m in thickness. However, in the overlying Late Cretaceous sediments of the Prince Creek Formation, the coals are typically much thinner, have a lower rank*, and a different overall composition. *Podozamites* leaf remains disappear both from the coals and the associated carbonaceous shales and are replaced by leafy shoots of *Parataxodium* and *Metasequoia*. The inorganic siliciclastic and overall ash content of these coals are higher as is the frequency and thickness of charcoal horizons.

---

* The rank of a coal refers to its degree of coalification, i.e. the chemical changes that occur with increasing temperature and pressure as the peat is buried. Low rank coals are those which have not been subjected to high temperatures and pressures; they include lignite.

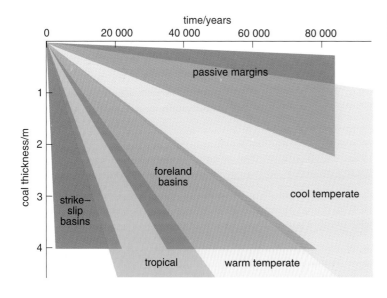

Figure 4.39 The relationship between climate and tectonic setting for the development of coals. *(McCabe and Parrish, 1992.)*

○ What factors could explain this reduction in coal formation in the Late Cretaceous?

● The increase in charcoal could suggest longer periods of dry conditions that allowed the vegetation to burn more freely when wildfires started. If increased drought frequency and intensity was the cause, it would also explain the reduced thickness of the coals because dryness would result in oxidative decay of a larger proportion of the peat than was previously the case. The higher inorganic clastic content of the Late Cretaceous coals suggests that they represent swamps rather than raised bogs and therefore had a higher nutrient status. This, in turn, would have led to more woody vegetation and more fuel to feed the wildfires. Other explanations for the higher siliciclastic content could be higher subsidence rates and/or lower growth rates for the peat.

To try to understand the cause for this change in coal formation, we need to look at the changing nature of the polar forests throughout the Late Cretaceous. Take a look at Table 4.1 comparing the floral diversity based on leaf fossils at a major plant group level between the Nanushuk Formation and overlying Prince Creek Formation.

Table 4.1 Nanushuk Formation and overlying Prince Creek Formation floral diversity comparison.

Flora	Nanushuk Formation	Prince Creek Formation
Conifers	12 shoot forms	2 shoot forms
Ferns	18	2
Sphenophytes	*Equisetites*	*Equisetites*
Ginkgophytes	4	0
Cycadophytes	2	0
Liverworts	1	0
Angiosperms	67	2
Totals	104	6

Figure 4.40 Field photograph of a tree embedded and preserved in floodplain shales of the Nanushuk Formation in northern Alaska. *(Bob Spicer, Open University.)*

It is clear from the data in Table 4.1 that there is a major reduction in floral diversity in the Upper Cretaceous. Diversity is often associated with growth conditions; benign conditions give rise to more diversity than less benign conditions. So, at first sight, we could conclude that conditions conducive to growth worsened during the Late Cretaceous on the Arctic Slope. However, there could be other causes such as less favourable circumstances for preservation. The Prince Creek Formation is composed of sandstones, mudstones and bentonites (clays produced from diagenesis of volcanic ash) as well as the thin coals. Overall, this sediment suite is similar to that of the Nanushuk Formation, and if anything, may have been deposited more rapidly if we interpret the higher siliciclastic content of the coals as an indication of inorganic deposition rate relative to peat growth. The fossil leaves of the Prince Creek Formation, like those of earlier deposits, show little evidence of biological degradation (there are very few leaves with evidence of pre- or post-mortem 'chewing' by invertebrates or skeletonization caused by bacterial and fungal activity). This suggests similar rapid burial after being shed from the parent plants.

If preservational biases are an unlikely explanation for the diversity decrease, we need to examine if our other hypothesis, that of environmental deterioration, is supported by other evidence.

4.3.8 Arctic woods

Fossil tree trunks are common in the Arctic Slope Cretaceous sediments either as *in situ* standing stumps (Figure 4.40) or as dispersed logs and branches. To date, no convincing vessels have been found in Arctic Slope Cretaceous woods, which could be taken to indicate that woody angiosperms were absent. However, angiosperm leaves that appear to have been deciduous are abundant.

The evidence for deciduousness includes a preservational style that suggests the leaves were thin, and possessed expanded petiole bases adapted for shedding rather than long-term retention. There are also presumed relationships to modern forms that are predominantly deciduous. Long-lived woody plants produce deciduous leaves, while herbaceous (non-woody) plants have leaves that wither on the plant instead of being shed. Consequently, leaves of herbaceous plants are almost never found in the fossil record. If woody angiosperms were so abundant, a more plausible explanation for the lack of wood with vessels is that vessels were not properly developed in the Arctic woods produced by angiosperms. This is not as far fetched as it might appear because vessels are not universal in all living woody angiosperms.

The absence of vessels is actually advantageous for palaeoclimatic reconstructions because the uniformity of woods composed mostly of tracheid cells means that it is easier to compare climate signals between different specimens.

Box 4.2 The growth of trees

Tree rings are only produced when there is some variation in the rate and type of cell production. Wood cells come in several forms. Those that make up most of the water conduction system and which are aligned up and down a tree are called tracheids (Figure 4.41). These are elongated cells whose contents have died, leaving an open tube. Because these cells are aligned along a trunk or branch axis, they are called axial tracheids. Other tracheids run radially from the centre of a tree out to the bark. These are called radial or ray tracheids which, together with other cells, make up the rays. Ray cells bind the axial tracheids together and limit the tendency for a tree to split lengthways as the branches are produced. They also allow fluid to flow radially, and not just axially, which is important if the tree is damaged and fluid flow has to go around the damaged area. As well as tracheids, there are a variety of other cell types such as packing cells, resin cells, and, in angiosperms, specialized large fluid-conducting cells called vessels (not shown in Figure 4.41).

As a tree grows, new cells are continuously formed by actively dividing cells in the cambium. The cambium lies just under the bark and produces new cells both inwards, towards the centre of the branch or trunk (the tracheids), and outwards. Those that are produced outwards remain alive and conduct the products of photosynthesis around the plant and are known as phloem cells. Phloem cells are rich in sugars and proteins. When the tree dies, they are rapidly attacked by bacteria and fungi and so they decay and are rarely preserved.

Because a single cambium cell divides repeatedly to form a new axial tracheid, these cells, when seen in a section cut across a trunk, are arranged in lines radiating out from the centre of the trunk. Under good growing conditions, a new ring of axial tracheids is produced every few days. When growing conditions deteriorate, the axial tracheid production rate drops and instead of cells with large hollow interiors (lumina, singular lumen) and thin walls that are good at conducting fluids, the cells formed have smaller lumina and thicker walls. These cells are structurally stronger, but are less efficient for water conduction than the cells produced during good growing conditions. Any variation in growing conditions can bring about these changes, but most often seasonal changes in light, temperature, or water availability produce repeated switching from large-lumened thin-walled tracheids to small-lumened, thick-walled tracheids. The large thin-walled cells tend to be produced in the spring when demand for fluid conduction is high (earlywood cells), whereas smaller, thicker-walled cells are produced late in the growing season when the demand for fluids is less intense (latewood cells).

Figure 4.41 (a) Drawing of a piece of branch or trunk of a conifer tree showing three faces or planes of section. The transverse face is that which is commonly referred to as a cross-section and is useful in showing the ring characteristics used for climate studies and dendrochronology. The radial face shows the detail of the ray in longitudinal section, while the tangential face is at a tangent to the ray and shows the ray in cross-section. (b) Drawing of the detail of the tracheids, both axial and radial. Other cell types are present but are not featured here for clarity. *(Eames and MacDaniels, 1947.)*

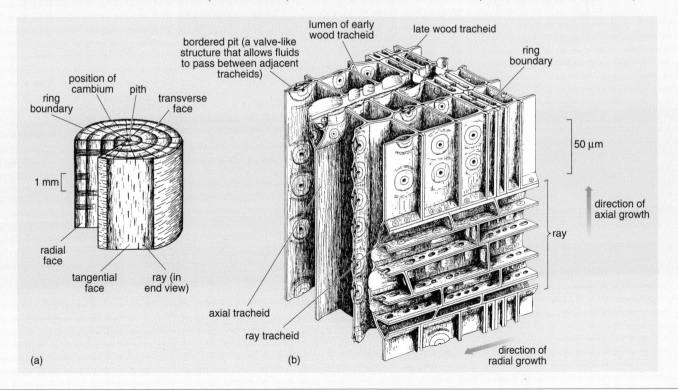

The woods preserved in the Nanushuk Formation typically have wide rings composed of several hundred earlywood cells. These have thin walls and large central spaces adapted for conducting fluid efficiently during the spring and summer when growth is rapid and transpirational demand is high. The amount of latewood cells is small in comparison to earlywood cells, but the ring boundaries are well defined. Latewood cells have thick walls and the cell-conducting space is comparatively small (Figure 4.42). This is not a disadvantage because they are produced late in the summer when fluid demand is low and the plant is entering dormancy. The advantage of the thick cell walls is that they afford structural strength. No cells are produced during the dormant period.

Figure 4.42 Light micrograph of a transverse section of wood from a Nanushuk Formation tree. *(Bob Spicer, Open University.)*

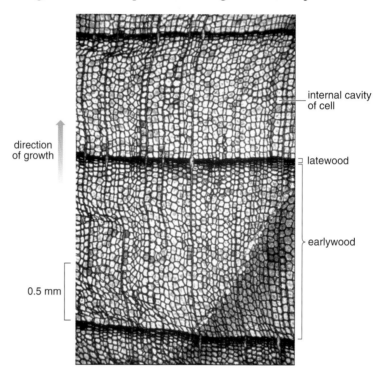

direction of growth

internal cavity of cell

latewood

earlywood

0.5 mm

○ What do you think the wide rings and large production of earlywood in Nanushuk woods indicates about the growth conditions?

● They suggest that growing conditions were benign during the spring and summer. Benign here means that temperatures had to be warm (usually between 10 °C and 40 °C), water had to be in abundant supply, but without drowning the root systems, and there had to be sufficient light for photosynthetic activity to significantly outweigh the respiratory drain on food reserves.

The small amount of latewood suggests the transition from full growth to full dormancy was rapid; that there was not a long period of growth slowdown before winter. This rapid seasonal change from summer to winter is exactly what one would expect in the high Arctic. A re-examination of Figure 4.6 shows that at the palaeolatitudes at which these trees were growing (approximately 75° N) the transition from the summer insolation regime of continuous daylight to continuous twilight or winter darkness takes place during a period of just over two months. The 'break-even' point for photosynthetic productivity over respiratory drain lies somewhere around halfway through that period.

○ What assumption are we making here about the Earth's obliquity in Nanushuk time?

● We are assuming the obliquity was more or less the same as now. In fact, this assumption appears justified because the rings show such small amounts of latewood implying a rapid metabolic shutdown at the end of the growing season. This feature is characteristic of the polar light regime and hence an obliquity that is likely to be within the normal Milankovich variations.

Another noteworthy feature of the Nanushuk woods is the overall rarity of what are called false rings. These are not formed at the end of the growing season but during it, and mark episodes when growth was, for some reason, restricted. As growth slows, the tracheid cells develop thicker walls and smaller cell spaces. In some instances of extreme cold, the cells might even suffer permanent frost damage if thin-walled cells are still being produced when the freeze occurs. To date, no well-documented frost damage has been seen in Nanushuk woods. Several different species of trees occur in the Nanushuk sediments and they all show the same ring features of wide earlywood and narrow latewood. This is an important observation because it suggests that the ring's features were environmentally induced and were not a feature of the growth characteristics of a particular species that could have been genetically coded.

Woods from the Prince Creek Formation (Figure 4.43) show a different set of characteristics from those typified by the wood cross-section shown in Figure 4.42. Here, the earlywood cells are fewer in number in each ring and the latewood correspondingly occupies a higher proportion of the ring width. Also the earlywood growth pattern is periodically interrupted by false rings. False rings may be distinguished from seasonal rings in the following way. In both ring types, the cell wall and lumen dimensions gradually change going into the ring formation phase. However, coming out of a true ring there is a sudden transition to large earlywood cell production, whereas in a false ring the recovery is again gradual.

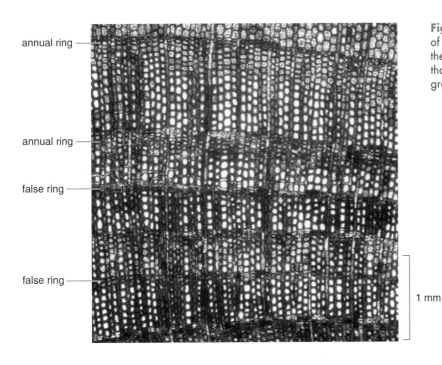

annual ring

annual ring

false ring

false ring

1 mm

Figure 4.43 Light micrograph of a transverse section of part of a log from the Prince Creek Formation. Note the narrower rings and the development of false rings that indicate an interruption of growth rate during the growing season. *(Bob Spicer, Open University.)*

Table 4.2 Comparison of conifer growth-ring characteristics from Late Cretaceous woods of the Nanushuk Formation, the Prince Creek Formation and the Early Tertiary woods from Sagwon.

Parameter	Nanushuk Formation (Albian/Cenomanian)	Prince Creek Formation (Campanian/Maastrichtian)	Sagwon (Palaeocene)
Ring width (mm)			
range	0.4–12.9	0.14–5.88	0.35–4.70
range of means	1.1–4.9	0.39–3.67	0.84–2.98
mean of means	2.81	1.77	1.84
Mean sensitivity			
range	0.28–0.79	0.10–0.77	0.14–0.38
mean	0.44	0.40	0.30
Latewood			
number of cells	1–15	1–>30	2–12
ratio to earlywood	max. 0.30	max. 0.83	max. 0.38
False rings	rare	abundant	common

Table 4.2 summarizes the ring characteristics of the Nanushuk and Prince Creek Formation woods but also includes features for Early Tertiary (Sagwon) woods from the Arctic Slope.

○ Looking at the mean ring widths given in Table 4.2, the ratio of numbers of latewood to earlywood cells and the false ring observations, how would you describe the changes in growing conditions from the mid-Cretaceous to the Early Tertiary on the Arctic Slope?

● Growing season conditions in Nanushuk time were generally benign, but then worsen as the Late Cretaceous progressed. The narrower rings of the Prince Creek woods suggest less benign growing conditions throughout the spring and summer and the frequent false rings indicate that even these relatively poor conditions were interrupted by even worse events. In the Early Tertiary, spring and summer conditions improved again.

There is one measure in Table 4.2 that needs explanation and that is mean sensitivity, a measure of interannual variability in ring width. It is calculated as follows:

$$\text{mean sensitivity} = \frac{1}{n-1} \sum_{t=1}^{t=n-1} \left| \frac{2\,(x_{t+1} - x_t)}{x_{t+1} + x_t} \right|$$

where x_t is the width of ring t and x_{t+1} is the width of the adjacent younger ring. Woods with mean sensitivities of <0.3 are termed 'complacent' and are interpreted to have grown under conditions that were stable from year to year. Woods with mean sensitivities >0.3 are termed 'sensitive', suggesting variable conditions from year to year. Sensitive trees of a particular taxon live at the edges of the range of that taxon, and the sensitivity is generally linked to climatic effects, although other factors, such as waterlogging of roots, also can affect sensitivity. Table 4.2 indicates that all the Arctic Slope woods can be termed 'sensitive' and are potentially good indicators of climate.

Although climate is most important in determining mean sensitivity, it is not the only factor, as the sensitivity of the Nanushuk woods may have been due to a shifting sedimentological and hydrological environment, rather than to climate directly. The woods generally lack other characteristics, such as significant latewood and false rings, which would have indicated a stronger climatic effect on the growth of the trees. The sensitivity of most of the Prince Creek woods could also have also been due to variations in sedimentology and hydrology, as they were deposited in similar environments to those of the Nanushuk Formation. However, the Prince Creek woods exhibit additional characteristics which indicate a somewhat more severe climate than that encountered by the woods from the Nanushuk Formation. Nevertheless, many of the individual woods in the Prince Creek were complacent, suggesting that, although climate might have been more severe overall, the interannual variability was not extreme.

If overall growth conditions in Prince Creek time were less benign than earlier or later, what were the likely causes of the false rings? False rings were abundant in the woods from the Prince Creek Formation, whereas only one specimen of wood from the Nanushuk Formation has been observed to have false rings. Multiple false rings within a single growth ring have not been observed in woods from the Nanushuk Formation. Insect attack (reducing the leaf area available for photosynthesis) as a cause of false rings in the Prince Creek woods is unlikely because there is no evidence for insect attack in either the woods or the associated leaves, and because many growth rings have more than one false ring. Insects in seasonal climates tend to have rigid life cycles and attacks on trees by a given species of insect will occur during a relatively constrained time period. So, if the false rings were formed during the stress of insect attack, several species of insects would have had to have been involved. There is no evidence for even temporary drought in the Prince Creek sedimentary deposits; indeed, the system was very wet. Fire and freezing are both plausible explanations for growth disruption in Prince Creek woods. Charcoal is abundant in Prince Creek sediments (the presence or absence of charcoal can be useful in distinguishing these rocks). However, none of the false rings so far examined shows any evidence of direct charring. Freezing, or at least temperatures dropping to +4 °C where biological activity is effectively stopped, is an equally likely explanation, given the adverse conditions suggested by the paucity of earlywood.

Clearly, there is evidence in the tree rings of a decline in Arctic Alaskan growing conditions as the Late Cretaceous progressed. This change could be the result of the continued approach by the region to the pole were it not for the fact that there is a clear reversal of the trend in the Early Tertiary. As there is no sudden southward movement of the region (at least not one sufficient to explain the tree ring changes), it seems likely that what is being detected here is a change in global climatic conditions. To test this we need to look elsewhere and ideally characterize and quantify the climatic change.

4.4 Russia

The north-eastern part of Russia, often mistakenly called Siberia, has a wealth of Cretaceous plant-bearing rocks, and coals are abundant. Figure 4.44 is a map of the area showing the geological regions as identified by Russian geologists. Of particular interest is the Okhotsk–Chukotka volcanogenic belt — a region of active volcanism from the Late Aptian through the Late Cretaceous. We are going to look at three areas in some detail (the Grebenka, Chauna and Vilui Basin areas) before providing a broader overview.

Figure 4.44 Map of Alaska and north-eastern Russia showing some significant geological regions. Marked regions are as follows: 1, Arctic Slope of Alaska; 2, Grebenka area; 3, Chauna area; 4, Vilui Basin. *(Kelley et al., 1999.)*

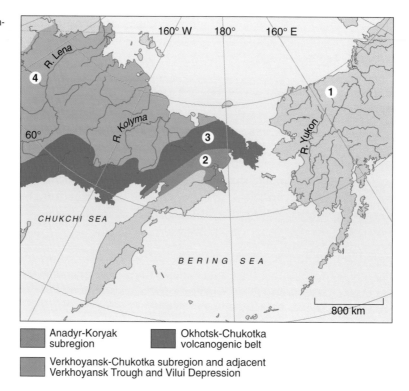

Anadyr-Koryak subregion

Okhotsk-Chukotka volcanogenic belt

Verkhoyansk-Chukotka subregion and adjacent Verkhoyansk Trough and Vilui Depression

4.4.1 The Grebenka flora

The first case study from Russia is that of the Grebenka flora. This flora has been regarded as the most important mid–Late Cretaceous palaeofloristic record in north-eastern Asia. During the past 40 years, every publication discussing north-eastern Asian regional phytostratigraphy and palaeobotany has referred to it. Moreover, the Grebenka flora seems to be one of the most diverse mid-Cretaceous floras of Eurasia, and possibly of the world, despite its position within the Cretaceous Arctic Circle.

The Grebenka flora is so-called because the plant remains come from a number of localities along the Grebenka River, a tributary of the Anadyr River (Figure 4.45).

The fossils occur in exposures of the Krivorechenskaya Formation and include specimens of *Nilssoniocladus* that, in combination with material from northern Alaska, were used to understand the biology of the Arctic cycadophytes. The main plant-bearing locality is situated approximately 50 km to the south and east of the Okhotsk–Chukotka volcanogenic belt and the succession was deposited in a fore-arc basin, during the accretion of the Kony–Murgal Arc onto the north-eastern Asian continental margin, i.e. the final closure of the Mongol–Okhotsk Ocean. Recent palaeomagnetic work from the plant-bearing outcrops themselves suggest a palaeolatitude of 72° N and the age of the beds, based on ^{40}Ar/^{39}Ar dating of volcanic ashes in the succession, is 96.5 ± 0.5 Ma. This is in agreement with a late Albian to early Cenomanian age based on marine fossils in correlative beds, palynological work and palaeomagnetostratigraphy as shown in Figure 4.46.

The ashes concerned appear to have been minimally reworked because the age spread of individual mineral grains is very small. This suggests they were all derived from the same short-lived set of eruption events. The Krivorechenskaya Formation reflects depositional environments of alluvial to coastal plains and

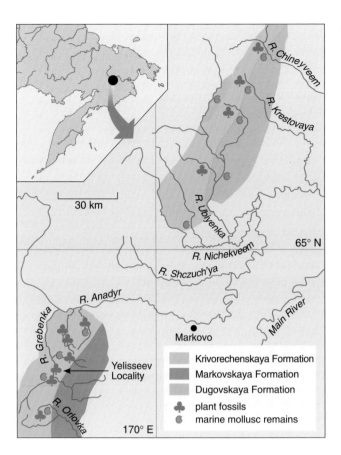

Figure 4.45 Map of Chukotka, north-eastern Russia, showing principal fossil localities and geological mapped units. *(Spicer et al., 2002.)*

Stage	International zonal standard (Pokhialainen, 1994)	Ammonite zones of NE Russia (Pokhialainen, 1994)		Inoceramid zones of NE Russia (Pokhialainen, 1994)		Stratigraphical position of the plant fossil localities (Shczepetov & Herman, 1990): According to ammonites and inoceramids	Stratigraphical position of the plant fossil localities (Shczepetov & Herman, 1990): According to palynology and palaeomagnetic data
Lower Turonian	*Mammites nodosoides*	*Marshalites tumefactus*		*Inoceramus labiatus*			
	Watinoceras coloradoensis					Orlovka Mountain Locality	
Cenomanian	*Neocardioceras juddi*			*Pergamentia reduncus*			
	Metoicoceras geslinianum						
	Calycoceras guerangeri						
	Alternacanthoceras jukesbrownei						
	Acanthoceras rhotomagense	*Turrilites costatus*		*Pergamentia pressulus*			
	Mantelliceras dixoni	*Neogastroplites americanus*	*Hypoturrilites gravesianus*	*Gnesioceramus comancheanus*	*Inoceramus dunveganensis ajensis*	Gornaya River Locality	
	Mantelliceras mantelli						Grebenka River Locality
Upper Albian	*Stoliczkaia dispar*		*Pseudhelicoceras mordax*		*Inoceramus concentricus sulcatus*	Malaya Grebenka River Locality	
	Mortoniceras inflatum						

Figure 4.46 Correlation chart demonstrating the stratigraphical position of the main Grebenka River plant fossil locality.

adjacent shallow marine basins. The geology of the region is dominated by the influence of the nearby volcanogenic belt, which was the source of the clastic material that entombs the fossil flora.

Figure 4.47 shows a graphic log of part of the primary plant-bearing locality. One thing that should be immediately obvious is the coarse-grained nature of many of the sedimentary deposits compared to those in the section we looked at on the Kukpowruk River. This suggests a much more energetic depositional regime. The framework clasts of the conglomerates are epiclastic, i.e. composed of reworked pre-existing volcanic rocks of the Okhotsk–Chukotka volcanogenic belt. Most sandstones are dominated by volcanic rock fragments that exhibit a plethora of shapes and compositions, but each grain is typically dominated by a fine-grained groundmass of plagioclase laths. In some horizons, they clearly outnumber the volcanic rock fragments, i.e. they form arkoses, which also are rich in hornblende and biotite. These accessory minerals are comparably fresh and only slightly abraded (angular to subangular), and were therefore regarded as suitable for radiometric dating. Quartz grains are only occasionally present. The recorded detrital compositions broadly correspond with that of the inferred andesitic sediment sources of the Okhotsk–Chukotka volcanogenic belt.

Most of the stream channel sandstones are made up of <1 m-thick irregular sheets, or isolated channel fills within gravelly facies, within the channel package. The sandstones range from medium- to very coarse-grained, and are typically poorly to moderately sorted. Some sandstone beds show normal grading. They are dominated by shallow, low-angle troughs, commonly with pebble lags, but do also include horizontally laminated sandstone beds. Scour-and-fill units also occur. The facies association includes very few tabular and trough cross-bedded co-sets and a notable lack of rippled beds. Dune foreset migration palaeodirections indicate broadly southward sediment transport when corrected for tectonic tilt. Reworked (size-sorted and abraded) fossil logs are abundant along some bedding planes, and constitute the only observed biota. This facies association commonly shows a sharp, slightly erosive base, and is interbedded with stream channel conglomerates.

Organic matter is present throughout the succession, as solitary detrital particles of silt and sand size, or as large wood fragments. *In situ* rootlets and pedogenic (soil-related) textures frequently occur, but almost all the palaeosols are immature. Siltstones are dominated by fragmented angular framework grains and are rich in dispersed organic matter throughout.

Unlike the Arctic Slope sediments, the Grebenka succession contains conglomerates. These are matrix-rich, clast-supported and are moderately to well sorted. The clasts are typically well rounded and of a high sphericity, whereas the matrix is uniformly composed of moderately to well-sorted coarse-grained sandstone which corresponds compositionally and texturally to most sandstones of the section. Overall, the sedimentary signatures indicate an active fluvial system. Channel fill, bar, levée and crevasse sheet sediments are all represented, as well as those accumulated in floodplain ponds. Floodplain interfluve palaeosol deposits are also present and these preserve plant remains from mature forests.

Clearly, this represents a more energetic system than the ones so far considered in Alaska. Coarse-grained sediments, deposited rapidly and at higher river energies, are dominant. Almost all of the palaeosols are immature, suggesting that plant communities colonizing new sediment surfaces had little time to become fully established before they were inundated by more sediment. This sediment created

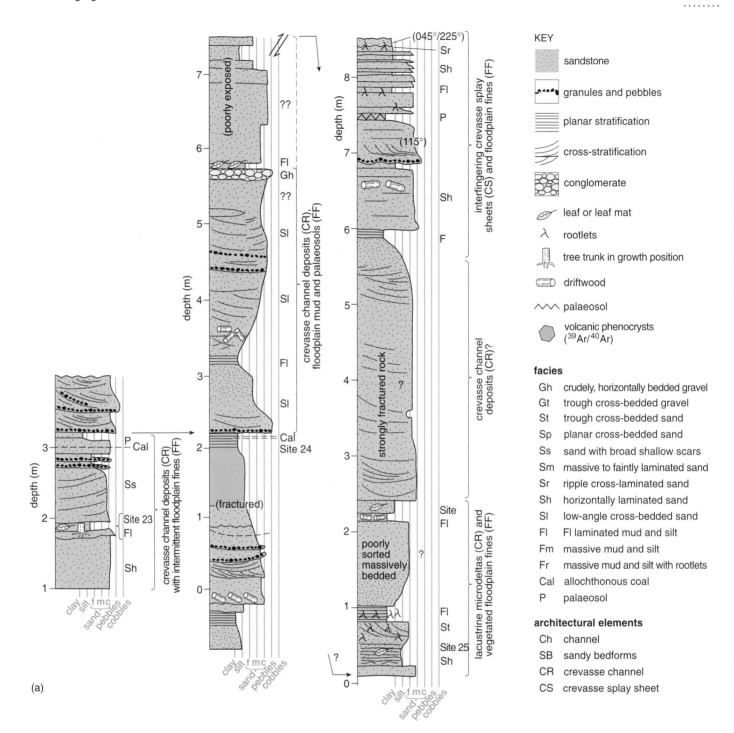

Figure 4.47 (a) Part of a graphic log describing the main exposure from which the Grebenka Flora was collected. *(Spicer et al., 2002.)*

yet newer sites for colonization. Overall, the channel system seems typical of a braided architecture rather than lower-energy meandering rivers. As with any braided river system, rapid choking of channels, channel switching, and variable flow regimes created a patchwork of environments for plant colonization. Equally high deposition rates enhanced the preservation potential of these communities, which explains the diversity of the preserved flora.

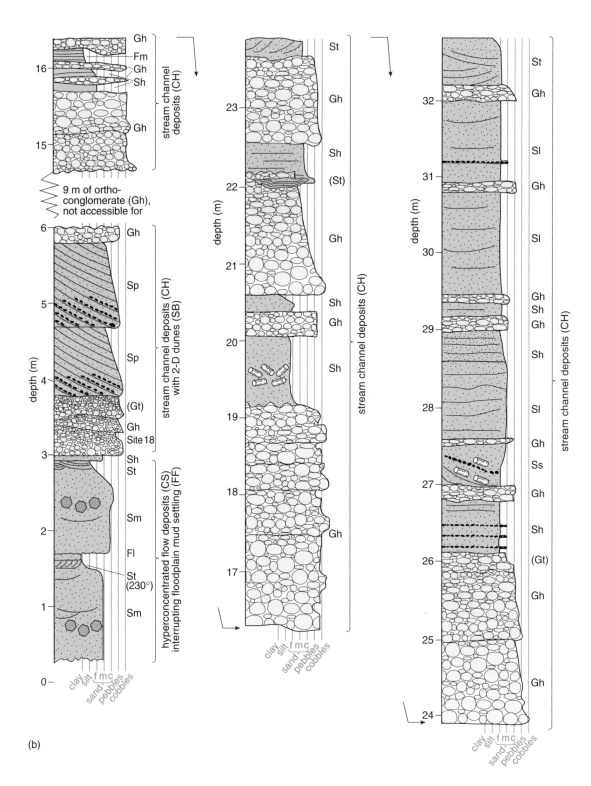

Figure 4.47 (b) Part of a graphic log describing the main exposure from which the Grebenka Flora was collected.

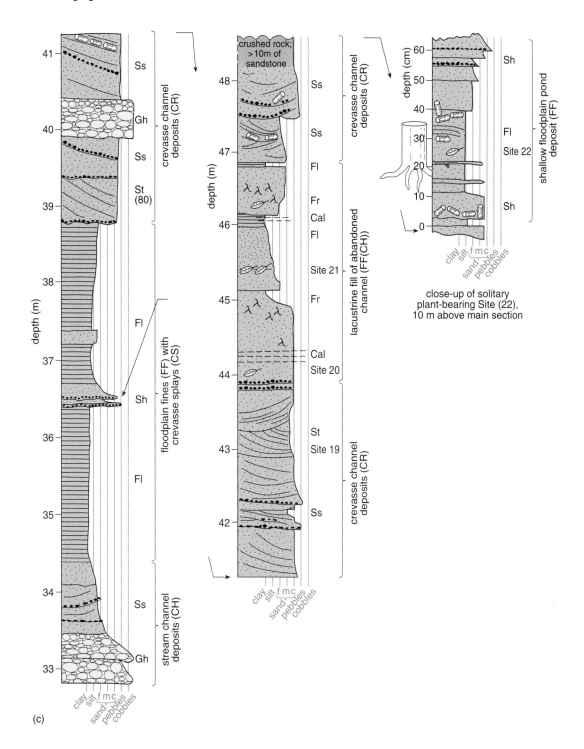

Figure 4.47 (c) Part of a graphic log describing the main exposure from which the Grebenka Flora was collected.

By looking for repeated associations between sedimentary facies and plant remains, it has been possible to reconstruct not only the community composition but also the dynamics of community development. Plant fossils occur in abundance, and particularly rich horizons have been studied in detail. These are identified as 'sites' in the sedimentary log and they are numbered. Using data from these sites, the vegetation dynamics of the region have been reconstructed (Figure 4.48).

Figure 4.48 Flow diagram describing the development of vegetation from the early pioneer stage through to mature forest communities in the early Cenomanian of north-eastern Russia, based on the assemblages preserved in the Grebenka River region. Genus names in bold indicate the most consistent and abundant components of the different associations.

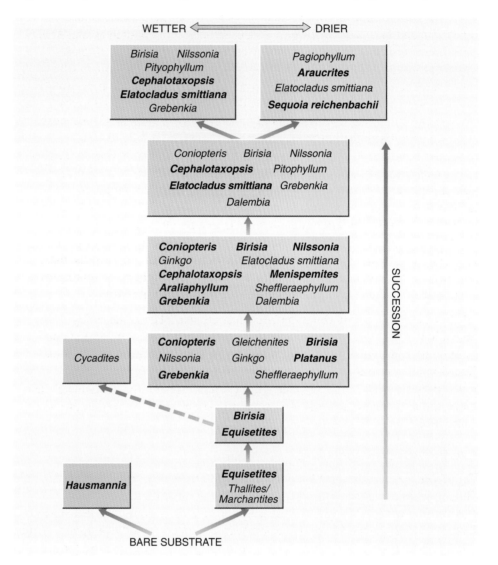

Starting with a bare substrate, an *Equisetites* and bryophyte (mosses and liverworts) pioneer community may develop if the surface is not permanently flooded. If this is not the case, the pond may become colonized by the water fern *Hausmannia*. Assuming a drier situation, the *Equisetites* community may be invaded by the fern *Birisia* forming the *Birisia* marsh that we saw in Alaska. However, in Grebenka, the next phase in the succession saw the arrival of angiosperms typically represented by *Platanus*, *Grebenkia*, and *Sheffleraephyllum*. These were probably woody bushes rather than trees and grew with the remnants of the *Birisia* community mixed with other ferns such as *Gleichenites* and a few ginkgos. As time went on, and if the area was not inundated by more sediment, more species would have invaded including other angiosperms (e.g. *Araliaphyllum*), climbers (*Menispermites*), the cycads (*Nilssonia*), and conifers (*Elatocladus* and *Cephalotaxopsis*). Eventually, the conifers became more dominant and, in drier sites, the more xerophytic forms with small leathery leaves such as *Pagiophyllum*, *Araucarites* and *Sequoia* took over while the damp-loving plants such as the ferns disappeared. Examples of these different leaf forms are shown in Figure 4.49 (overleaf).

There are some important similarities between the Alaskan Cenomanian vegetation and that of north-eastern Russia. Both were deposited at similar high latitudes in subsiding basins. However, sedimentation rates in north-eastern Russia were much higher because of the abundant supply of volcaniclastic material from the Okhotsk–Chukotka volcanogenic belt. This high rate of siliciclastic sedimentation prevented the development of long-lived mires because any low-lying land was quickly inundated by sediment. Consequently, there are no coals in the Grebenka sediments. However, what the high sedimentation rates did was to preserve the leaves representing a wide range of plant communities growing on the sediment surfaces. This is the reason for the high species diversity of the flora. Moreover, by analogy with modern sedimentary systems, the hundred or so metres of sediment exposed at the main plant locality probably represents no more than a few thousand years of accumulation. If this is so, then the plant remains represent conditions that existed for a fraction of the shortest Milankovich cycle. In other words, there is little time-averaging and any climatic information we might glean represents a reliable snapshot of conditions that existed at a single point in time.

What does all this tell us about the Grebenka climate? The modern-day vegetation of the Grebenka region is rather like that of much of Alaska: open tundra with some small thickets of trees and bushes and overall a low species diversity. In contrast, the Cenomanian vegetation was much richer and communities in stable sites were heavily forested. Apparently, the climate was warmer, but by how much it is difficult to say at this stage. We will return to this issue shortly.

4.4.2   The Chauna flora

The second case study from Russia is from within the heart of the Okhotsk–Chukotka volcanogenic belt. The eruption of vast quantities of ash over a sustained period beginning in the Late Aptian and lasting until the very Early Tertiary not only aided the preservation of fossil material on the periphery of the belt, but also provides us with a record of the vegetation within it. So far, we have been considering vegetation growing at or near the Cretaceous sea-level, but in the case of this second Russian flora, the Chauna flora, the heavy ash falls on captured communities that were living in intermontane basins within the volcanic highlands.

The Chauna Group, within which the flora is found, consists of large-volume ignimbrites and andesitic basalts and tuffs (Figure 4.50, overleaf). Determining the relative age of the flora based on comparative floral biostratigraphy proved problematic because the Chauna flora had a composition quite unlike any other fossil flora from north-eastern Russia.

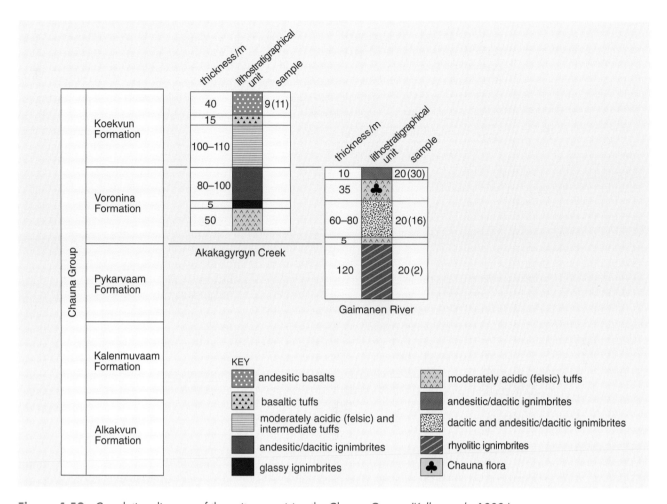

Figure 4.50 Correlation diagram of the units comprising the Chauna Group. *(Kelley et al., 1999.)*

○ Why do you think the composition of the flora might have been so different?

● There are two possible reasons. First, the flora represents plant communities which are living at higher elevations than most of those preserved as fossils in coastal plane settings. Secondly, large thickness and frequent ash falls produce substrates that are both well-drained and nutrient-rich. These soil conditions lead to the development of specialized communities that are difficult to compare with those on more 'normal' substrates.

The Chauna Group consists of five formations made up of over 4000 m of rhyolitic tuffs, andesite–dacite tuffs, ignimbrites and lavas, and tuffaceous sandstones and conglomerates. Not only did the type of erupted material change with time but also great thicknesses of volcanic material were produced. The five formations of the Chauna Group cannot be distinguished on the basis of their floral assemblages that collectively represent a single floral type typified by the Chauna flora (Figures 4.51, 4.52). The flora is characterized by the presence of

Figure 4.49 Photographs of some of the different leaf forms from the Grebenka flora: A, *Hausmannia*; E and H, *Araucarites*; B, *Pagiophyllum*; I, *Elatocladus*; F, *Sequoia*; G, *Cephalotaxopsis*; M, *Nilssonia*; K, *Scheffleraephyllum*; C and D, *Gleichenites*; J, *Menispermites*; L, *Platanus*; N, *Grebenkia*. Note that specimens F, G and I all have distinct clusters of small leaves at their bases suggesting these leafy shoots were discrete entities that became detached near these bases instead of being fragments of larger shoots that just happen to have broken off. This may well indicate that the parent plant shed these shoots regularly and may have been deciduous. Such features are not seen in plants such as *Araucarites* (E and H). *(Bob Spicer, Open University.)*

endemic (i.e. native to that area) ferns (*Kolymella* and *Tchaunia*, Figure 4.51b, e), large-leaved cycadophytes (*Heilungia* and *Ctenis*, Figure 4.51c, d, h, j), a group of uncertain affinity, the czekanowskialeans (*Phoenicopsis*, Figure 4.51g), numerous conifers more common in post-Cretaceous rocks, and an extreme rarity of angiosperms. The same flora persisted throughout the period of deposition of the Chauna Group and relatively advanced forms (such as *Quereuxia* and *Trochodendroides*, Figure 4.52f, h, i) occur even in the lower formations, indicating that the Chauna sediments were deposited quite quickly in geological terms.

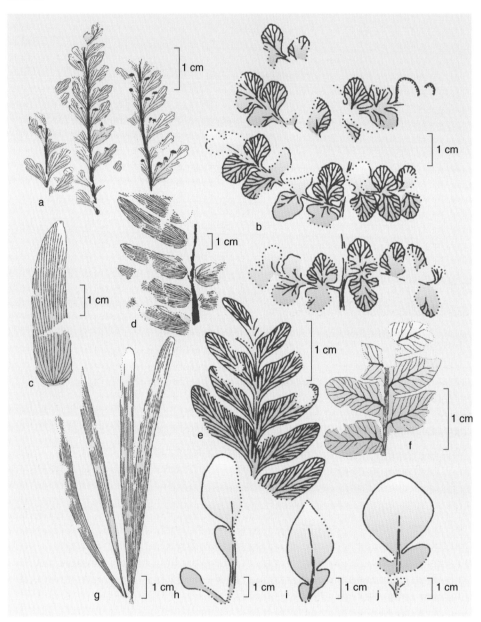

Figure 4.51 Some characteristic leaf forms from the Chauna flora: a, *Coniopteris*; b, *Kolymella*; c and d, *Heilungia*; e, *Tchaunia*; f, *Cladophlebis*; g, *Phoenicopsis*; h, i and j, *Ctenis*. (Kelley et al., 1999.)

Estimates of the age of the flora based on the peculiar mix of old and relatively modern species have ranged from the Albian through to the Turonian. However, recent ^{40}Ar/^{39}Ar ages for the Chauna Group show that over a stratigraphical

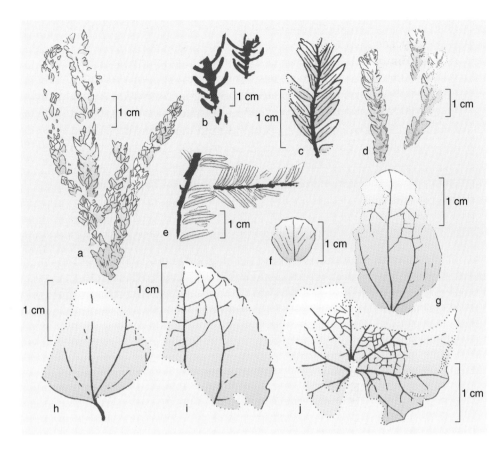

Figure 4.52 Some characteristic leaf forms of the Chauna flora: a, *Pagiophyllum*; b, *Araucarites*; c, *Metasequoia*; d, *Pagiophyllum*; e, *Taxus*; f, *Quereuxia*; g, *Zizyphoides*; h and i, *Trochodendroides*; j, *Menispermites*. (Kelley et al., 1999.)

sampling interval of 350 m all sample measurements indicate an age of 88 Ma and have uncertainties that overlap, suggesting the eruption of a large volume of material over a short interval of time. This is consistent with the observation that the floral assemblages are indistinguishable over an even greater stratigraphical thickness. By combining the radiometric and floral biostratigraphical data from rocks above and below the Chauna Group volcanics, we can say that the formation of the northern Okhotsk–Chukotka volcanogenic belt must have taken place during the Turonian–Campanian interval and not the late Albian–Turonian as had previously been thought.

This has important implications not only for our understanding of the evolution of the Arctic floras, but also the overall Arctic environment during the Late Cretaceous. We know from evidence such as the Grebenka sediments that volcanism in north-eastern Russia was in full swing by the early Cenomanian and, in fact, other radiometric data confirm an early Albian start for large-scale volcanism in the Okhotsk–Chukotka volcanogenic belt. The Chauna Group data reveal that during the mid–Late Cretaceous the volcanism was if anything even more vigorous. Plants colonizing this highly disturbed, edaphically (i.e. determined by the soil) dry but nutient-rich, landscape were a mix of relictual mid-Cretaceous conifers and ferns together with evolutionarily advanced conifers and rare angiosperms.

The persistent large-scale volcanism must have had a profound effect on the Arctic environment. The provenance of the abundant Late Cretaceous bentonites in northern Alaska has long been a contentious issue because of the dearth of proximal sources. The first thin Cretaceous bentonites occur in the Albian in the western North Slope, e.g. the Kukpowruk River exposures of the Nanushuk Formation, while thick ashes occur later in the Cretaceous and are found further east. For the tuffs in northern Alaska, the $^{40}Ar/^{39}Ar$ ages on biotite and feldspar range from 100 Ma to 67 Ma, but most are 93–85 Ma. It is possible then that these Alaskan bentonites were sourced in north-eastern Russia, and the eastward penetration reflects greater volumes of ash, ejected into the atmosphere to greater heights, as a result of increasing intensity of eruptive activity. The ages of the Chauna Group tephras match the ages of the most abundant ash deposits in northern Alaska.

If there were extensive ash plumes blowing to the east, is there any evidence for winds in the opposite direction? The answer to this question is a clear 'no'. To the west of the Okhotsk–Chukotka volcanogenic belt there is a large Cretaceous Basin, the Vilui basin, close to the Yakutian city of Viluisk, but mid- to Late Cretaceous rocks are totally devoid of any ashes or diagenetic products of them. This shows clearly that, just as today, the predominant wind patterns in the Arctic circled the pole from west to east. The Vilui basin is our third and final Russian case study. Here we will examine what the vegetation and climate was like in a Cretaceous continental interior.

4.4.3 The Vilui flora and the continental interior environment

A simplified geology of the Vilui basin is shown in Figure 4.53. The broadly synclinal basin is filled with non-marine Cretaceous sediments overlying Jurassic marine rocks. Stream channel sandstones, interbedded with tabular floodplain deposits, dominate the Early Cretaceous strata exposed in the basin. The floodplain units include autochthonous coal seams and stacked palaeosols. Mid–Late Cretaceous successions show strong and upwardly increasing channel cannibalism and reworking of floodplain and levée deposits. Basal channel deposits often include mudstone and peat balls, slumped tree bases, driftwood and log jams. The fossil soils are mostly rare and are immature when they do occur. Deposits that accumulated in abandoned channels often contain well-preserved leaf assemblages. This is particularly so where there are rhythmical variations in grain size indicative of periodic (seasonal?) reactivation of channel stream flow. Preservation of delicate leaves suggests limited downstream transport prior to deposition.

In the central parts of the Vilui basin, the Timerdyakh Formation is overlain by the Linde Formation. This is a pebbly to sandy cross-bedded siliciclastic unit dominated by minimally transported kaolinized basement material, which probably resulted from local tectonic rejuvenation. The Early Cretaceous detrital mudstones and palaeosols of the Vilui basin have kaolinite and smectite in approximately equal amounts whereas kaolinite dominates, sometimes together with illite, but without notable smectite content, in the mid–Late Cretaceous Timerdyakh and Linde formations. This apparent change in kaolinite to smectite ratio most likely reflects a change from generally more arid to humid weathering conditions in the hinterland (see Box 4.3 for an explanation of the relationship between clay minerals and climate), from mid-Cretaceous times onwards, but overall there is no evidence for pronounced aridity.

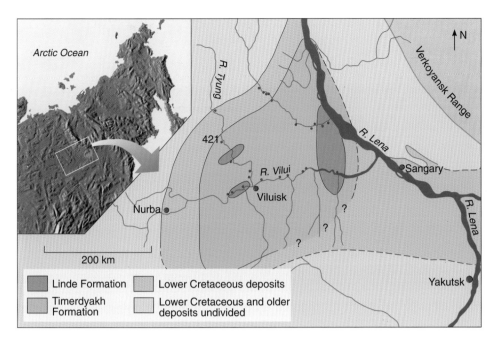

Figure 4.53 Map of the Vilui basin showing the principal geological units. Fossil sample sites are shown as numbered red dots.

Box 4.3 Clay minerals and their relationship with climate

The weathering sequence for rocks in various climatic conditions provides the basis for using clay minerals as climatic indicators. A typical sequence might be as follows:

rock → illite (from feldspars) → chlorite (from Fe–Mg minerals) → montmorillonite → kaolinite (in warm, humid conditions).

Other clay minerals such as vermiculite form under cool climates, whereas smectite forms under warm climates with seasonal rainfall (like bentonite, smectite can be formed from the breakdown of volcanic ashes).

It is important to note that clay minerals indicative of a given climatic regime need not dominate the clay assemblage in any sample to have climatic significance, particularly as some clay types (e.g. sepiolite, attapulgite and palygorskite) are indicative of warm, arid conditions and are rare.

Fossil pollen is locally abundant in the Vilui sediments. In the Timerdyakh Formation, 190 taxa have so far been recovered. Of these, 61 are spores (at least 33 represent ferns), 16 are gymnosperm pollen, 10 are monocot pollen of uncertain affinities (including two probable palm species), and 87 are dicot pollen. Not more than 10% of the assemblage shows evidence of reworking. The remarkable diversity of palynomorphs, together with presence of thermophilic (warmth-loving) taxa, indicate warm and humid climatic conditions, similar to those of today's warm temperate regions. However, palaeomagnetic studies of exposures along the Vilui River suggest, like Grebenka, a palaeolatitude of 72° N. The majority of the pollen grains are well preserved (Figure 4.54) and are often still found in the clusters in which they formed. Such clusters suggest that at least some pollen was minimally transported but, in general, pollen and spores tend to be carried further than leaves by wind or water before they become degraded.

Figure 4.54 Scanning electron microscope images of examples of the well-preserved Late Cretaceous pollen grains and spores from the Vilui basin. The bottom-right image shows a cluster of four of the grains shown in the image in the middle of the bottom row. *(Unpublished scanning electron micrographs courtesy of Dr. Christa-Charlotte Hofmann.)*

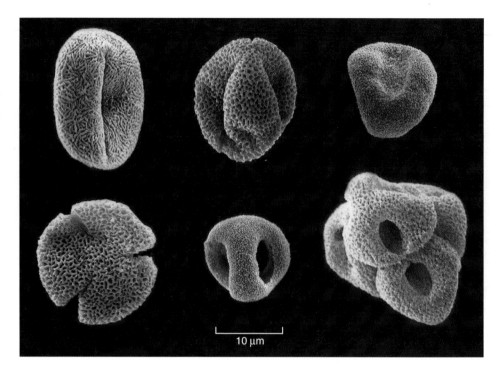

10 µm

Such palynofloras therefore give us more of a regional picture of the ancient vegetation than leaf assemblages do.

○ What are the possible disadvantages in using pollen and spore data for vegetation and climate reconstructions?

● Pollen and spores, being more robust than leaves, can be easily reworked from older sediments into younger sediments. Also, different plants produce spores and pollen in different amounts, so giving a biased record of the relative frequencies of the source plants.

Use of reworked pollen and spore data 'smears' any temporal resolution because the palynological assemblage that includes reworked grains represents a 'time-averaged' view of vegetation. Moreover, each time reworking takes place, grains tend to be moved greater distances from their source, thus blurring spatial resolution. Reworking may be detected by detailed examination of the spore wall, in particular by using the scanning electron microscope (SEM). Any erosion or other alteration of the grain wall pattern suggests reworking. It is therefore significant that the pollen from the Vilui basin shows only a minor reworked component.

Pollen grains are also useful in providing a relative age for the sediments. The lack of volcanic ash means that we cannot use radiometric-dating methods for these deposits. Instead, we have to use palynostratigraphy despite the obvious potential pitfalls introduced as a result of reworking. In this case, because reworking is minimal, the palynostratigraphical information is robust.

Leaf fossils from the Timerdyakh Formation consist of conifer leafy shoots, some ferns, and an abundance of angiosperm leaves. Some leaves are quite large (>15 cm across) and in friable clays are very difficult to collect. However, there are also many small leaves so the overall size range of the leaves is quite large.

Figure 4.55 A selection of angiosperm leaves from the Timerdyakh Formation, Vilui basin. Note the range of leaf size and the presence of both smooth and toothed margined forms (see Section 4.5). *(Bob Spicer, Open University.)*

At least 25 different morphotypes (broadly equivalent to species) have been recovered and, although this is but a fraction of the diversity indicated by the pollen, this is sufficiently diverse to determine quite a lot about the palaeoclimate of the region. Qualitatively, the leaf types (examples of which are shown in Figure 4.55) indicate a temperate, quite warm, climate with abundant water supply. If the Vilui basin climate had been arid, the leaves would all have been small; large leaves cannot grow in dry climates because they lose too much water to the surrounding atmosphere.

○ Does this suggest high rainfall in the Vilui basin when the leaves were alive?

● Not necessarily, because we are sampling leaves that accumulated in a large river floodplain. The water available to the plants might have fallen on the surrounding highlands at the basin margin while the climate in the basin itself might have been drier.

To recap, the sedimentary deposits suggest that the Vilui basin floodplain experienced fluctuating river discharge, and these fluctuations occurred regularly as evidenced by rhythmical variations in sediment grain size. Such regular fluctuations could have been produced seasonally. The preservation of large and otherwise delicate leaves suggests minimal downstream transport before burial so the leaf flora at least represents plants growing close to the site of final deposition. The large size of some leaves suggests a wet regime but this could have been due to proximity of rivers rather than within-basin precipitation. However, clay mineralogy also suggests humid conditions, and both floral composition and clay mineralogy indicate a degree of warmth.

As useful as these observations are, they are only qualitative. What we need is a method of determining ancient air temperatures in a quantitative manner. This is what we will be looking at in Section 4.5.

4.5   Nature's ancient meteorologists

We have already seen that some aspects of plant fossils carry an important climatic signal. Tree rings yield qualitative data on growing conditions during the period of active cell production while the overall appearance or physiognomy of the vegetation allows us to determine thermal conditions such as mean annual temperature or temperature range even though precision is often low. Similarly, we can say something about relative wetness or dryness, e.g. from the abundance of coal, but this can be complicated by sediment supply rates and rates of subsidence.

We are now going to examine how the leaves of woody dicotyledonous angiosperms can provide much greater precision for determining both ancient temperatures and, with less precision, estimates of the moisture regime. In the case of the Grebenka flora, these estimates can be very precise because:

- the diversity of the woody dicots (84 species or morphotypes) is high;
- the rates of deposition were high (resulting in good preservation and the 'capture' of a climate signal averaged over only a short time interval);
- a wide variety of communities are represented.

This last point is important because it means that we are not dealing with an assemblage that represents just a local microclimate, nor one that is highly disturbed and so not in equilibrium with the environment. The Grebenka flora does contain assemblages from disturbed settings, but it also contains many that represent well-established communities. Overall, the greatest angiosperm species diversity is in mature communities even though, in this case, conifers often dominate the climax or mature 'steady-state' vegetation.

4.5.1   Appearance is everything

The relationship between leaf physiognomy and climate was first noticed early in the 20th century, but before the advent of computers this relationship was restricted to single characters such as leaf margin type. Bailey and Sinnot in a 1915 paper in *Science* noted that in warm climates woody dicots tend to have leaves with entire (smooth) margins, while in cool climates there are more species with toothed leaves. The relationship is represented in Figure 4.56 from a paper by Jack Wolfe (1993), then of the US Geological Survey, who examined the relationship in the humid forests of south-east Asia.

Figure 4.56 Plot of mean annual temperature (MAT) versus percentage of leaf margination. The relationship is good and by examining the ratio of toothed to non-toothed woody dicot leaves in a fossil assemblage it should be possible to estimate the MAT when those leaves were growing. There is some scatter around the red regression line and this provides a measure of the statistical uncertainty in the estimate.

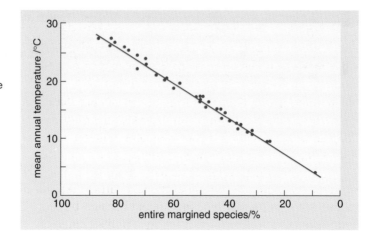

This is an impressive relationship, particularly for biological material that usually has considerable variability. Why should there be such a strong relationship with climate?

The fact that a relationship exists between vegetation and climate is obvious even to someone without specialist knowledge. The overall appearance of vegetation, i.e. its physiognomy, but not necessarily its species composition, tends to be characteristic of the particular climatic regime in which the vegetation grows. Anyone looking at a picture of rainforest vegetation knows it represents warm and wet conditions, while desert vegetation suggests a much drier climate. Such casual observations have been developed into a number of schemes for relating the physiognomy of vegetation to climate in more quantitative ways. An example of this is the life zones map of Holdridge (Figure 4.57). Vegetation can therefore be classified physiognomically, and climatic signals can be derived from it independently of taxonomic considerations. This clear relationship between physiognomy and climate arises because only those plants with a genome that can generate particular appropriate morphological features (phenotypes) are able to survive in a particular climatic regime. Inappropriate phenotypes fail to survive, either through direct environmental elimination (in which case the physical environment takes its toll directly), or by competition with better-suited plants (eventually leading to the demise of the less well-suited, and therefore less competitively successful, individuals).

Figure 4.57 Global vegetation categorized on the basis of plant architecture instead of the more usual taxonomy (e.g. species or major plant group composition). A key has not been provided here as the basis for vegetation classification is complex. However, note that the different vegetation types reflect latitude (a proxy for warmth) as well as altitude. *(Valdes et al., 1999.)*

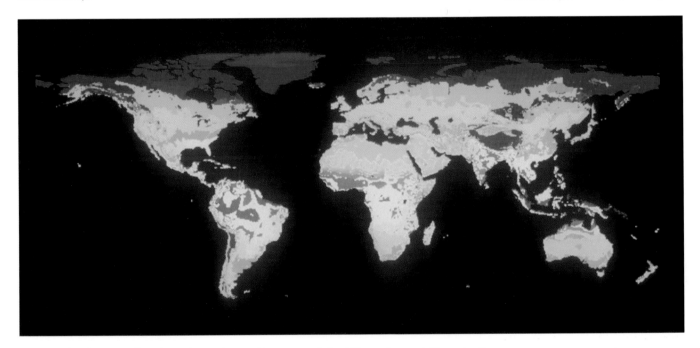

Although all parts of a plant throughout all stages of the life cycle contribute to the overall success or failure of the whole plant, the organ that plays the most critical role in environmental adaptation is the leaf. The photosynthetic function of a leaf demands that it is efficient at intercepting light and exchanging gases with the atmosphere, while affording the minimum water loss concomitant with maintaining water flow within the plant and evaporative cooling of the leaf surface. All this must be achieved with the minimum of structural tissue investment because building leaf tissue costs energy and food resources. Thus, there are only a limited range of engineering solutions that can satisfy the often-conflicting constraints that exist in a given set of environmental conditions. Because the unchanging laws of gas diffusion, fluid flow and mechanics impose such constraints, these solutions are time-stable and independent of taxonomic affinity.

So pervasive is the premium on selection for leaf efficiency that leaf physiognomic 'tuning' to the environment can be seen to vary not only between plant lineages, and between individual plants within lineages (ecophenotypes), but also within individual plants. For example, leaves at the top of a tree crown are exposed to high light levels and wind energies and tend to be small and thick (sun leaves). In contrast, those in a darker, more sheltered, humid subcanopy tend to be larger and thinner (shade leaves). If the environment around a tree changes with time (e.g. through removal of surrounding trees), then subsequent cohorts of leaves will display different appropriate morphologies.

The capability for leaf morphological plasticity must be genetically coded, but how independent of genetic predetermination are character states that might have some correlation with climate? On initial consideration, leaf size, because of its variability on even a single individual plant, or variability during different growth stages, might be thought to be subject to minimal genetic predetermination, but this cannot be so. Plants must produce leaves that are adaptive to the entire growing season, and this production must, therefore, be genetically predetermined. Plants in Mediterranean-type climatic regions, e.g. California and Chile, receive abundant rainfall during the spring when the new leaf crop is produced, followed by extreme summer drought. If the plants responded only to the spring rains and produced large leaves, these leaves would be non-adaptive during the summer. Clearly, selection favours genotypes that are tuned to overall climatic conditions and not just to those experienced by the plant during leaf development. Vegetation in recently glaciated and non-glaciated parts of the Northern Hemisphere also demonstrates a good correlation with climate. This shows that genetic, and phenotypic, tuning of foliar physiognomy takes place over geologically short time-scales (< 1 Ma) as a result of taxonomic elimination and migration, as well as selection for novel genotypes that arise as the result of chance mutation or hybridization.

Clearly, in leaf architecture, there is the potential to obtain much more information than just mean annual temperature. In 1993, Jack Wolfe examined leaf physiognomy using many more characters than leaf margin type. He began investigating the relationship between leaf physiognomy and climate using computer techniques that are capable of seeking out relationships in more than two dimensions (e.g., the two axes of Figure 4.56). He called the technique 'Climate Leaf Analysis Multivariate Program' or 'CLAMP'.

4.5.2 CLAMP

CLAMP uses multivariate statistics to position modern vegetation samples in multidimensional space based only on the characters found in the leaves of at least 20 woody dicots growing in those communities. The position of any sample relative to the others is a measure of how similar (or dissimilar) the samples are in terms of leaf physiognomy. The closer together two points are, the more similar is their leaf architecture. The plot in Figure 4.58 shows modern vegetation samples (shown as spheres) in two-dimensional space. The number of leaf characters used is 31, which includes leaf size, leaf shape, apex and base characters.

The samples are colour-coded: red representing samples growing in warm climates and blue for samples in cold regimes. You can see that there is a trend in one dimension that reflects temperature even though the ordering is based only on leaf physiognomy. This trend is summarized by the red arrow, which represents the mean annual temperature (MAT) vector.

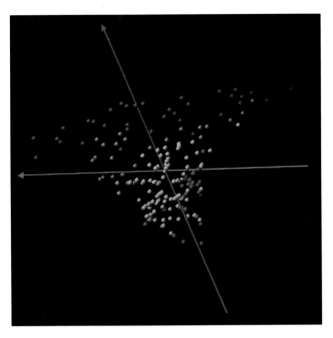

Figure 4.58   CLAMP site physiognomic space. This two-dimensional plot of vegetation samples shows coloured spheres arranged relative to one another based on the physiognomic characteristics of at least 20 species of woody angiosperms that occur in each vegetation sample. The balls are colour-coded based on the mean annual temperature observed at the site where the samples were growing: cold (blue) and warm (red). Only leaf physiognomic data are used and the data are ordered from cold to warm. The trend can be summarized by a mean annual temperature vector (red arrow). The blue arrow represents a similar vector for precipitation during the growing season.

There is other information in the ordering of the samples. The blue arrow represents a vector for precipitation during the growing season. Note that the plots shown here are in only two dimensions because we are limited by the two-dimensional surface of the paper. However, in CLAMP the relationships between climate of leaf physiognomy is examined in 31 dimensions and the positions of many more climate vectors can be plotted running through that space.

Because we know the climate under which our modern reference samples are growing, these vectors can be calibrated. Figure 4.59 shows the regression curve for the observed mean annual temperature plotted against the position along the MAT vector.

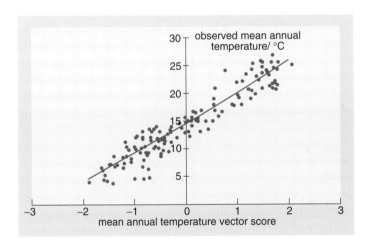

Figure 4.59   Regression plot for MAT using CLAMP versus the vector score for the MAT vector for 143 vegetation samples.

Note that there is some scatter around the regression curve but that the r^2 value for the goodness-of-fit is high (0.95). The scatter again provides us with a measure of the statistical uncertainty of the technique. There are several ways by which uncertainty may be measured but here we will use one of the most straightforward: standard deviation (deviation from the mean value on either side). In this case, the standard deviation of the mean of the residuals is 1.17 °C. In a population exhibiting a normal (Gaussian or bell curve) distribution, one standard deviation (σ) should encompass 68% of the samples, while two standard deviations (2σ) includes 98% of the samples. It is usual to express uncertainty in terms of two standard deviations. In this case, two standard deviations equate to an uncertainty of ± 2.34 °C.

The uncertainties associated with other climate vectors are given in Table 4.3. You can see that measures of precipitation are rarely precise. Figure 4.60 shows the regression curve for the observed precipitation during the growing season plotted against the growing season precipitation vector score. Note that the spread of points about the regression line is not uniform along its length.

Table 4.3   CLAMP analysis of Grebenka floras showing results for mean annual temperature (MAT, °C), warm month mean temperature (WMMT, °C), cold month mean temperature (CMMT, °C), length of growing season (LGS, months), growing season precipitation (GSP, ×10 mm), mean monthly growing season precipitation (MMGSP, ×10 mm), precipitation during the three wettest months (3-WET, ×10 mm), precipitation during the three driest months (3-DRY, ×10 mm).

		Grebenka:		Standard deviation
	All sites	**Site 22**	**Sites 19–26**	**Residuals**
MAT	13.19	11.99	11.33	1.17
WMMT	20.99	19.39	19.35	1.58
CMMT	6.23	5.14	3.85	1.88
LGS	7.45	6.91	6.56	0.70
GSP	74.09	72.90	56.10	33.57
MMGSP	9.04	8.84	8.07	3.69
3-WET	38.91	38.10	32.36	14.02
3-DRY	13.90	13.23	14.08	9.29

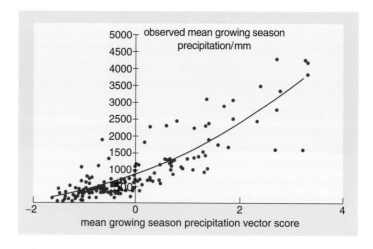

Figure 4.60 Observed growing season precipitation (MGSP) plotted against the growing season precipitation vector score.

○ Why do you think this is?

● In dry climates, there is very little spread showing that leaf physiognomy reflects very strongly these dry conditions. However, where water is plentiful, leaf architecture is not constrained by the necessity to conserve water, and a wide variety of architectural forms can exist. This translates to a broad spread about the regression line and a loss of precision.

The characters of at least 20 species of woody dicot fossil leaves are scored in the same way as fossil leaf assemblages, and can be positioned relative to the modern samples. Their positions along the calibrated vectors can be determined. In this way, we can estimate the climate under which the fossil leaves were once growing.

Table 4.3 also shows the results of a CLAMP analysis on the Grebenka flora. Note that Grebenka site 22 has been separately processed. This was done to test how much variation might be produced by examining an assemblage from an individual depositional setting (broadly equivalent to a separate community) as compared to the flora as a whole. The overall flora gave a MAT of 13 °C, which is 3 °C warmer than that of London today. Site 22 (Figure 4.47c), however, gave a slightly cooler result. Sedimentological information from Site 22 suggests that it is the remains of a shallow floodplain pond and this could explain the lower temperature estimates. However, there is little difference in precipitation estimates suggesting the whole system was water-rich. Note, however, that the uncertainties for temperature and precipitation estimates are greater than the differences between the whole flora and Site 22. The same phenomenon is seen in the CLAMP results for Sites 19–26. These sites all represented well-developed interfluve vegetation. Again, there is no significant difference in the CLAMP results and those of the other samples but what difference there is can be explained in terms of the microclimate in a cool moist forest.

As well as the MAT being warm for a position of 72° N, the warm month mean temperature (WMMT) is relatively cool at around 21 °C, while the cold month mean temperature (CMMT) of +6 °C suggests some winter frosts, but no hard freezes. This low annual temperature range is perhaps surprising given the strong seasonality imposed by the light regime at 72° N and suggests some mechanism other than direct sunlight for keeping the region warm during the winter. We will return to this issue later.

As we saw earlier, the angiosperms really only started to become ecologically important on a global scale at the end of the Albian, and yet in Grebenka we are using angiosperm leaves to determine palaeoclimate. How do we know that angiosperm leaves are carrying an accurate climatic signal this early in their evolution? To test this, we can plot climate estimates derived from leaves against some other proxy climate indicator. The simplest method is to plot MAT against latitude where we would always expect there to be a temperature gradient from warm at low latitudes to cool at high latitudes. Figure 4.61 shows a plot of

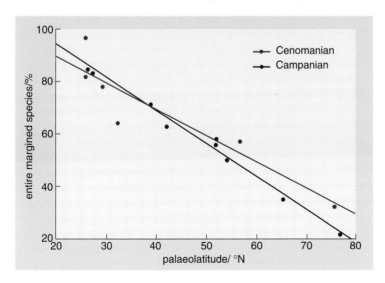

Figure 4.61 Plot of percentage of species with leaves with entire margins versus palaeolatitude for the Cenomanian and Campanian.

percentage of leaf margination against palaeolatitude for both the Cenomanian and the Campanian. There is clearly a trend from warm to cool with increasing latitude even though we know from other data such as oxygen isotopes that the equator-to-pole temperature gradient at this time was shallower than now. The Cenomanian plot shows more scatter but a relationship is clearly established even this early in angiosperm evolution. Of particular interest is that there is no evidence of a change in slope in the regression line as we approach high latitudes. This is a good indication that the polar light regime did not unduly affect the foliar physiognomy temperature calibration.

How accurate is this calibration? So far, all we have is a correlation with palaeolatitude without any idea as to the accuracy in terms of temperature measurement. We do not have any Cenomanian thermometers of course, so we have to use other temperature proxies, ones that are quantitative.

○ What temperature proxy could we use?

● We have just mentioned oxygen isotopes and these are likely to provide our best proxy despite the fact that most data are marine based (and therefore only indirectly comparable to the air temperatures provided by CLAMP or simple leaf margin analysis). There are some drawbacks to using the oxygen-isotope method, such as correcting for imprecisely known ancient ice volumes and salinities, and problems associated with diagenetic changes that have altered the isotopic composition of the fossils. However, this method is widely accepted and it exploits principles (isotope fractionation) that are entirely unconnected with foliar physiognomy. This last point is important as it avoids circularity.

To test CLAMP against oxygen-isotope data, we need to analyse a flora that was growing on a coastal plain, or better still an island, the climate of which was under strong maritime influence. Moreover, we need to know the palaeo-sea-surface temperatures near to our flora as measured by oxygen-isotope analysis.

One of the most intensively studied regions of the Cretaceous open ocean system is the Tethys Ocean and within that area there are several candidate floras. The most recently analysed is the Peruc flora near Prague in the Czech Republic. Here, CLAMP analysis reveals a climate summarized in Table 4.4.

Table 4.4 Summary of CLAMP data from the Cenomanian floras of the Czech Republic. Climate parameter abbreviations and uncertainties are defined in Table 4.3.

	Floras		
	Chucle	**Vyshehorovice**	**Peruc**
MAT	19.86	18.38	17.06
WMMT	28.33	27.30	26.83
CMMT	11.23	9.73	7.92
LGS	11.01	10.34	9.67
GSP	204.44	256.86	175.61
MMGSP	21.37	28.22	19.44
3-WET	91.91	115.32	81.84
3-DRY	41.43	62.96	38.55

Sedimentological evidence suggests the Chucle flora represents an assemblage of leaves derived from vegetation that was on a better-drained substrate than Vyshehorovice which was from a low-lying interfluve swamp. Like Grebenka, Peruc yielded a fossil assemblage derived from vegetation growing in a range of environments. One thing to note immediately is the warmer MAT at these lower latitudes and the low mean annual temperature range associated with the maritime climate. Moreover, the temperature signatures from the three sites are similar and, more importantly are similar to Tethyan $\delta^{18}O$ data. This can be seen by comparing the CLAMP results with the mean annual sea-surface temperature (SST) in the climate model result in Figure 4.62. The SSTs were provided to the model using oxygen-isotope data. Vyshehorovice yields by far the wettest signature even though all values are within the large uncertainties associated with the precipitation predictions.

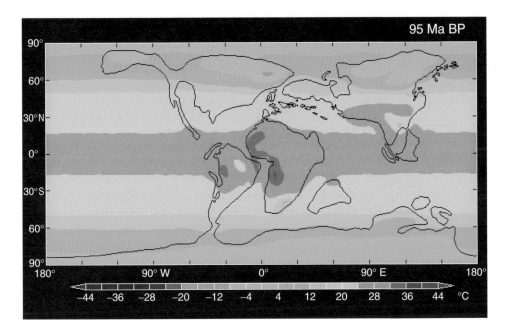

Figure 4.62   Map of mean annual sea-surface temperature during the Cenomanian. (Valdes et al., 1999.)

This congruence between the CLAMP results and oxygen-isotope data is reassuring because it indirectly confirms that the method is not strongly influenced by the higher atmospheric CO_2 values that are believed to have existed in the Cenomanian, and indeed throughout the Cretaceous. The fact that the calibration based on modern leaf physiognomy decodes climate signals consistently with oxygen-isotope data suggests that this calibration is robust through time and the method can be used with some confidence from the early Cenomanian through to the present day — a time-span of almost 100 Ma.

In continental interiors, such as the Vilui basin, we have no such independent checks on the validity of the CLAMP-derived temperature estimates. Using oxygen isotopes in freshwater systems, while not impossible, is not routinely carried out because of the complexities involved in applying necessary correction factors. However, there is no reason to suppose that plants would reflect the climate any differently to the way they reflect the climate in continental margin sites. In fact, CLAMP is so far the only widely available method for quantitatively determining the palaeoclimates of continental interiors.

When CLAMP is applied to the Timerdyakh Formation leaf flora, the results are congruent with the qualitative interpretation of the palynoflora and the clay mineralogy. The CLAMP analysis yielded a MAT of 13.1 ± 13.51 °C (2σ), a WMMT of 21.1 ± 3.6 °C, and a CMMT of 5.8 ± 5.1 °C. Note that compared to today's mean annual temperature range in the Viluisk region of around 67 °C (WMMT–CMMT), the Cretaceous was remarkably equable. The length of the growing season (7.4 ± 1.7 months) is more a function of the high palaeolatitude (72° N) light regime than temperature. The GSP (827 ± 632 mm), the MMGSP (133 ± 73 mm), the 3-WET (495 ± 274 mm), and the 3-DRY (293 ± 176 mm) precipitation all suggest a moderately wet regime all year round. This is also reflected in the mean annual relative humidity ($77.5 \pm 16.3\%$). However, it is worth remembering that the CLAMP figures for precipitation may be spurious in that the plants were growing near to large river systems collecting moisture from the margins of the basin.

4.6 Polar ecosystems and climate near the end of the Cretaceous

Both qualitative and quantitative data relating to terrestrial Arctic climate suggest a gradual decline in MATs as the Cretaceous drew to a close. We have already seen that Alaska tree ring data suggest less-benign growing seasons during the Campanian and Maastrichtian near-polar environments, and declines in floral diversity suggest cooler conditions.

The leaf diversity is so low that neither leaf margin analysis nor CLAMP can be used reliably. Trees from the Prince Creek Formation were much smaller in girth as evidenced by the fact that no logs with a diameter of more than 25 cm have been found. In contrast, the Nanushuk Formation logs often exceed 50 cm in diameter. Despite the small size of the Campanian and Maastrichtian trees, their presence does tell us that during part of the year temperatures must have been greater than +10 °C because, below this temperature, tree growth is virtually non-existent. In today's world, the rate at which important enzymes operate is curtailed below this temperature and there is no reason to suppose Cretaceous plants had a different biochemistry.

We can obtain a better idea of Maastrichtian temperatures by looking at foliar physiognomic data from more diverse floras both before and after this time period. Both CLAMP and leaf margin analysis of Coniacian floras from the Arctic Slope of Alaska suggest an MAT of 13–14 °C with a CMMT of 9 °C. Early Tertiary MATs using leaf margin analysis appear to be around 6–7 °C (Figure 4.63).

Figure 4.63 Mean annual air temperatures for northern Alaska from the mid-Cretaceous to the Early Tertiary based on plant fossils. Error bars represent uncertainty in estimating MAT with small sample sizes. (*Spicer and Parrish, 1990.*)

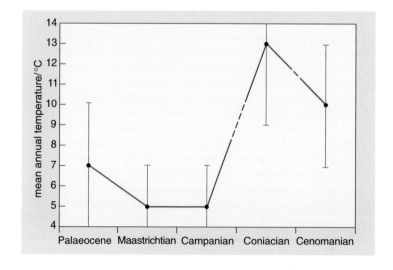

If the Maastrichtian sediments are providing a reliable sample of the vegetation, then we can say that MATs are likely to have been lower than the Early Tertiary estimates. This means at best they may have been around 5 °C. However, there is no evidence of ice signatures in the sediments from this time, and tillites and dropstones that could be interpreted as signatures of glaciation are absent. We may therefore conclude that CMMT are unlikely to have been very low, although occasional winter freezes may have occurred during the long winter darkness.

If occasional freezing occurred at sea-level, and MATs were in the region of 5 °C, then we can also say something about the conditions at altitude in the newly uplifted Brooks Range. It is difficult to measure precisely the altitude of ancient mountain belts such as the Brooks Range, but it is reasonable to suppose that by analogy with similar recently uplifted mountains today, like the Andes, an altitude of 1500 m would be easy to achieve. The general lack of coarse-grained siliciclastics in the Umiat and Corwin delta plain sedimentary deposits indicates that we are not dealing with extremely high mountains and 1500 m again seems reasonable in that context. To estimate the temperatures at the top of these mountains, we need to use what is known as the lapse rate, i.e. the rate at which mean annual temperature decreases with increasing atmospheric height. This latter rate is partly dependent on atmospheric composition, but given the uncertainties in our knowledge of the topography its effect is small. The best we can do is apply what is known as the global average free air lapse rate (the lapse rate in the open atmosphere away from any mountains) and for the present day this is about 5.5 °C km^{-1}. With a MAT of 5 °C at sea-level and such a lapse rate, we get a MAT of –0.5 °C at 1000 m. At 1500 m, this decreases even further. Therefore, it is likely that the peaks of Brooks Range could have supported small permanent ice fields in latest Cretaceous times.

The fossil trees from the Arctic Slope during the Maastrichtian are all of the conifer type (lacking vessels) and are associated with abundant conifer shoots, e.g *Parataxodium* and *Metasequoia*. There are very few angiosperm leaves, which suggests that long-lived woody angiosperms were rare. Examination of the fossil pollen reveals very rich angiosperm diversity and the stratigraphical longevity of any particular angiosperm pollen species is very short. This, coupled with a high spatial heterogeneity, indicates a rich herbaceous angiosperm flora with a high evolutionary turnover which, in turn, may indicate short life cycles such as are associated with opportunistic annual weeds. We know that ferns were also present on the floodplains because they have been found in life position, rooted in a palaeosol, within thick volcanic ash falls.

Maastrichtian Arctic Slope sediments are also very rich in dispersed fossil charcoal but almost no well-developed mires that produced coal are preserved. At best, organic accumulations are in the form of transported leaf mats or organic-rich palaeosols. The lack of mire development may be due to high rates of inorganic sediment deposition but relative drought (as evidenced by frequent occurrence of dispersed charcoal) is also a possible explanation.

○ What simple climatic explanation might account for the drying under progressively cooler polar conditions?

● The development of a progressively stronger polar high-pressure cell. Normally, cooling leads to higher relative humidities and lower evaporation, so producing apparently wetter conditions. However, near the pole, cooler global conditions produce a stronger high-pressure cell where descending air has a low relative humidity. Such a strong polar 'high' is developed in today's 'icehouse' world and is responsible for the low precipitation observed on the Arctic Slope.

The overall picture of the vegetation is one of relatively small deciduous conifers with a ground cover of ferns and herbaceous short-lived angiosperms. Frequent wildfires probably maintained the vegetation in a rather open state, with clumps of trees separated by open areas colonized by opportunistic angiosperm 'weeds'. In fact, it probably looked a lot like modern Taiga and is shown in the reconstruction during our introductory 'tour' (Section 1.1.2; Figure 1.9).

The foliar physiognomic data provide us with invaluable evidence for just how warm the Cretaceous was over land, and what the variation in temperature was throughout the year. When we can compare these data with other independent temperature estimates, such as those from oxygen-isotopic data, there is agreement which gives us confidence in the calibration. However, CLAMP also throws up some unexpected results. In the Arctic, for example, not only were temperatures warmer than they are today, but there was surprisingly little seasonal variation in temperature, particularly as temperatures barely fell below freezing even in the depths of the dark polar winter. We may need to look for some feature of the Cretaceous world that maintained winter warmth even in the absence of sunlight. A similar phenomenon is seen in the continental interiors. Here again, winter temperatures do not appear to fall to the very low values we see in today's world. In Part 2 of this book, we will look at the broader picture of the Cretaceous environment to see if we can begin to understand how the Cretaceous world worked, but first we shall take a brief look at the fauna that inhabited the strange world of the Cretaceous Arctic.

4.7   The fauna of the polar forests

The Nanushuk Formation sedimentary deposits not only contain plants but also yield some clues as to the animal life in the polar forests. The floodplain lakes undoubtedly supported fish, but the small size of the bones and generally acidic conditions arising from the decay of the litter shed by the luxuriant deciduous forests, have left few traces. However, freshwater mussels also inhabited these waters and one, in particular, has left us an interesting record of its death. Figure 4.64 shows this specimen, which was recovered from an exposure of upper Nanushuk (Cenomanian) rocks along the Colville River.

Figure 4.64 Cast of a freshwater mussel shell from the Nanushuk Formation showing a series of puncture wounds made by a small toothed vertebrate. (Bob Spicer, Open University.)

At first, it looks like a poor specimen of a bivalve preserved as a cast. That is to say the shell material has dissolved away, leaving just the rock that hardened around the shell before it was lost. Looking more carefully, you will see it has a number of holes in it. By taking a rubber impression to reveal the geometry of the

holes, Mike Parrish, a palaeontologist from the University of Michigan, was able
to show they were the holes left by teeth. Moreover, the teeth patterns showed
that the animal that did the biting had an elongated snout and actually took three
bites, slightly rotating the shell between each bite, before either dropping or
spitting out the partly chewed shell. We cannot say for certain what the animal
was that attacked the shell but one possible candidate was the toothed bird
Hesperornis, some bones of which have been found from Cenomanian-age rocks
in the area.

Other shelled animals are found as casts. Among them, again from the Nanushuk
Formation, is a turtle whose nearest living relatives are native to Asia. Today,
turtles can survive winter low temperatures of $-14\,^{\circ}$C by burrowing in the mud at
lake bottoms and aestivating (hibernating), so the fact that turtles were living on
the Arctic perhaps provides us with a lower limit on winter temperatures. This
limit, however, is well below that indicated by foliar physiognomy.

By far the most spectacular terrestrial animal fossils from Alaska are the dinosaurs.
Dinosaur footprints are known from the Nanushuk Formation (Figure 4.65b), but
by far the greatest number of bones have come from the Prince Creek Formation
along the Colville River where it nears the present-day Arctic coast. In the
sediments of ancient floodplain ponds and river channels, filled with inwashed
Parataxodium conifer needles from forest floor litter, we find abundant bones and
even some skin impressions of the duck-billed dinosaurs (Figures 1.8 and 4.65a).

(a)   (b)

Figure 4.65   (a) Hadrosaur skin impression from the Alaskan Arctic Slope. (b) Dinosaur footprints from the Nanushuk Formation.
(Gil Mull, State of Alaska Geological Survey.)

The hadrosaurs generally belong to the genus *Edmontosaurus* which, as its name
suggests, is particularly well known from Alberta. This herbivore (Figure 4.66)
grew up to 18 m in length and is known to have moved in large herds because of
the great numbers of skeletons often found together. Often, these bone aggregations
are found in river channels and we might envisage a Maastrichtian scene similar
to that which can be seen today when huge herds of wildebeest migrate across
swollen African rivers and inevitably many individuals drown. Migration
behaviour is essential to the survival of herbivores that live in large herds for
mutual protection because grazing/browsing pressure is locally intense due to the
sheer numbers of individuals and 'fresh pastures' need to be constantly sought.

Figure 4.66 Reconstruction of *Edmontosaurus* on the Late Cretaceous Arctic Slope of Alaska. The newly uplifted Brooks Range is depicted in the background. *(John Watson, Open University.)*

○ What do you suppose happened to the hadrosaurs during the long polar winter darkness?

● The most likely scenario is that they migrated south down the margins of the Western Interior Seaway to their winter feeding grounds near or below the Maastrichtian Arctic Circle where there was sufficient light to support evergreens.

Calculations show that such a journey to, for example, southern Alberta where, incidentally, hadrosaur nest sites are found in some abundance, could be accomplished over a three-month period travelling for 12 hours each day at a steady directional 'browse' rate of no more than $2 \, \text{km} \, \text{hr}^{-1}$. For a large animal, this would not have been taxing and it is interesting to note that so far no very small infants have been found in Alaska, only semi-mature or mature individuals.

The hadrosaurs were not alone. The horns from the neck frills of styracosaurs, and the bones of *Pachycephalosaurus*, have also been found. Even skin impressions have been recovered, showing that these Arctic dinosaurs had no special adaptations to cold in the form of fur or feathers.

As well as the herbivores, we find the teeth of carnivores such as troodonts and the infamous *Tyrannosaurus*. So far, no bones of these have been found in the Arctic and it is just possible that the teeth travelled to the Arctic embedded in the flesh of the wounded migrating herbivores. However, top predators are always few in number compared to herbivores and the lack of bones is not evidence for their absence from the Arctic polar ecosystem. Indeed, it is likely that the carnivores hunted the hadrosaur herds just as wolves hunt the migrating caribou herds today, picking off the young, sick, or elderly individuals.

○ Why is it unlikely that the hadrosaurs probably did not survive over winter on the Arctic Slope by hibernating?

- There would be little shelter on the Corwin and Umiat delta floodplains for great numbers of large dinosaurs. Some shelter would have been necessary to protect them from both predation and the cold. The carnivores would only have to remain 'awake' for a little longer than the herbivores for them to have a feast and decimate the herbivore populations. Moreover, as the winter cold deepened towards freezing, staying warm during torpor, particularly at body extremities, would have been problematic. If hadrosaur physiology had been anything other than warm-blooded, then this problem would have been even more acute.

We do not have surface rocks in northern Alaska marking the very end of the Cretaceous, and the last hadrosaurs are well below the likely stratigraphical position of the boundary. Exactly how far below we are not sure because of extensive faulting. What is certain though is that within the last million years or so of the Cretaceous, Arctic dinosaurs spent their summers eating the luxuriant plant growth by the light of the midnight Sun at 85° N.

4.8   Summary

- Global temperature changes are more strongly expressed at high than at low latitudes, and fossil remains of the late Cretaceous polar forests provide a wealth of data relevant to understanding the Earth system at times of global warmth.

- The Cretaceous polar forest ecosystem thrived under a combination of light and temperature regimes that does not exist in our present world. When reconstructing and interpreting this ecosystem, the principle of uniformitarianism can only be applied with considerable care.

- The sedimentary and fossil record of the late Cretaceous Arctic environment is particularly well captured in the foreland basin setting of Northern Alaska. The terrestrial environmental record is preserved in the floodplain sediments of fluvio-marine deltaic successions. These deltas formed at palaeolatitudes of between 75° and 85° N.

- By recording the repeated associations between plant fossil remains such as leaves and wood and their entombing sediments, a detailed picture of the structure and dynamics of the polar forests can be reconstructed.

- Forests in the Early Cretaceous, before the arrival of flowering plants, were conifer-dominated with fern and scouring rush understories. Minor elements included deciduous cycadophytes with ginkgophytes growing along disturbed river margins. Abundant, thick, ash-free coal beds suggest the development of raised mires. These in turn indicate a year-round wet climate regime in contrast to the low precipitation experienced at similar latitudes today.

- The large quantities of Arctic Cretaceous coal represent an extremely effective carbon sequestering system. Virtually wholesale deciduousness or winter die-back of vegetation, coupled with winter temperatures low enough to limit decay, high precipitation and delta sediment compaction, led to preservation of large amounts of terrestrial organic matter.

- When the angiosperms migrated from lower latitudes into the polar regions around the middle of the Cretaceous, they initially displaced the ginkgophytes from riparian environments but never penetrated the stable mire environments. The vegetation remained conifer-dominated, but as the late Cretaceous progressed plant diversity and tree stature declined.

- This decline is interpreted as a deterioration of growing conditions associated with a local and global cooling trend. As the polar high-pressure cell strengthened as a result of this cooling, so the Arctic became drier and raised mire development waned. Fire frequency also increased.

- These conditions are not restricted to Arctic Alaska but are seen across Russia. Generally, the climatic data recovered from plant fossils and sediments are in agreement with computer climate modelling but serious discrepancies exist between the models and data in continental interiors. In particular, the models suggest drier conditions with a larger annual range of temperature and lower mean annual temperature than the geological record suggests. The cause of this model/data mismatch is, as yet, not understood.

4.9 References

BAILEY, I. W. AND SINNOT, E. W. (1915) 'A botanical index of Cretaceous and Tertiary climates', *Science*, **41**, 831–4.

DOYLE, J. A. AND HICKEY, L. J. (1976) 'Pollen leaves from the mid-Cretaceous Potomac group and their bearing on early angiosperm evolution', in BECK, C. B. (ed.) *Origin and Early Evolution of Angiosperms*, Columbia University Press, pp. 139–206.

HICKEY, L. J. AND DOYLE, J. A. (1997) 'Early Cretaceous fossil evidence for angiosperm species', *Botanical Review*, **43**, 3–104.

MULL, C. G., HOUSEKNECHT, D. W. AND BIRD, K. J. (2003) *Revised Cretaceous and Tertiary Stratigraphic Nomenclature in the Central and Western Colville Basin, Northern Alaska,* Alaska Department of Natural Resources/US Geological Survey.

PARRISH, J. M., PARRISH, J. T. , HUTCHINSON, J. M. AND SPIER, R. A. (1987) 'Late Cretaceous vertebrate fossils from the North Slope of Alaska and implications for dinosaur ecology', *Palaios*, **2**, 377–89.

POKHIALAINEN, V. P. (1994) *Mel Severo-Vostoka Rossii (The Cretaceous of Northeastern Russia)* Magadan: NEISRI, Far Eastern Branch, Russian Academy of Sciences [in Russian].

SHCZEPETOV, S. V. AND HERMAN, A. B. (1990) 'Cretaceous flora of the Anadyr River right bank' [in Russian], *Izv. Acad. Nauk SSSR. Ser. Geol.* **19**, 16–24.

WOLFE, J. A. (1993) 'A method of obtaining climate parameters from leaf assemblages', *US Geological Survey Bulletin*, 2040.

5 Changing climate and biota — the marine record

Peter W. Skelton

As with the terrestrial record, which you explored in Chapter 4, clues to past conditions in the seas and oceans include diagnostic deposits, fossil evidence and isotopic data (especially carbon-isotope and oxygen-isotope ratios), inferences from all of which can be compared with the results of computer modelling. Indeed, the marine record of the Cretaceous is relatively more complete and extensive than its terrestrial counterpart, and so provides rather more comprehensive coverage of climatic and associated changes through the period, especially those of shorter duration. Although high latitudes registered the largest-amplitude signals of climate change (as shown in Chapter 4), we also need to pay particular attention to the marine record from low latitudes, in order to investigate the oceanic contribution to polar heat transport during the period.

○ Thinking back to the controls on oceanic circulation discussed in Chapter 2, why might the low latitude record be of special interest for the Cretaceous?

● In Section 2.1, you saw that the thermohaline circulation of the oceans today depends upon the extensive formation of sea-ice at high latitudes. The likely absence of polar ice at sea-level for much of the period, especially during the Late Cretaceous, means that we must look elsewhere for possible sources of relatively dense brines that, by sinking, might have helped to drive oceanic circulation. It was noted in Section 2.3 that intense evaporation in warm, arid regions could have provided such an alternative source. Partially enclosed basins in low latitudes (e.g. the early Atlantic) would thus have been obvious candidate sites.

Besides their possible role in ocean circulation, the low latitudes of the Cretaceous world also hosted the vast carbonate platforms, which, together with the Chalk of mid-latitudes, constituted a significant component of the global geological carbon cycle, as noted in Tour stop 3 of Section 1.1.3. Hence, in this Chapter, we will review the marine record of the period starting from low latitudes and working our way back up towards the poles.

5.1 Carbonate platforms

Two aspects of the carbonate platforms that developed around the Cretaceous tropical belt are remarkable — first, their exceptional extent (e.g. Figure 2.16) and secondly their episodic development (Figure 5.1). The latter aspect hints at the temperamental nature of the Cretaceous marine realm, although its instability was expressed rather differently from the climatic fluctuations of the Quaternary, as you saw in Tour stop 4 of Section 1.1.4.

In Figure 5.1, the pattern of Cretaceous Tethyan/Atlantic carbonate platform development is represented in a very generalized way that ignores the many small differences between the histories of individual platforms attributable to more regional factors. However, the overall patterns for the New (southern USA, Mexico and Caribbean regions) and the Old (mainly around the Mediterranean and Middle East regions) Worlds are shown separately. Though initially similar, the platforms in the two areas developed somewhat differently from late Aptian times onwards.

Figure 5.1 Generalized history of carbonate platform development during the Cretaceous in the Tethyan/Atlantic oceanic realm, for (left) the New World (i.e. the Americas) and (right) the Old World (the other continental masses). Major crises in the growth of platforms are indicated by bold horizontal bars, and inferred oceanic anoxic events (OAEs) are shown on the right.

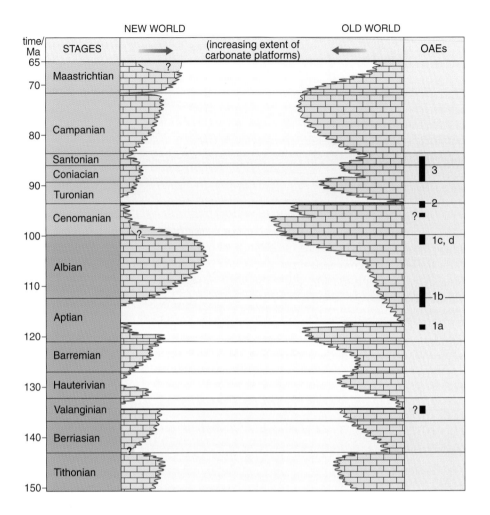

○ Which major palaeogeographical change was associated with this regional divergence in platform history?

● The opening of the northern Atlantic Ocean (Section 2.2.1).

Notwithstanding these differences in development, there were some interruptions of global extent. Foremost among these was that at the close of the Cenomanian, although crises of nearly comparable scale occurred in the mid-Valanginian and mid-Aptian. All three episodes were accompanied by major oceanographic perturbations, as you will see shortly. The Maastrichtian decline was somewhat different in character, however, and its causation remains a subject of debate. Before dwelling on the shared misfortunes of the platforms however, let us first look at the patterns of growth of these extraordinary 'biosedimentary' systems when they were in good health. Numerous examples have been studied both from outcrops, and in the subsurface in connection with hydrocarbon exploitation (e.g. Simo *et al.*, 1993; Philip and Skelton, 1995; and Alsharhan and Scott, 2000).

The platforms had a prodigious capacity for rapid growth and expansion, thanks to the distinctive benthic shelly biota that constituted their main carbonate factories. On the (much smaller) tropical carbonate platforms of today, the most productive carbonate factories are the upstanding marginal reefs built up by corals and other framework-building organisms. Yet, compared to the Quaternary, the development of reef frameworks was extremely limited in the Cretaceous.

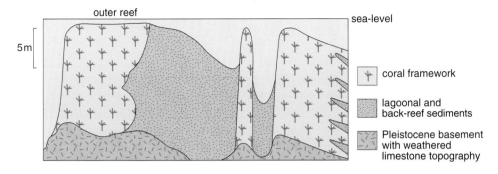

Figure 5.2   Diagrammatic cross-section across the margin of One Tree Reef in the southern Great Barrier Reef of Australia, demonstrating how Holocene reefs built up from pre-existing topographical irregularities. Vertical exaggeration ×10. *(Marshall and Davies, 1982.)*

Much of the classic reefal topography of today grew up during the rapid eustatic sea-level rise of the Holocene, upon prominent foundations that were sculpted by karstic weathering of older platform deposits exposed by the previous large-amplitude eustatic fall in sea-level (e.g. Figure 5.2). The development of antecedent karstic surfaces on such a scale, though not unknown (e.g. Figure 1.17a), was also much less frequent in the Cretaceous platforms.

○   From the discussion of the controls on sea-level change in Chapter 3, how would you explain this difference in the importance of antecedent karst development between Cretaceous and Quaternary platforms?

●   The platforms of the Quaternary world were subject to large-amplitude glacio-eustatic oscillations of sea-level, hence repeated emergence and submergence, which would not have occurred — at least on anything like the same scale — in the largely greenhouse world of the Cretaceous (Figure 3.11).

On the Cretaceous platforms, the most productive carbonate factories were constituted by vast congregations of shelly organisms that spread, like submarine 'meadows', across the shallow tops and often gently sloping flanks of the platforms. Various kinds of organisms were involved, ranging from algae with calcareous skeletons, via benthic foraminifers — some types achieving relatively 'giant' (centimetre-scale) sizes — to larger, highly gregarious shelly animals, especially bivalve and gastropod molluscs, as well as corals and other colonial forms. Among the most abundant contributors, especially in the late Cretaceous, were the rudists. These bizarrely shaped, sessile bivalves were variously adapted as sediment-dwellers, either lying prone upon current-swept surfaces (hence termed recumbents), or implanted in accumulating sediment (so termed elevators) (Figure 5.3a), although some closely overgrew their substrates with large basal surfaces (and termed clingers). Meanwhile, coral colonies, mostly of platy to domal form, often carpeted the sea-floor in slightly deeper zones — especially downslope on ramps flanking the platforms (Figure 5.3b).

Although these fossil associations have often been referred to as fossil 'reefs' in the literature — in a very loose sense — rudist researchers Eulàlia Gili, Jean-Pierre Masse and Peter Skelton (1995) have argued that this description is misleading, so is best avoided. The bulk of evidence suggests that the clustered rudists were largely sediment-supported in life and did not produce skeletal

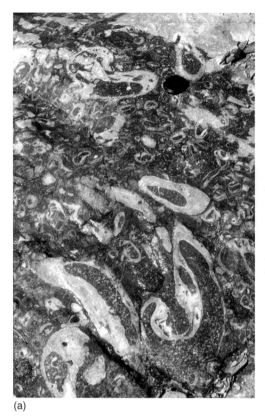

Figure 5.3 (a) Rudist association in a bed of Lower Aptian limestone in SW Mexico (Huetamo area), viewed from above, showing transverse sections of small tubular elevators coming up from beneath, together with large shells of a recumbent form, again seen in section, at the top of the bed. (b) Vertical section of a bed of Barremian limestone packed with platy, plus some branching corals, from SW Mexico (San Lucas area). Scale: lens cap 5.5 cm across. *(Peter Skelton, Open University.)*

(a)   (b)

frameworks projecting significantly (>tens of centimetres) above the sediment surface. Nor, in most instances, did their accumulation generate more than modest relief on the sea-floor relative to surrounding (usually bioclastic) facies on the platform top, although some examples of mounds with an original relief of several metres have been documented. Hence the rudist congregations would usually have presented a somewhat monotonous appearance to any passing observer, perhaps resembling oyster beds carpeting the platform top as far as the eye could see (Figure 1.13). The limited increments of accommodation space usually provided by greenhouse-type oscillations of sea-level (Figure 3.11b), gave rise only to tabular parasequences within which the corals and/or rudists formed thin (up to a few metres) but laterally extensive biostromes or lensoid bodies (collectively termed coral lithosomes or rudist lithosomes) (Figure 5.4).

Figure 5.4 Generalized model for the depositional structure of Cretaceous Tethyan carbonate platforms. Columns show minor depositional cycles (parasequences) in outer and inner platform settings, with rudist lithosomes in pale blue (one example in each column is extended laterally to show its cross-sectional form). Expanded logs show idealized cycles. Typical platform dimensions are shown below (note vertical exaggeration).

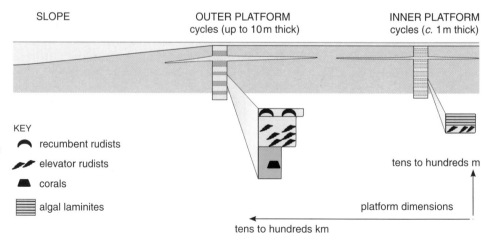

Parasequences in the outer platform typically might commence with coral beds, followed by rudist lithosomes, perhaps overlain by transported bioclastic limestones, reflecting deposition in progressively shallower conditions (Figures 5.4, 5.5). In some cases, however, corals were rare or absent, and the parasequences comprised just rudist lithosomes and associated bioclastic deposits. Parasequences produced in the more restricted conditions of the inner platform were usually thinner, lacked corals, and were sometimes capped by algal laminites formed in intertidal conditions (Figure 5.4).

(a)

Figure 5.5 Example of a shallowing cycle (parasequence) from an outer platform-top sequence in the Santonian of the southern Central Pyrenees, Spain (another section of the cycle illustrated in Figure 3.12b; see Skelton *et al.* (1995)). (a) Section through the cycle showing a biostrome of platy to domal corals with scattered large elevator rudists in its lower part (1), followed by a biostrome of clustered elevator rudists in the middle part (2), and culminating with bioclastic beds in its upper part (3). (b) Detail of slender elevator rudists in unit 2, here largely toppled. (c) Photomicrograph of the fine matrix of unit 2, consisting largely of tiny chips of rudist shell produced by boring sponges. (d) Photomicrograph of the current-swept and coarser bioclastic sediment of unit 3. *(Peter Skelton, Open University.)*

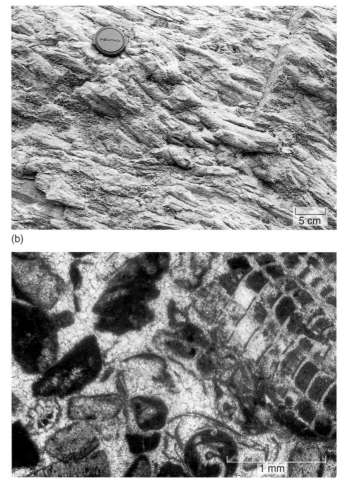

(b)

(c)

(d)

This characteristic geometry of deposition had major feedback effects on the ecology of the platform biota. Despite not building reefs, as such, the rudist associations seem to have been capable of producing carbonate at rates to match any modern reef. The German rudist researcher Thomas Steuber (2000) has identified periodic oscillations of oxygen and carbon isotopes in sectioned rudist shells, which are most plausibly interpreted as annual growth increments (i.e. reflecting seasonal variations in water temperature and/or composition). From these, he has been able to estimate the mean annual rate of carbonate production in individual shells, and hence bulk rates of carbonate production in the congregations, assuming typical population densities on the sea-floor. His calculations for various Late Cretaceous rudists gave values in a range of 2.2–35.7 kg $CaCO_3$ m^{-2} year^{-1}, which compares closely with estimates of carbonate production in modern coral reefs (1–35 kg $CaCO_3$ m^{-2} year^{-1}). However, the crucial difference is that rudist congregations expanded over large parts of the platform tops, rather than being mainly limited to their outer margins, as modern reefs are. When washed out (e.g. by storms), the shells could be broken down both by boring organisms (Figure 5.5c) and current action. Thus the congregations provided a vast carbonate factory area, from which bioclastic debris could be harvested and laterally redistributed by currents. Indeed, the very lack of reefal barriers had the feedback effect of promoting such redistribution by opening up large areas of the platform tops to the influence of storm waves and currents.

Work by Italian geologist Gabriele Carannante and his colleagues (1999) has highlighted a further, compositional factor that specifically relates to post-Cenomanian rudist associations. They have drawn attention to the relatively high proportion of calcitic, as opposed to aragonitic, bioclasts that were derived from these later rudist/foraminifer-dominated associations (a theme to which we will return in Chapter 8). As calcite is less soluble than aragonite, such sediments were less prone to early cementation (by reprecipitation of material dissolved near the sediment surface) than aragonite-rich sediments, so they remained susceptible to current reworking for long periods.

Repeated colonization of the platform surfaces by rudists and associated shelly biota, in concert with factors promoting redistribution of shell rubble and sand derived from them, as outlined above, allowed rapid filling of the thin but laterally extensive increments of accommodation space furnished over the platforms. Short-term rises in relative sea-level were thus matched by rapid aggradation (vertical build-up), yielding the characteristically tabular platform-top sedimentary cycles (e.g. Figure 5.5). Meanwhile, copious export of bioclastic material to neighbouring depressions also allowed frequent progradation (lateral building out of inclined beds) around the platform margins (e.g. Figure 1.12).

Once initiated, the Cretaceous carbonate platforms thus had great potential for rapid, self-sustaining growth and expansion, making them capacious and responsive sinks for carbon — so long as environmental conditions for their growth remained favourable.

5.2 Platform crises and oceanic anoxic events

As noted in the previous Section, brief phases of platform demise coincided with major oceanographic perturbations, the chief expression of which was the widespread deposition of organic-rich sediments. Some of these perturbations were of global effect, which has proved useful for stratigraphical correlation. The

Figure 5.6 The Vispi Quarry in the Contessa Gorge, near Gubbio in northern Italy, showing the '*livello* Bonarelli' as a thin dark band running obliquely across the quarry face on the left, separating the pale grey-white *Scaglia Bianca* from the pink *Scaglia Rossa* slope deposits. (*Iain Gilmour, Open University.*)

best-studied example fell at the close of the Cenomanian (Section 1.1.4). In the region around Gubbio in northern Italy, the corresponding level is marked within fine-grained slope sediments by a 1 m-thick dark band of organic-rich shale and radiolarian sands known as the '*livello* Bonarelli' (Figure 5.6).

Associated with the Bonarelli horizon is a pronounced positive shift of δ^{13}C values (Box 1.1). This isotopic signature can also be detected well outside the Tethyan region, as, for example, in the English Chalk (Figure 5.7), where it coincides with a distinct, clay-rich layer known as the Plenus Marls (named after the belemnite, *Actinocamax plenus*, which is common at that level, Figure 5.8).

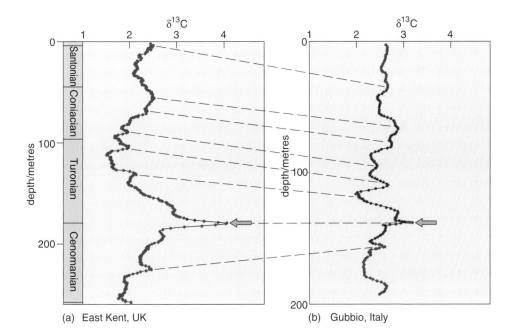

Figure 5.7 Smoothed composite curves of δ^{13}C values (‰ deviation from PDB standard) for carbonates in the Upper Cretaceous sections of (a) the Chalk of East Kent, SE England (arrow indicates position of the Plenus Marls) and (b) the Gubbio area in northern Italy (arrow indicates position of the *livello* Bonarelli). Correspondences between the two curves are shown by dashed tie-lines. (*Jenkyns et al., 1994.*)

Figure 5.8 Outcrop of the Plenus Marls (indicated by the seated man) in a Chalk quarry in Buckinghamshire, England. *(Peter Skelton, Open University.)*

○ What can you infer concerning the relative rates of burial of organic carbon (C_{org}) and carbonate carbon (C_{carb}) from a positive shift in seawater $\delta^{13}C$ values?

● C_{org} is preferentially enriched in the light carbon isotope, ^{12}C (Box 1.1). Hence, for the $^{13}C/^{12}C$ ratio of the carbon left in solution (in carbonate and bicarbonate ions) to have risen — yielding more positive $\delta^{13}C$ values — the rate of burial of C_{org} relative to C_{carb} must have increased.

The interpretation of these stratigraphical levels has been much debated. Competing explanations for the organic-rich deposits involve either enhanced planktonic productivity, or merely increased transmission of organic detritus through the water column to the sea-floor thanks to anoxia in the latter (preventing oxidative breakdown, so enhancing preservation), or a combination of the two. An additional contributory factor could have been a decrease in the rate of inorganic sediment input from the land because of high sea-levels. However, the global extent of the $\delta^{13}C$ signature implies drastically increased burial of organic carbon relative to carbonate carbon, overall, which would be hard to explain merely in terms of regional variation in productivity, and extensive anoxia seems likely in at least some parts of the oceans. Such episodes were thus termed oceanic anoxic events (OAEs) by the American oceanographer Seymour Schlanger and British geologist Hugh Jenkyns (1976). Three main OAE intervals have been identified within the Cretaceous, with OAE 2 corresponding to the terminal Cenomanian episode (see Figure 5.1), though OAE 1 has since been subdivided into distinct shorter episodes and other, more minor events recognized as well.

The close of the Cenomanian was attended by the drowning of carbonate platforms from Mexico to the Middle East (although those in the former region were already substantially reduced from their Albian acme, Figure 5.1). In particular, all the recumbent rudists of the time, which tended to thrive on the outermost platform zones, became extinct, though numbers of species appear to have declined progressively through the late Cenomanian. Benthic foraminifers, especially the 'giant forms', showed a similar pattern. Exactly how the OAE was linked with the mass extinction remains unresolved. Some authors have suggested that the expanding zone of anoxia combined with rising sea-level might have effectively 'sterilized' large parts of the platforms. However, the possibility of earlier initiation of the wave of extinctions hinted at above means that the OAE could just as well have been a symptom, rather than a cause, of environmental crisis.

Earlier such crises occurred in the mid-Valanginian and over the latter part of the early Aptian (Figure 5.1), both likewise associated with positive global $\delta^{13}C$ excursions, though in neither case is there an unambiguous eustatic signal. For example, early Aptian platforms in the Adriatic and eastern Arabian regions became emergent, whereas new platforms became established with flooding of previously continental areas along the North African margin in the late Aptian. There does seem to be some evidence for associated climate change, however, from arid to wetter conditions in many areas in both the mid-Valanginian and the mid-Aptian (e.g. Figure 5.9).

The reasons for the late Albian/early Cenomanian decline of carbonate platforms in the New World, and contrasting expansion in the Old World (Figure 5.1), remain unclear. We can but note several major tectonic changes at the time (such as the opening of the North–South Atlantic connection), besides at least regional emergence (e.g. Allison Guyot, Section 1.1.5) as well as minor OAE events registered elsewhere. A regional explanation for the demise of the Pacific examples, based on excessive heating of low latitude waters, has also been suggested by Wilson *et al.* (1998).

(a)

Figure 5.9 Mid-Aptian changes of facies in (a) the Lusitanian Basin, Portugal, where platform limestones of early Aptian age (middle and below right) are overlain by fluviatile siliciclastic sandstones of mid-Aptian age (above left); and (b) Istria, where the early Aptian platform limestones show a shallow karstic weathering surface, overlain by green, clay-rich fossil soils of late Aptian age followed by Albian limestones. (c) Close-up of the karstic surface and fossil soil (below) in (b). *(Peter Skelton, Open University.)*

(b)

(c)

Following the recovery of platforms in the Turonian, OAE 3 struck in the Coniacian/Santonian (Figure 5.1), this time coinciding with phosphate-rich, black shale deposition in relatively shallow waters over large areas of NE Africa, Sinai and Israel. Yet there seems to have been somewhat less-marked extinction among the platform biotas than on previous occasions, though there was a major turnover in planktonic foraminifers in the Santonian. With this final perturbation, the Cretaceous series of major OAEs ended and marine anoxia became an occasional sideshow thereafter, recurring only in restricted basins and under special conditions.

5.3 The sedimentary record in mid- to high latitudes

In mid-latitudes, the most remarkable feature of the Late Cretaceous world, in particular, was the massive expansion of Chalk deposition on continental crust (Section 1.1.1), associated with both global transgression and evolutionary radiation of the calcareous plankton. It is possible that the effect of the eustatic rise in sea-level went beyond merely flooding large areas of the continents with seawater. Initially, with the continental margin usually quite shallow (like today), the confrontation of circulation patterns on the shelf and in the ocean probably created a shelf-break front — a sort of wall of circulating water that would have served as a partial barrier between the two water masses (Figure 5.10a). A key difference for the planktonic ecosystems inhabiting the two water masses would have been the rhythm and rate of nutrient supply. Continuing sea-level rise, however, is likely to have led to the shelf-break front eventually being overstepped by oceanic currents, bringing their planktonic ecosystems with them (Figure 5.10b). Thus, what we would regard as planktonic denizens of the open ocean today, such as coccolithophores, would, during the late Cretaceous, have gained free access to the broad shelf and intracontinental seas of the time. Moreover, mid-latitude aridity together with the subdued relief of the neighbouring hinterlands would also have contributed to the 'oceanic' character of Chalk deposition by severely reducing siliciclastic input. Although the fossil record of the calcareous plankton has now been traced as far back as the late Triassic, when they were apparently restricted to low latitudes, they underwent extensive evolutionary radiation and spread up into higher latitudes in the late Cretaceous. The exceptional oceanographic conditions outlined above no doubt contributed to their success.

Despite its superficial monotony, the Chalk again testifies to a dynamic history of environmental change. The effect of the terminal Cenomanian perturbation has already been noted (Figures 5.7, 5.8), though whether anoxia, as such (as opposed to just the global carbon-isotope signature), impinged on the Chalk seas has been questioned. At a much finer scale, we also have the background depositional oscillations attributed to orbital forcing of climate (Section 1.3.1).

At high latitudes, deposition of the Chalk gave way to more siliciclastic-rich sediments, as seen, for example in the Arctic Basin and the Lower Saxony Basin in northern Germany. This change of facies is consistent with the continental evidence for relatively humid polar climates (Chapter 4). In addition, some evidence from Lower Cretaceous marine deposits is suggestive of colder, perhaps even briefly glacial, episodes. Dropstones are isolated boulders encased in finer-

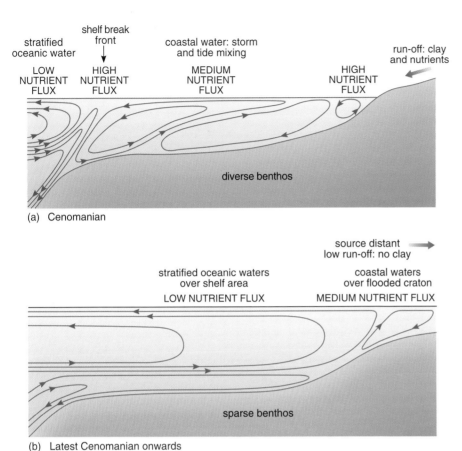

(a) Cenomanian

(b) Latest Cenomanian onwards

Figure 5.10 Reconstructed shelf profile in southern England in (a) Cenomanian, and (b) latest Cenomanian/early Turonian times. (Gale et al., 2000.)

grained marine deposits, which can be interpreted to have fallen from melting ice rafts, either calved from glaciers or merely derived from break-up of seasonal shore-ice. They are common, for example, in mudrocks of Valanginian to Aptian age that were deposited between 53° and 72° S palaeolatitude, in Central Australia, as well as in Arctic Basin deposits in Alaska and northern Siberia. Another, rather more specific indicator is the presence of glendonites, which are calcite replacements of the mineral ikaite, a low temperature/high pressure form of hydrated $CaCO_3$. Ikaite crystallizes to form distinctive stellate aggregates at around 0 °C, above which it soon alters to calcite. Modern examples have been recovered from the Antarctic shelf at nearly 2 km depth, and from much deeper (~ 4 km) levels at lower latitudes, where it remains stable (under the high pressures) up to a few °C. In the Arctic Basin, beds containing glendonites imply brief cold episodes during the Valanginian and late Aptian. Moreover, the German geologist Edwin Kemper (1995) has argued that alternating bands of light, carbonate-rich, and dark, carbonate-poor mudrocks of mid–late Aptian age in the Lower Saxony Basin reflect the onset of short-term climatic oscillations between warm/arid and cool, wet conditions, respectively. The regressive tendencies of the dark bands suggest glacio-eustatic falls in sea-level during the cold phases.

It is intriguing that the main early Cretaceous intervals of inferred cooling at high latitudes, in the Valanginian and Aptian, appear to have coincided broadly with platform extinction events, isotopic indications of OAEs and other hints of climatic change in low latitudes, as discussed in the previous Section. How they might have been linked causally requires further investigation, especially better-resolved stratigraphical correlation in order to elucidate the exact sequence of events.

5.4   The pattern of climatic change

Potentially one of the most useful keys to past temperatures is the measurement of oxygen-isotope ratios in minerals that crystallized in the environments in question (including skeletal minerals). The principles are straightforward enough, but because of the different possible influences on the results, interpretation of them is often fraught with difficulties. Two distinct fractionation processes are relevant. First, when, for example, a mineral such as calcite crystallizes, it preferentially takes carbonate ions containing the heavier isotope of oxygen, ^{18}O, into its crystal lattice, so the ratio of ^{18}O to the commoner, light isotope, ^{16}O, is greater than that in the surrounding water. The strength of this fractionation process declines with increasing temperature. Thus, if the seawater's isotopic ratio is known, its original temperature can be inferred from the isotopic ratio observed in the calcite. The ratio of $^{18}O/^{16}O$ is usually recorded as a deviation from a standard, $\delta^{18}O$, which is calculated in the same manner as $\delta^{13}C$ for carbon isotopes (see Box 1.1). Accordingly, proportional enrichment in ^{18}O, relative to the standard, is signified by positive values of $\delta^{18}O$, while negative values of the latter indicate relative depletion of the heavier isotope.

○   Other things being equal, would you expect increasing values of $\delta^{18}O$ in calcite to indicate increasing or decreasing temperatures during crystallization?

●   As the effect of the fractionation declines with warming, increasing enrichment in the heavy isotope (yielding greater $\delta^{18}O$ values) should correspond to *decreasing* temperature.

No doubt the phrase 'if the seawater's isotopic ratio is known' did not escape your notice earlier, since we are concerned with Cretaceous seawater that we cannot directly sample: that is where the second fractionation process becomes relevant. When water evaporates, molecules containing the light oxygen isotope can escape (as vapour) slightly more easily than those containing the heavy isotope, so the vapour tends to have a lower $\delta^{18}O$ value than the water. The total amount of water vapour in the atmosphere, and even that returning to the sea via rain and rivers, is tiny compared to the volume of water in the oceans, so changes in the former are unlikely to have much impact on the isotopic ratio of the latter.

○   Can you think of another, larger reservoir to which water vapour from the oceans contributes, changes in the size of which might thus have a detectable effect on the oceanic ratio of oxygen isotopes?

●   The continental ice-caps, which are fed by snow ultimately derived from oceanic vapour, represent sizeable reservoirs of isotopically 'light' water (note, however, that sea-ice, which forms directly from seawater without intervening evaporation, is not relevant here). Hence, variations in the amount of water locked up in the ice-caps can be expected to modify the isotopic ratio observed in the oceans.

Hence, as ice-caps expand, drawing off isotopically light water from the oceans, so the $\delta^{18}O$ value of the remaining seawater increases. In order to calculate an estimate for the original temperature of seawater in which a given sample of calcite crystallized, it is therefore necessary to correct for the difference in volume of the ice-caps then and now. A check also has to be kept on salinity. If there is excessive evaporation from a more or less restricted water mass, the remaining water, as well as showing increased salinity, will also be biased towards a heavier isotope ratio. Lowered salinity due to meteoric influx, by contrast, would show the opposite bias. And before such considerations, we also need to be certain that the calcite is original and has not suffered subsequent modification of its isotopic ratio. Diagenetic calcite precipitated from meteoric water and/or at high temperatures, for example, will have appreciably lighter $\delta^{18}O$ values than pristine calcite precipitated in the sea. So, oxygen-isotope data need to be handled with considerable caution. It is worth remembering at all times that they do not provide a direct measure of palaeotemperature (despite frequent loose talk of 'palaeotemperature measurements' in the literature), but only the basis for calculating estimates, which depend upon assumptions concerning diagenesis, ice volumes and salinity.

Numerous oxygen-isotope measurements have been carried out over the last few decades, mostly on calcitic shell material, but also on phosphatic skeletal remains from fish, for example. A large number of palaeotemperature estimates from such data were plotted against the palaeolatitudes of the sample sources by the Australian palaeoclimatologist Larry Frakes, in 1999, in what he termed a grossplot (Figure 5.11). Although the data are distributed very unevenly (the early Cretaceous being seriously under-represented, for example), and do not allow for longitudinal regional variation, they do portray a general pattern of warming into late Cretaceous times, though with considerable fluctuations.

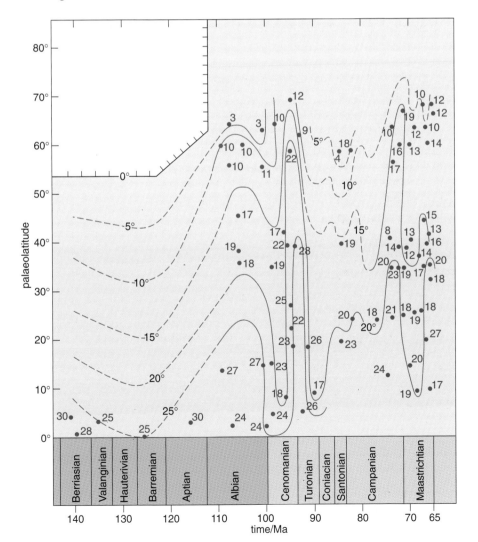

Figure 5.11 Frakes' (1999) grossplot of estimated marine palaeotemperatures (°C), against palaeolatitude and time. Smoothed contours shown at 5 °C intervals (dashed where uncertain or conjectural); note that there is some scatter of data points around these.

○ What, do you suppose, was the basis for the 0 °C limit shown in the top left part of Figure 5.11, despite an absence of pertinent isotope data there?

● The Valanginian to Aptian dropstones of central Australia, mentioned in the previous Section, would imply at least seasonal cooling to below 0 °C, to account for the formation of the ice that is inferred to have carried them.

One curious feature of the grossplot for the Maastrichtian, in particular, is the apparent inversion of the temperature gradient within the tropical belt: some low-latitude values seem to be anomalously low. Other authors, on the basis of similar data (not included here), had taken the estimated palaeotemperatures at face value, and referred to the apparent pattern as the 'cool tropic paradox'.

○ Why should the situation (if correctly diagnosed) have been regarded as a paradox?

● From the overall energy balance of the Earth, we should expect a general pattern of polar heat transport at all times (Chapter 4), and a reversed latitudinal gradient of temperature in the tropics would be inconsistent with that.

Various ingenious alternative explanations were offered, such as the effects of intense evaporation on low-latitude surface waters (elevating their $\delta^{18}O$ values, so yielding low estimates of temperature). However, a much simpler explanation emerged in 2001 from work by a British expert on foraminifers, Paul Pearson and co-workers: most of the planktonic foraminifers on which earlier results were based had evidently suffered sufficient diagenetic alteration to have distorted the palaeotemperature estimates. This conclusion resulted from their study of exceptionally well-preserved planktonic foraminifers recovered from Maastrichtian shelf-sea claystones in Tanzania. Unlike previously studied material, which had mainly come from deep-sea drilling cores, and which may thus have suffered some early diagenetic alteration on sinking through colder waters to the ocean floor, the Tanzanian material was revealed through scanning electron photomicrographs to be pristine. Collected from an area that would have been situated at palaeolatitude 19° S, the specimens yielded much higher estimates for surface water palaeotemperatures (in the range 28–32 °C) than those previously obtained from oceanic planktonic foraminifers. This result agreed remarkably well with the smaller number of estimates derived from well-preserved macrofossils. For example, aragonite in rudist shells recovered from Maastrichtian deposits on an originally equatorial Pacific guyot gave estimates in the range 27–32 °C. Today, by contrast, sea-surface temperatures rarely exceed 30 °C, except in restricted basins. The lesson from these findings is that many of the earlier data, at least those derived from planktonic foraminifers, must be regarded as suspect. Hence, the finer detail, though probably not the broad pattern, of Figure 5.11 should be taken with a pinch of salt.

The general picture that has so far emerged from oxygen-isotope studies is of relatively cool (in Mesozoic terms) though variable climates in the earliest part of the Cretaceous, with the possibility of some occasional sea-level ice at high latitudes. A broadly warming trend seems to have set in from the Aptian, though with some distinct, relatively brief, cool/wet phases, until a maximum was reached around the close of the Cenomanian or in the early Turonian. Further fluctuations then seem to have ensued until the Maastrichtian, when climatic deterioration set in (as noted in Section 4.6). With a more critical approach to sampling (especially in relation to diagenetic effects) in future work, we can expect a lot more of the detail to be fleshed out over the next few years.

Thus far, we have only dwelt on apparent changes in sea-surface temperatures over the Cretaceous, without really considering the third dimension of changes of temperature with depth. The latter aspect can be tackled, however, by sampling the skeletal remains of organisms whose depth-habitats are known or can be inferred. For example, the German geochemist Silke Voigt has used this approach in a study of the Cenomanian to Turonian Chalk of Europe (Figure 5.12). She was able to infer temperature/depth relations using data from brachiopods and bivalves that had grown in shallow (tens of metres) shelf seas, and sharks and belemnites that evidently dwelt in deeper (hundreds of metres), cooler water masses further offshore. It is of interest to note that the estimated temperatures of the latter, though substantially less than those of the contemporaneous shallow waters, are still significantly warmer than the 1–2 °C typical of such deeper waters today. The Cretaceous oceans were evidently warmer throughout than those of today.

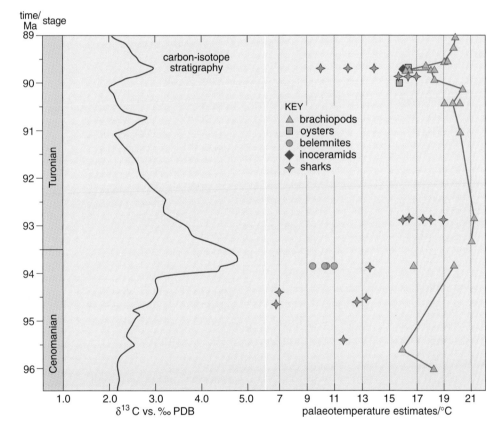

Figure 5.12 Palaeotemperature estimates derived from different kinds of fossil in a Cenomanian–Turonian shelf transect in the Chalk of northern Europe. A $\delta^{13}C$ curve for the succession is shown alongside (note again the pronounced positive shift in the uppermost Cenomanian). *(Unpublished data courtesy of Dr. Silke Voigt, Universität Köln.)*

With the confirmation of high tropical temperatures, it is worth also considering the likely effects on salinity. General circulation models for the Cretaceous yielding comparably high equatorial temperatures (up to 34 °C) predict associated high salinities (>40‰*) in arid zones, because of intense evaporation, especially in the western Central Atlantic and Tethyan regions. Convincing evidence for hypersaline surface waters in the Cretaceous oceans themselves is not yet forthcoming, though abundant evaporites in the Albian of northern Mexico, for example, confirm their existence at least in neighbouring restricted shelf areas. Whether the bulk of hypersaline water formed at the ocean surface or

* The symbol ‰ represents per mil or parts per thousand. In the case of seawater salinities, this is often abbreviated in the literature to a dimensionless value (e.g. 40) without the ‰ symbol.

seeped out from surrounding semi-restricted shelf seas, the next issue to be considered is under what circumstances it might then have sunk into the deep ocean, so contributing to oceanic circulation. The relationship of seawater density to salinity and temperature is complex (Figure 5.13). Basically, water density (measured at the sea-surface) increases with increasing salinity, but decreases with increasing temperature (except within a few °C above freezing point at lower salinities — which need not concern us here). In the modern world, brines produced by the formation of sea-ice readily sink because they are already cold. In the case of brines produced by evaporation, however, some cooling is usually necessary for sinking to occur.

Figure 5.13 Diagram relating density of seawater (curved contours for values in kg m^{-3}) to temperature and salinity. Potential values of temperature and density are those observed at the sea-surface (both show slight increases with depth for any given sample of water, due to compression). *(Hay and DeConto, 1999.)*

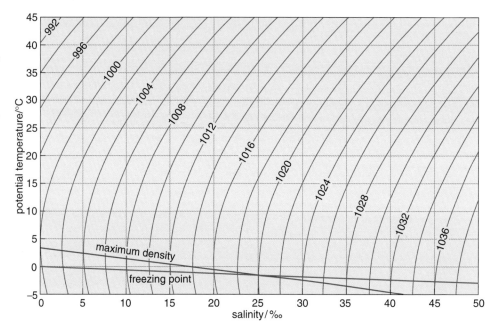

○ Suppose that surface currents introduce water of salinity 35‰ and temperature 14 °C to an area of intense heat and evaporation, such that the temperature of the water rises to 30 °C and the salinity to 40. By at least how much would the hypersaline brine then have to cool before it started to become denser than the incoming water, and so begin to sink?

● On Figure 5.13, one of the curved lines of equal density (1026 kg m^{-3}) passes through a salinity of 35 and temperature of 14 °C, and then on through 40 and about 28.5 °C, so the brine would have to cool by at least 1.5 °C (from 30 °C) before its density began to exceed that of the incoming water, allowing it to sink.

Thus, hypersaline brines produced in the Cretaceous tropics would probably have sunk as they cooled during the winter months, perhaps at slightly higher latitudes, thereby driving a so-called 'halothermal' circulation, as a sort of reversed version of the thermohaline circulation of today's icehouse world (Figure 2.4). It is not yet possible to determine the route that the deep circulation might have taken from its tropical source areas, as we do not have sufficient information on the topography of the Cretaceous ocean floors.

High sea-surface temperatures at low latitudes would have had one further crucial effect on the climate system. With each 10 °C rise in temperature, the amount of water vapour that the atmosphere can hold approximately doubles. Besides eventually delivering a lot more rain (somewhere else), the increased vapour content would also have invested the atmosphere with a great deal more transportable energy, in the form of latent heat of evaporation (the heat required to turn water into vapour, which is released again when the vapour condenses). Calculations by Bill Hay and Robert DeConto (1999) suggest that this last aspect may have provided the extra boost to polar heat transport that ensured the high-latitude warmth of the Cretaceous, an idea that we will explore further in Chapter 8.

If Cretaceous oceanography is difficult to reconstruct, then atmospheric weather patterns might seem even more intractable, yet Ito *et al.* (2001) have suggested an ingenious way to monitor past storm intensities. Assuming that the wavelength of hummocky cross-stratification (HCS) provides a measure of the size (orbital diameters) of storm waves, they surveyed HCS wavelengths from numerous Mesozoic–Cainozoic inner shelf successions as proxies for past storm intensities (Figure 5.14). A clear change through time was observed, with, once again, a major peak in the mid-Cretaceous, implying a maximum of storm intensity then. This conclusion is certainly consistent with the relatively warmer sea-surface temperatures discussed above. Given the intensity of some of the hurricanes that afflict the Caribbean/Gulf of Mexico area today, for example, as the ocean warms up each summer, their counterparts in the late Cretaceous must have been truly ferocious.

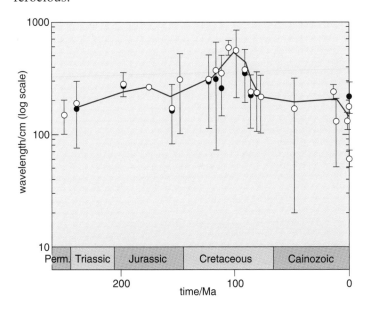

Figure 5.14 Variation in the wavelength of hummocky cross-stratification through the Mesozoic and Cainozoic, based on measurements from inner shelf deposits from Japan and other areas around the world. The wavelength data are plotted on a logarithmic scale: vertical bars show ranges of values; open and filled circles show midpoints and means, respectively. *(Ito et al., 2001.)*

5.5 Marine biodiversity

Whereas sea-level rise, marine shelf-space, mean temperature, storminess and oceanic volcanism all peaked in late Cretaceous times and declined thereafter, the fossil record indicates that marine (like terrestrial) biodiversity continued to increase, overall, to the present day (Figure 1.24). This sustained rise, only briefly reversed by the K/T mass extinction, appears to have been driven by a combination of intrinsic (biotic) as well as extrinsic (environmental) factors.

Extrinsic factors. One of the commonest instigators of speciation (hence increase in biodiversity) is the geographical separation of formerly interbreeding populations enforced by the development of some barrier to dispersal (vicariance), such as the expansion of an ocean between two continents. Regions that thereby acquire marked differences from one another in their complements of species are referred to as faunal or floral provinces. Provinces can be quantitatively defined in various ways, all somehow relating the numbers of taxa (e.g. species) that are restricted to given regions (endemic taxa) with those that are shared between regions (cosmopolitan taxa).

○ From consideration of the palaeogeographical history discussed in Chapters 1 (Section 1.2) and 2, how would you expect provinciality to have changed in general terms from the Mesozoic to the Cainozoic?

● The history of progressive continental break-up (Pangaea to Laurasia and Gondwana, and thereafter to today's continents) should have led to an increase in provinciality.

The effects of Atlantic opening, for example, have been investigated for various different groups of fossil organisms (e.g. Figure 5.15), confirming the tendency for increasing provincial distinction.

Figure 5.15 Effect on ammonite genera of Atlantic opening. Curves show percentages of genera endemic to the New and Old Worlds, respectively. (Kennedy and Cobban, 1976.)

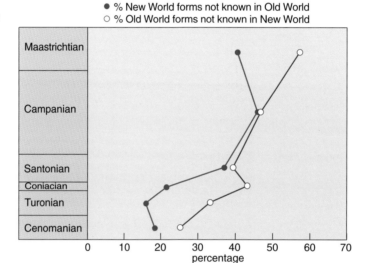

Cainozoic cooling added a further dimension to diversification by imposing more extreme latitudinal gradients of environmental conditions, which led to a finer subdivision of provinces, especially along the N–S oriented coasts bordering the Atlantic and Pacific Oceans. Together, these palaeogeographical changes contributed to the sustained rise in global biodiversity.

Intrinsic factors. After the mass extinction at the close of the Permian (Section 1.2), a remarkable set of linked changes in the nature of ecological relationships began to emerge among the newly radiating marine fauna. Chief among these was a significant increase in the intensity of predation enacted by various new groups of predators specializing on shelly prey. The new predators included shell-crushing crabs, lobsters, fish and marine reptiles, as well as starfish able to prise open their prey and evert their stomachs within their shells, so as to digest them in place, and snails equipped to drill through shells to gain access to their occupants. The prey in turn responded with an increasingly diverse array of defensive adaptations. For example, snails evolved more robust, compactly coiled shells, with fortified rims and obstructive protrusions, and several groups of bivalves began to burrow more deeply into the sediment, while others (such as oysters) adopted a cemented habit, which inhibited preparatory manipulation by the predators. These changes have sometimes been graphically described as an 'evolutionary arms race'. Meanwhile, surface-grazing by new stocks of roving sea-urchins, for example, besides the enhanced extent of burrowing, increasingly disturbed the sea-floor, and the benthos accordingly evolved various new compensatory strategies. Among crinoids, for example, stalkless, free-living taxa proliferated at the expense of sessile, stalked forms. These and other associated ecological changes have collectively been termed the Mesozoic marine revolution (MMR) by the Dutch palaeontologist Geerat Vermeij (1977). Although the MMR had started already in the early Mesozoic, elements continued to be added thereafter. Shell-drilling by gastropods, for example, seems to have taken off from Albian times, judging from the fossil record of their distinctive drill-holes.

The root causes of the MMR remain unclear, but the net effect of the proliferation of specialist predators and grazers, and the multiplicity of defensive adaptations evolved in response to them, was an increase in numbers of species within communities, and hence, again, a sustained increase in global biodiversity. That increase indeed continued in the Cainozoic, as yet more sub-plots (including molluscivorous birds and marine mammals, besides numerous new groups of predatory gastropods, for example) were added to the ecological drama.

Interaction of the two factors. Also unclear is the extent to which the exceptional conditions of the Cretaceous contributed (if at all) to the progress of the MMR. It has been argued, for example, that the vastly expanded shallow seas of the period accelerated it by providing an enlarged arena for the evolution of adaptive novelties, though we are far from being able to test such a speculative proposal. Certain evolutionary patterns, however, throw some doubt on it. The Cretaceous fossil record of predatory gastropods (including their trace fossils) is most diverse in regions that originally pertained to mid-palaeolatitudes. Only in the Cainozoic record does the peak of their diversity shift to the tropics, where we still see it today. Though the possibility of biased sampling cannot yet be excluded, the current evidence does seem to suggest that perhaps more temperate regions of the Cretaceous world were the main crucible for the rich tropical marine ecosystems of today. The biota that had dominated the Cretaceous carbonate platforms largely disappeared along with the Maastrichtian demise of the latter (Figure 5.1), which is why they appear as unfamiliar (even bizarre) to our eyes as the dinosaurs.

5.6   Summary

- Cretaceous tropical carbonate platforms underwent rapid, episodic development, with some regional variation, and major global interruptions in the mid-Valanginian, mid-Aptian and at the close of the Cenomanian, before their final decline in the late Maastrichtian.

- The main carbonate factories of the platforms were constituted by a distinctive shelly benthos, which, unlike that of today's tropical carbonate platforms, rarely developed reef frameworks, instead forming extensive 'meadows' spread across sedimentary substrates of low relief. Antecedent karstic surfaces were also rarely developed in Cretaceous carbonate platforms, in contrast to Quaternary platforms, in which they have frequently been produced by glacio-eustatic oscillations, so providing a template for modern reef topography.

- Rudist bivalves were among the most important contributors to the platform carbonate factories, growing variously as recumbents on current-swept surfaces, elevators implanted in accumulating sediment, or clingers closely overgrowing the substrate. Corals, mainly of domal to platy form, often carpeted the sea-floor downslope from the rudists.

- The typically limited increments of accommodation space that were provided by greenhouse oscillations of sea-level tended to yield tabular parasequences containing thin but laterally extensive coral and rudist lithosomes. Inner platform parasequences are generally thinner than those of the outer platform and usually lack corals, but may be capped by algal laminites.

- Bulk carbonate production rates by Late Cretaceous rudists matched those of modern reefs, per unit area, but the rudist congregations were spread across large parts of the platform tops, unlike modern reefs, yielding vast carbonate factories from which bioclastic debris could be harvested and redistributed by storm currents. The prevalence of calcite in such debris from post-Cenomanian associations limited early cementation, so further enhancing its susceptibility to current reworking.

- Major crises in platform growth coincided with widespread deposition of organic-rich sediments, associated with positive shifts in $\delta^{13}C$ values, observed world-wide. The most likely explanation is the spread of anoxia in at least some parts of the oceans (oceanic anoxic events, or OAEs), allowing increased transmission of organic detritus to the sea-floor. Following an (unnumbered) event in the mid-Valanginian, OAE 1 comprised several episodes in the Aptian/Albian, OAE 2 coincided with the close of the Cenomanian, and OAE 3 occurred in the Coniacian/Santonian. How the platform crises and OAEs were connected remains unresolved, though climatic change, from arid to wetter conditions, appears also to have been involved in some instances.

- A divergence in history between the New and Old Worlds occurred in the late Albian/early Cenomanian, when platforms in the former region began to decline, just as those in the latter were expanding, though again the reasons remain unclear.

- The spread of Chalk facies over large areas of continental crust in mid-latitudes during the late Cretaceous may have been triggered when the exceptional eustatic sea-level rise allowed ocean currents to overstep shelf-break fronts. The calcareous plankton, which dated back to at least the late Triassic, diversified greatly at this time.

- Cretaceous high-latitude successions are more siliciclastic-rich, reflecting the more humid conditions there. Dropstones and glendonites (calcite replacements of ikaite) in Lower Cretaceous deposits imply brief cold phases, especially in the Valanginian and late Aptian, the latter possibly also accompanied by glacio-eustatic oscillations. Better-resolved correlation is needed, however, to determine how these phases were connected with events at low palaeolatitudes.

- Estimates of seawater palaeotemperature can be derived from oxygen-isotope ratios in, for example, calcitic shells and the phosphatic skeletons of fish, though corrections must be made for continental ice volumes and salinity, as well as checks for diagenetic alteration. Plotting of palaeotemperature estimates against latitude and time ('grossplots') confirm a general pattern of warming into late Cretaceous times. Earlier concerns about anomalously cool tropical estimates (the 'cool tropic paradox') have since been attributed to diagenetic alteration: it now seems likely that tropical sea-surface temperatures in the late Cretaceous extended into the low 30s °C. The thermal maximum for the period appears to have been around the close of the Cenomanian or early Turonian.

- Comparisons of data from shallow, and deep-water organisms indicate cooler palaeotemperatures for the latter, though still significantly warmer than the 1–2 °C typical of deep waters today.

- High sea-surface temperatures were probably associated with raised salinities in arid zones, especially in the western central Atlantic and Tethyan regions. On cooling, these brines could have sunk, so driving a 'halothermal' circulation in the opposite sense from the thermohaline circulation of today's icehouse world.

- Low-latitude heat and evaporation would also have furnished the atmosphere with a much greater vapour content, the latent heat of evaporation of which could have been a significant factor in the enhanced polar heat transport of the period. A survey of changes in the wavelengths of hummocky cross-stratification through time imply a Mesozoic–Cainozoic peak in storm intensity in the mid-Cretaceous, which is consistent with the data on sea-surface palaeotemperatures.

- In contrast to the patterns of environmental change that peaked in late Cretaceous times, marine biodiversity continued to increase after the Cretaceous, due to a combination of increasing provinciality and the effects on community diversity of the Mesozoic marine revolution (MMR), principally involving more specialist predator/prey interactions. It is not clear what influence the exceptional conditions of the Cretaceous had on the MMR, though it should be noted that many of the ancestors of today's tropical marine ecosystems evolved outside the Cretaceous tropical seas.

5.7 References

ALSHARHAN, A. S. AND SCOTT, R. W. (eds) (2000) *Middle East models of Jurassic/ Cretaceous carbonate systems*, SEPM (Society for Sedimentary Geology), Special Publication, **69**.

CARANNANTE, G., GRAZIANO, R., PAPPONE, G., RUBERTI, D. AND SIMONE, L. (1999) 'Depositional system and response to sea level oscillations of the Senonian rudist-bearing carbonate shelves. Examples from Central Mediterranean areas', *Facies*, **40**, 1–24.

FRAKES, L. A. (1999) 'Estimating the global thermal state from Cretaceous sea surface and continental temperature data', pp. 49–57 in BARRERA, E. AND JOHNSON, C. C. (eds) *Evolution of the Cretaceous Ocean-Climate System*, Geological Society of America Special Paper **332**.

GILI, E., MASSE, J.-P. AND SKELTON, P. W. (1995) 'Rudists as gregarious sediment-dwellers, not reef-builders, on Cretaceous carbonate platforms', *Palaeogeogr., Palaeoclim., Palaeoecol.*, **118**, 245–267.

HAY, W. H. AND DeCONTO, R. M. (1999) 'Comparison of modern and Late Cretaceous meridional energy transport and oceanology', pp. 283–300 in BARRERA, E. AND JOHNSON, C. C. (eds) *Evolution of the Cretaceous Ocean-Climate System*, Geological Society of America Special Paper **332**.

ITO, M., ISHIGAKI, A., NISHIKAWA, T. AND SAITO, T. (2001) 'Temporal variation in the wavelength of hummocky cross-stratification: implications for storm intensity through the Mesozoic and Cenozoic', *Geology*, **29**, 87–89.

KEMPER, E. (1995) 'Changes in the marine environment around the Barremian/Aptian boundary', *Geol. Jb.*, **A141**, 587–607.

PEARSON, P. N., DITCHFIELD, P. W., SINGANO, J., HARCOURT-BROWN, K. G., NICHOLAS, C. J., OLSSON, R. K., SHACKLETON, N. J. AND HALL, M. A. (2001) 'Warm tropical sea surface temperatures in the Late Cretaceous and Eocene epochs', *Nature*, **413**, 481–487.

PHILIP, J. AND SKELTON, P. W. (eds) (1995) *Special Issue — Palaeoenvironmental models for the benthic associations of Cretaceous carbonate platforms in the Tethyan realm*, *Palaeogeography, Palaeoclimatology, Palaeoecology*, **119**.

SCHLANGER, S. O. AND JENKYNS, H. C. (1976) 'Cretaceous oceanic anoxic events: causes and consequences', *Geologie en Mijnbouw*, **55**, 179–84.

SIMO, J. A., SCOTT, R. W. AND MASSE, J.-P. (eds) (1993) *Cretaceous Carbonate Platforms*, American Association of Petroleum Geologists, Memoir **56**.

SKELTON, P. W., GILI, E., VICENS, E. AND OBRADOR, A. (1995) 'The growth fabric of gregarious rudist elevators (hippuritids) in a Santonian carbonate platform in the southern Central Pyrenees', *Palaeogeography, Palaeoclimatology, Palaeoecology*, **119**, 107–126.

STEUBER, T. (2000) 'Skeletal growth rates of Upper Cretaceous rudist bivalves: implications for carbonate production and organism-environment feedbacks', pp. 21–32 in INSALACO, E., SKELTON, P. W. AND PALMER, T. J. (eds) *Carbonate platform systems: components and interaction*, *Geological Society, London, Special Publications*, **178**.

VERMEIJ, G. (1977) 'The Mesozoic marine revolution: evidence from snails, predators and grazers', *Paleobiology*, **2**, 245–258.

WILSON, P., JENKYNS, H. C., ELDERFIELD, H. AND LARSON, R. L. (1998) 'The paradox of drowned carbonate platforms and the origin of Cretaceous Pacific guyots', *Nature*, **392**, 889–894.

PART 2   THE WORKINGS OF THE CRETACEOUS WORLD

6   Biogeochemical cycles

Simon P. Kelley

In Part 1, we described how the Earth's climate system was very different from today's during the Cretaceous, with dense forests and duck-billed dinosaurs roaming the northern slope of Alaska. Wide seas rich in calcareous plankton covered many of the world's continental shelves and carbonate platforms grew in the Tethys/Atlantic seaway and covered extensive lines of oceanic volcanoes in the Pacific. Deep ocean water was not near to freezing as it is today, but as warm as 15 °C and mean land surface temperatures were almost 10 °C higher than today. So why were environmental conditions so different in the Cretaceous? Was there one overriding cause or did many things combine to reinforce global warming? In this Chapter, we will look at how the carbon cycle works over geological time-scales and how it affects the climate and interacts with other cycles such as those for sulfur, phosphorus and iron. These questions are critical in order to understand how climate worked during the Cretaceous and what the future holds for us in our own warming environment.

6.1   The carbon cycle

To find out why the Earth's climate operated in such a different way during most of the Cretaceous Period, we need to do a little ground work. What separates the Earth from other planets is our equable life-supporting climate. The surface temperature of Venus, the planet next closest to the Sun, is around 450 °C. Mars, the planet next furthest from the Sun, has a surface temperature of around –50 °C. Theoretically, the Earth should have a surface temperature of around –18 °C. However, greenhouse gases such as CO_2 (carbon dioxide), O_3 (ozone), N_2O (nitrous oxide), H_2O (water vapour), SO_2 (sulfur dioxide) and CH_4 (methane) absorb light from the Sun at certain wavelengths (Figure 6.1) and re-radiate heat towards the Earth's surface, thus keeping the surface at an average temperature of +15 °C.

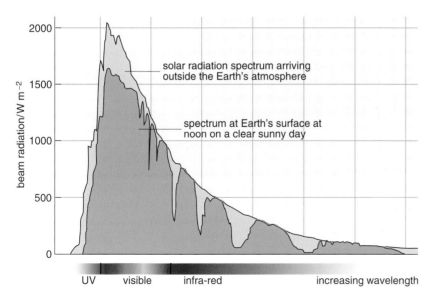

Figure 6.1   Comparison of the solar radiation arriving outside the Earth's atmosphere (yellow) with the superimposed solar radiation at the Earth's surface (red) at noon on a clear sunny day. Note that the amount of radiation arriving at the surface varies with the wavelength of the light. Several broad bands are strongly absorbed in the atmosphere, particularly at long, infra-red wavelengths. These absorption bands are characteristic of the greenhouse gases, in particular, N_2O, CO_2, H_2O and CH_4. *(Image by Dr. Andrew Marsh of Square One Research PTY Ltd.)*

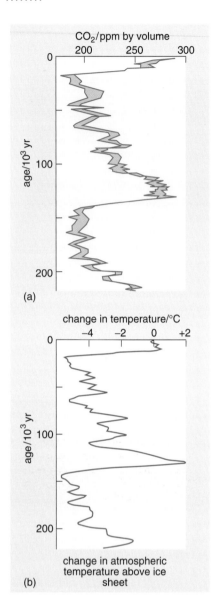

CO$_2$/ppm by volume

(a)

change in temperature/°C

change in atmospheric temperature above ice sheet

(b)

Figure 6.2 Record of CO$_2$ from the Vostok ice core (a), which tracked changes in global temperatures (b), from a generally warm period around 140 000 years ago to the last glacial maximum, just 18 000 years ago.

Without the greenhouse effect, the surface of the Earth would be covered in ice. In fact, records from the 2-km-deep Vostok core, drilled into the ice in Antarctica, close to the South Pole, show that changes in temperature tracked changes in atmospheric CO$_2$ (Figure 6.2). Although CH$_4$ also tracks temperature changes on a short time-scale, CO$_2$ is the only greenhouse gas with a long residence time in the atmosphere and thus affects long-term climate change. The relationship between atmospheric CO$_2$ and the biosphere (all plant and animal life) and geosphere (rocks and the deep Earth) mean that changes in CO$_2$ are linked to the operation of the carbon cycle. Life, therefore, has had an important impact upon the long-term changes in the Earth's climate.

The atmospheric CO$_2$ level has risen from 315 ppm (parts per million) to around 360 ppm over the last 40 years, a far faster rate of rise than any during the recent glacial events (Figure 6.2). It is this increase, and the concurrent rise in global temperatures, which is currently worrying scientists and governments alike. However, to make sense of the extreme greenhouse world of the Cretaceous, we need to take a longer-term view of atmospheric CO$_2$. During the last 200 000 years, the concentration of CO$_2$ in the atmosphere has varied between around 200 and 300 ppm but, in contrast, levels of CO$_2$ in the atmosphere during the Cretaceous Period probably reached three to six times the present-day value. This is what makes the Cretaceous Period so important: it provides us with a model of the Earth with a much higher concentration of atmospheric CO$_2$, little or no polar ice and higher temperatures across the globe — a 'Greenhouse world'.

The level of CO$_2$ in the atmosphere is regulated by the carbon cycle, but sub-cycles operating over different time-scales make its operation complex. During a single day, atmospheric CO$_2$ interacts with plants as they photosynthesize and respire, but this is too short a time-scale to equilibrate with CO$_2$ in the soil. Over a few weeks to months, atmospheric CO$_2$ can react to changes in season and the decay of plant matter in the soil. Over years and millennia, the oceans interchange with the atmosphere, but there is an even bigger reservoir of carbon which dominates the system over millions of years. This largest and dominant reservoir is rock, and we must look to the rocks to understand how CO$_2$ levels in the atmosphere and the climate change on a geological time-scale.

Box 6.1 explains in more detail how carbon is exchanged over a short time-scale between living organisms, the atmosphere, soil (including detritus or dead organic matter) and oceans. However, in order to understand how the Cretaceous climate system worked, and why it was so different to the present day, we must focus on the long-term carbon cycle. By far the largest carbon reservoir coming into contact with the atmosphere is rocks at the surface, such as the carbonates and organic-rich mudrocks we saw on the tour in Chapter 1. Carbon has a typical lifetime in rock of 100–200 Ma, but it can be stored for thousands of millions of years as we can see from the carbon-rich rocks found in ancient Precambrian rocks. CO$_2$ is released into the atmosphere from rocks by geological processes such as weathering or deep burial, which causes metamorphism of carbon-rich rocks releasing CO$_2$. Although we can quantify the processes controlling modern short-term CO$_2$ variations by studying the biosphere, this evidence is not available for the Cretaceous.

○ What evidence do we have for the interaction between the carbon cycle and climate during the Cretaceous?

● The evidence for the state of the carbon cycle during the Cretaceous is recorded in rocks, either in the form of fossils which we can use to determine conditions, or preserved as chemical signatures.

We have to apply the principle of uniformitarianism to infer the workings of the Cretaceous carbon cycle from evidence presented by rocks. This is a fairly robust principle since the processes involved in the carbon cycle were much the same as for the present day. We can make reasoned guesses at the ability of Cretaceous plant biomass to convert atmospheric CO_2 into organic carbon, or the rates at which shelly organisms convert atmospheric CO_2 to $CaCO_3$ (calcium carbonate). However, in order to quantify all this we would require knowledge of the relative contributions of the different pathways through the carbon cycle during the Cretaceous. We cannot measure these parameters directly, but by creating a model for the workings of the Cretaceous, we can investigate the possibilities and place some constraints upon what may actually have happened, as we will see later in this Chapter.

○ From the evidence of the rocks which we saw in Sections 1.1.1–1.1.6, which forms of life helped to sequester carbon from the atmosphere during the Cretaceous, storing it for many millions of years?

● Both marine and terrestrial forms of life left huge carbon deposits during the Cretaceous. Chalk deposits and the huge limestone deposits of the Tethyan/Atlantic seaway are the remains of marine life. Thick coal deposits at high latitudes are the remains of terrestrial forests.

Box 6.1 Short-term and intermediate-term carbon cycles

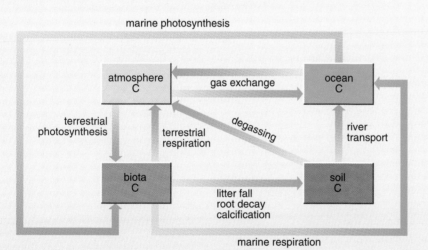

Figure 6.3 A box model for the short-term carbon cycles showing the processes that control exchange of carbon between the main reservoirs of ocean, atmosphere and soil. *(Berner, 1999.)*

Let us take a closer look at the main reservoirs for carbon and their interactions on the modern Earth. We can use this as a model against which to compare the carbon cycle in the Cretaceous world. There are two intertwined biogeochemical cycles: the terrestrial carbon cycle and the marine carbon cycle (Figure 6.3). The terrestrial carbon cycle involves the daily photosynthesis and respiration of plants, the yearly cycle of growth and decay, and involves mainly organic carbon. This cycle responds on a daily basis to changes in climate and atmospheric CO_2. The marine carbon cycle acts at a slightly longer time-scale of perhaps 100 000 ka. This cycle is dominated by transport of carbon as microscopic particles or as carbonate ions dissolved in water. The size of the one dominating reservoir, the deep oceans, means that the residence times for carbon are much longer than in the terrestrial carbon cycle.

The terrestrial carbon cycle

The processes of photosynthesis and plant respiration are responsible for the short-term terrestrial carbon cycle. The carbon reservoirs are CO_2 in the atmosphere, plant biomass, and soils. Animal biomass makes up such a small proportion of this (around 0.01%) that we can effectively ignore it — quite a sobering thought, although the effect that certain animals have upon atmospheric CH_4 and CO_2 cannot be ignored! Plants sequester around 120×10^{12} kg of carbon from the atmosphere each year but return half of this amount (60×10^{12} kg) via respiration.

○ Where do you think most of the carbon is stored in the biosphere?

● The remaining carbon is stored mainly in soils rather than in the more visible plant material, but most is returned to the atmosphere by decomposers such as bacteria; only a small proportion is preserved. Soils are an extremely important reservoir in the short-term carbon cycle because they contain more carbon than the atmosphere and plant biomass combined (see Table 6.1).

Since the short-term cycle is dominated by biomass, there is therefore a huge variation in local carbon cycles across the globe, each distinct ecosystem (e.g. tropical rainforest, tundra, savanna or boreal forest) storing different amounts of carbon as plant material and producing different atmospheric fluxes. As you might expect, ecosystems such as tundra or desert scrub produce only small amounts of plant material per year (<0.1 kgC m^{-2} yr^{-1}). Temperate and tropical forests produce much larger amounts of carbon (generally >0.5 kgC m^{-2} yr^{-1}). Tropical rainforest and swamps produce the most plant biomass per square metre at 0.9 kgC m^{-2} yr^{-1} and 1.125 kgC m^{-2} yr^{-1}, respectively. The current global total biomass amounts to roughly 560×10^{12} kgC. You will not be surprised to learn that there are considerable uncertainties in these biomass numbers and thus there are several estimates. For example, estimates of the total current biomass (not including soils) range from 420×10^{12} kgC to 830×10^{12} kgC. The mean plant biomass production rate is 48.3×10^{12} kgC yr^{-1}, a figure dominated by tropical rainforests with around 15×10^{12} kgC yr^{-1}, although this is partially due to their large area of coverage (roughly 11.4% of the land surface of the Earth).

○ From your knowledge of the Cretaceous, what do you think would be the main difference between modern biomass production, and biomass production during the Cretaceous?

● The sequestration of CO_2 by biomass during the Cretaceous cannot have been dominated by tropical rainforest as it is today, because the Equator was much drier and most of the primary biomass production was concentrated at higher latitudes. In fact, as we saw in Chapter 4 the rock record shows that huge forests in areas as far north as Alaska were the major producers of biomass during the Cretaceous.

Table 6.1   Sizes of the Earth's major carbon reservoirs.

Carbon reservoir	Size ($\times 10^{15}$ kg)
Oceans	38
Marine sediments	3
Soil and organic detritus	1.50
Atmosphere	0.76
Biosphere	0.56
Carbonate rocks (limestone, chalk, dolomite)	42 000
Organic-rich rocks (oil, coal etc.)	10 500
Methane hydrates	~80*

* Includes both marine and continental deposits.

The marine carbon cycle

In the oceans, CO_2 exchange with the atmosphere is regulated by equilibrium exchange across the air/ocean interface. Thus, unlike the terrestrial carbon cycle, which is dominated by organic carbon, the marine cycle is strongly influenced by the chemistry of inorganic carbon, in the form of dissolved carbonate ions. Nevertheless, the upper layers of the oceans are teeming with life and chemical transformations in the marine carbon cycle are regulated by biological interactions. The cycle is also controlled by physical factors such as temperature, which in turn affect life. Most biological production is in the photic zone (the upper 200 m of the oceans) and although this part of the ocean does not produce biomass at anything like the rate of the rainforest ($0.057\,kgC\,m^{-2}\,yr^{-1}$ compared to $0.9\,kgC$ $m^{-2}\,yr^{-1}$), the huge area of the oceans ($332 \times 10^6\,km^2$) makes them the largest net producer of carbon on the planet, at around $18.9 \times 10^{12}\,kgC\,yr^{-1}$. The only other important producer of carbon in the oceans in global terms is the continental shelves, where nutrients supplied by runoff from the continents raise net production levels to around $0.16\,kgC\,m^{-2}\,yr^{-1}$, and account for some $4.3 \times 10^{12}\,kgC\,yr^{-1}$ of biomass production, dominated by bottom-dwelling forms such as kelp forests. The total biomass within the oceans, concentrated in the upper photic zone is around $1.76 \times 10^{12}\,kgC$ and net production (the total after two-way exchange has been taken into account) in the oceans is around $24.9 \times 10^{12}\,kgC\,yr^{-1}$.

○ Compare these figures for the oceans with the total biomass and net production of land biomass. What do you notice about the differences between them?

● Although the biomass in the oceans is only 0.2% of that on land, the net production is over 50% of the land-based biomass. The oceans are a far more efficient carbon pump than land plants.

In the oceans, primary production by phytoplankton involves photosynthesis which produces O_2, and respiration which releases CO_2 into solution (as HCO_3^-) when they die or are eaten by zooplankton which are, in turn, eaten by larger animals. Dead tissue is colonized by bacteria which break it down to material that is dissolved in the water. Thus both inorganic and organic carbon is returned to the oceans ready to be reused by other organisms. In general, this cycle is around 90% efficient. Some small particles of organic carbon (mainly dead bacteria, faecal pellets and algae) sink slowly below the photic zone where they are recycled by other animals in the water column. Almost all debris will be recycled before it reaches the sea-bed, except in times of high production,

such as the spring plankton bloom in the North Atlantic, when many particles reach the sea-bed, and ocean bottom dwellers feast on the falling debris. However, this is rare and normally less than 1% of the total carbon production in the upper photic zone is preserved as sediment on the ocean floor.

There is one final process controlling the way in which carbon is cycled in the oceans. Many oceanic organisms such as planktonic algae called coccolithophores (Figure 1.4b) and zooplanktonic foraminiferans as well as benthic organisms precipitate carbonate shells and skeletons as a defence. This stage is critical to the long-term carbon cycle because the burial and preservation of these carbonate body parts also sequesters CO_2 from the atmosphere.

When these animals die, their skeletons sink and begin to dissolve because carbonate is more soluble at greater pressures. This process returns the inorganic carbonate ions to solution and if the particles sink slowly, they are often dissolved completely before they can reach the sea-bed. However, many particles become coated in organic debris, in the form of faecal pellets (the ratio of organic to inorganic debris is around 4 : 1), and are preserved as they sink to the ocean floor where they are dissolved more slowly in the sediment. The depth at which the rate of dissolution of calcite or aragonite shells equals the rate of supply (the depths for each mineral are different) is known as the carbonate compensation depth (CCD — defined in Section 1.1.1), around 3000–6000 m deep in most present-day oceans. At times of heightened production, the rain of particles is so great that the CCD can become deeper. This also leads to a much higher preservation rate of carbonate on the continental shelves where the ocean is less deep; a factor which is of particular relevance to the preservation of carbonate in the oceans of the Cretaceous.

The geosphere is not important in short-term carbon cycles since processes such as weathering or formation of carbonates and coal only generate significant effects over millions of years. The exceptions to this rule are volcanic eruptions which currently eject around $1.8 \times 10^{10}\,kgC\,yr^{-1}$ as CO_2 into the atmosphere, but have delivered as much as $3 \times 10^{10}\,kgC$ of CO_2 in a single historical eruption over a few days. The volcanic contribution will be discussed in some detail in Chapter 7. The geosphere also contributes to the short-term carbon cycle by accelerated recycling of carbon into the atmosphere via anthropogenic burning of carbon stored in rocks (coal, oil and gas) which amounts to as much as $680 \times 10^{10}\,kgC\,yr^{-1}$. However, the difference between present-day CO_2 levels and those extant during the Cretaceous cannot be explained by this particular route since dinosaurs did not, as far as we can tell, burn coal or drive oil-burning cars.

6.2 The long-term carbon cycle

The long-term or geochemical carbon cycle operates over a multi-million-year time-scale. It is fundamentally different to the shorter time-scale carbon cycles in that it includes interactions between the ocean–atmosphere–biosphere system and the geosphere.

The long-term carbon cycle thus encompasses the short-term cycles but the overwhelming size of the carbon reservoirs in the geosphere dominates processes on a long time-scale. Currently, around $42\,000 \times 10^{15}$ kgC are locked up in buried carbonates (limestone, chalk and dolomite) and $10\,500 \times 10^{15}$ kgC are preserved as organic matter (coal, oil, organic mudstones). Compare these reservoirs with the largest reservoir in the biosphere (land plants containing 0.56×10^{15} kgC), the atmosphere (containing 0.76×10^{15} kgC) and the world's oceans (containing 38×10^{15} kgC). The geosphere thus contains around 1200 times the combined carbon in all surface reservoirs (Table 6.1, Figure 6.4). Many processes are involved, including silicate weathering, long-term CO_2 output from ocean ridges and volcanoes, metamorphism and diagenesis, and carbon burial as organic or carbonate rocks. These all have an impact on climate because the amounts of carbon in the geosphere are much larger than those in other reservoirs. Two other processes have long-term effects on atmospheric CO_2 but were more important in the early Palaeozoic: the evolution of land plants and the gradual increase in the intensity of radiation from the Sun. However, none of these processes on their own could be said to 'control' climate; it is the interaction between them that dictates climate variations. Although we cannot hope to constrain all of the different forcing mechanisms of the long-term carbon cycle, by understanding some of its pathways and feedbacks, we can understand how it works and begin to discover which effects might have tipped the balance into a greenhouse world during the Cretaceous Period.

6.2.1 Cause and effect in the long-term carbon cycle

The two largest reservoirs of carbon in the Earth's crust are calcareous sediments and carbonaceous sediments. Calcareous sediments form mainly in marine environments and are largely composed of fine-grained particles and shells or skeletons created by organisms from CO_2 dissolved in seawater (as HCO_3^- ions). Carbonaceous sediments in the marine realm are rich in organic carbon, derived from soft plant or animal tissue, which was not oxidized prior to burial. These sedimentary deposits are formed when organic productivity is high and matter is buried rapidly or bottom waters are anoxic, and they react to form a partially degraded residuum known as kerogen. Petroleum forms from kerogen by heating during burial and can become trapped as oil and natural gas deposits. In terrestrial sediments, rotting plant debris accumulates either due to rapid burial or swampy conditions. The peat layers formed in this way are progressively transformed into coal by heating and loss of volatiles during burial.

We have seen (Figure 6.1) that the amount of CO_2 in the atmosphere has a major forcing effect on the climate and global temperatures. Over a long time-scale, the CO_2 content of the atmosphere is controlled by geological processes and reservoirs. We can track the interaction of the different reservoirs and fluxes by referring to a diagram like Figure 6.4. This shows the sizes of reservoirs (shown as blue boxes) and fluxes between reservoirs (green arrows), but it is very difficult to see how the processes interact and how the various feedbacks work. However, it does emphasize the importance of the large reservoirs.

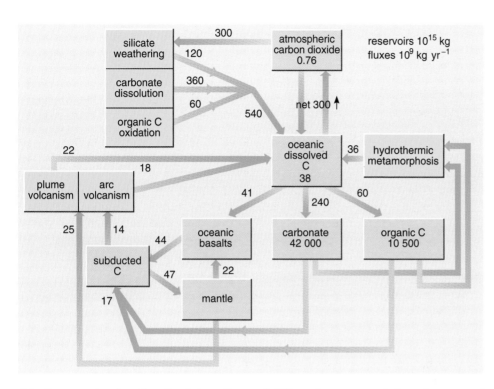

Figure 6.4 Box model for the long-term carbon cycle, showing the reservoirs (blue boxes, values given as $\times 10^{15}$ kg) or supplied by processes (yellow boxes), and estimated fluxes (green arrows, values as $\times 10^9$ kg yr^{-1}) of carbon between the reservoirs. Note the large carbon fluxes into the oceans, between the ocean and atmosphere and between the ocean and carbonate reservoirs, and the overwhelming size of the carbonate and organic carbon geological reservoirs. *(Arthur, 2001.)*

A better way to visualize the interactions of different processes is by a 'cause and effect' diagram such as the one Bob Berner of Yale University (1999) constructed for the carbon cycle (Figure 6.5c). We will see how this diagram works in the following pages, building upon the simplified cycles in Figure 6.5a,b. These diagrams emphasize the interconnectedness, and the cyclical nature of many of the processes. The ellipses represent ingredients (processes and reservoirs) and the arrowed lines between them represent the effect each has upon the next. Let us start with the cycle which leads to the build-up of calcareous sediments and the removal of CO_2 from the atmosphere: the silicate–carbonate sub-cycle.

Figure 6.5 Cause and effect diagrams for the long-term carbon cycle: (a) silicate–carbonate sub-cycle; (b) adding the organic matter sub-cycle; (c) the long-term carbon cycle. Each ellipse indicates a process or reservoir that feeds into other processes and is affected by other processes. The two square boxes indicate inputs to the system that are not affected by the other processes. Closed arrows indicate feedbacks that tend to accelerate the next process in the loop (positive feedbacks), and open arrows indicate feedbacks that tend to decelerate the next process in the loop (negative feedbacks).

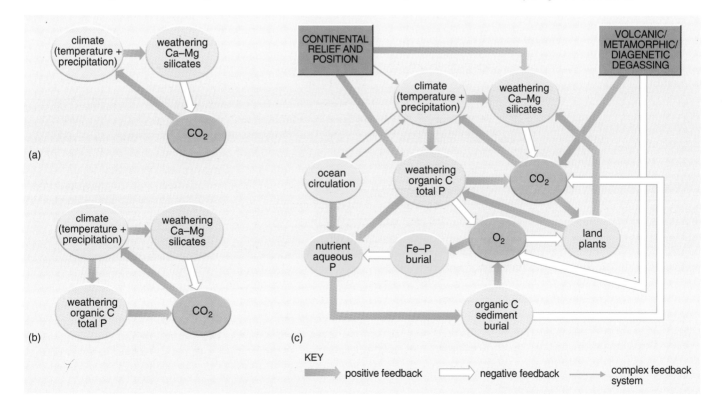

6.2.2 The silicate–carbonate sub-cycle

The silicate–carbonate sub-cycle describes the process of CO_2 removal from the atmosphere by weathering, a process that is accelerated by organic acids from decayed plant matter. CO_2 removal from the atmosphere occurs during weathering of Ca-rich and Mg-rich silicate minerals and can be represented in a simplified form by the equations:

$$2CO_2 + H_2O + MgSiO_3 \rightarrow Mg^{++} + 2HCO_3^- + SiO_2$$

and

$$2CO_2 + H_2O + CaSiO_3 \rightarrow Ca^{++} + 2HCO_3^- + SiO_2$$

This process occurs at rock surfaces and in soils. After dissolution, HCO_3^-, Ca^{++} and Mg^{++} ions are carried by rivers to the sea, where they are precipitated as shells and skeletons via biological action, and are finally deposited as calcareous sediments via the following reaction:

$$Ca^{++} + 2HCO_3^- \rightarrow CaCO_3 + CO_2 + H_2O$$

The overall weathering reaction is thus

$$CO_2 + CaSiO_3 \rightarrow CaCO_3 + SiO_2$$

Mg ions are precipitated as dolomite or exchanged within warm mid-ocean ridge basalts to release Ca ions which are, in turn, precipitated as carbonates. Weathering of other rocks such as limestones ($CaCO_3$) or dolomites ($Mg,Ca(CO_3)_2$) has no overall effect on atmospheric CO_2 levels. HCO_3^- ions produced during weathering find their way to the oceans in the same way as ions from silicate weathering, where they are precipitated as carbonates. However, the net result of this process does not affect the CO_2 level in the atmosphere.

However, CO_2 is returned to the atmosphere by another important process, which occurs when sediments are buried. CO_2 is released during heating as a result of thermal breakdown of carbonates during metamorphism and diagenesis, a process that can be represented simply by the equation:

$$CaCO_3 + SiO_2 \rightarrow CO_2 + CaSiO_3$$

which is the reverse of the weathering reaction above. In addition, smaller amounts of CO_2 from the Earth's mantle are released directly into the atmosphere by volcanoes. The three ingredients of the silicate–carbonate sub-cycle form a loop and each has a forcing effect on the next (Figure 6.5a). Here, the three ingredients (climate, weathering and atmospheric CO_2) are connected by arrows indicating the effect of each on the next. Follow the arrows around the loop in Figure 6.5a to see how each process affects the next. First, increased levels of atmospheric CO_2 lead to increased temperatures. This is a positive forcing mechanism, and one which would run away if left unchecked, destroying the delicate balance of life on Earth. If we continue around the loop, the next arrow also indicates a positive forcing mechanism, since a warmer climate also leads to faster weathering (most chemical processes are faster at higher temperatures). However, faster weathering reduces atmospheric CO_2 as shown in the equations above, and the arrow between weathering and atmospheric CO_2 thus indicates a negative feedback. The fact that this sub-cycle has one negative feedback means that the loop or sub-cycle as a whole cannot run away; it is a 'negative feedback' loop. A full circuit of the loop from any start point will tend to dampen the effects on the Earth's climate. However, there is one ingredient which is independent of the loop: volcanic degassing of CO_2, which can act to force the pace of climate change. We will see how this forcing may have affected atmospheric CO_2 in the Cretaceous in Chapter 7.

6.2.3   The organic matter sub-cycle

The organic carbon sub-cycle affects atmospheric levels of both CO_2 and O_2. Burial of organic matter or kerogen represents a net excess of photosynthesis products compared to respiration products, and can be represented by the reaction normally used to describe photosynthesis resulting in a net release of oxygen into the atmosphere:

$$CO_2 + H_2O \rightarrow CH_2O + O_2$$

To complete the organic sub-cycle, O_2 is consumed and CO_2 is produced during the weathering of organic-rich sediments such as coal, or the return of oil and gas to the surface and their subsequent oxidation by O_2 in the atmosphere:

$$CH_2O + O_2 \rightarrow CO_2 + H_2O$$

This reaction also represents the breakdown of organic matter during burial and metamorphism. Methane (CH_4) can be released as the result of near-surface processes such as biogenic action and maturation of kerogen, or deeper processes such as diagenesis and metamorphism. Methane (CH_4) is rapidly oxidized to CO_2 if it reaches the surface, but it may also become frozen in methane clathrates (ice containing very high concentrations of methane) which are stored in sediments on the continental shelves or in permafrost. Methane may also become trapped deeper in the crust, leading to natural gas deposits such as those in the southern North Sea. Let us add a new loop to the cause and effect feedback diagram involving the weathering of organic carbon (Figure 6.5b). As seen in Figure 6.5a, increased atmospheric CO_2 leads to a warmer climate and this, in turn, causes faster weathering. However, the final part of the organic matter sub-cycle is another positive process, because weathering of old sedimentary organic carbon-rich matter also tends to increase atmospheric CO_2. If we considered only this loop, it would spiral out of control, heating the climate to unbearable levels because it is a 'positive feedback' loop. All the effects in this loop are positive, tending to accelerate heating of the climate.

○ Can you think of how the effects of carbon-rich sediment weathering are limited in the natural environment? What might be the consequences of circumventing such limits?

● In real life, this loop is tempered by the fact that organic carbon-rich sedimentary deposits below the surface are not available to be weathered, in other words weathering of organic carbon-rich sedimentary deposits is limited by the amount of such deposits that are exposed at the Earth's surface. However, by bringing coal and oil to the surface and burning to oxidize them to CO_2, we accelerate this feedback loop and cause global warming.

6.2.4   Other loops in the long-term carbon cycle

Although this cycle seems to have catastrophic consequences, we will see countering feedback cycles in Section 6.5. Berner developed the diagram (shown in Figure 6.5c) including many known cycles that affect the Earth's climate. The positive and negative feedback effects can be deduced by following the arrows around the loops on this Figure. Without needing to refer to the box model of reservoirs and fluxes, we can see how the different processes amplify or damp each other in leading to climate change. In fact, we can determine the overall effect of any path through this diagram because any complete path with an even number of open arrows (including zero) is a positive feedback loop, whereas any loop with an odd number is a negative feedback loop. So far, we have encountered one negative and one positive loop, but another example might be the loop including atmospheric CO_2, land plants and silicate weathering (Figure 6.5c). Heightened CO_2 concentrations encourage more rapid growth of land plants which, in turn, enhance weathering as their decay products release humic acids into the soil and their roots break up soil particles. The importance of this mechanism has not been fully quantified in geological environments but its power is clear to anyone who has seen weeds force their way between cracks in a path and slowly break down the concrete — that may take a few years but then I don't weed very often! Smaller particles weather more easily and thus accelerate CO_2 drawdown from the atmosphere. A geological example of this effect was the first appearance of rooting plants and then trees during the Devonian, which resulted in significantly faster weathering and led to a drawdown of atmospheric CO_2.

Finally, consider the loop in Figure 6.5c that includes O_2, land plants, and weathering of organic C which contains two open arrows. Heightened O_2 concentration in the atmosphere has a negative effect on land plants, mainly by increasing rates of burning (wild fires). Reduction of plant biomass, particularly in highland areas, will tend to reduce weathering rates; however, reduced weathering of carbon-rich organic matter reduces O_2 uptake, allowing atmospheric O_2 levels to increase. Thus, despite two open arrows indicating negative forcing mechanisms, this is a positive feedback loop overall that tends to accelerate climate change. Note that phosphorus appears in several loops toward the bottom of the diagram; the importance of phosphorus is discussed in Section 6.5.

Can the interactions between the cause and effect loops in the carbon cycle account for present-day climate warming? Look again at Figure 6.5c, and particularly the organic carbon loop which includes climate, CO_2 and weathering of organic carbon. In pre-industrial times, this would have been dominated by weathering of exposed organic-rich deposits (including coals and organic-rich mudstones) or petroleum seepage. However, over the past 100 years, burning of fossil fuels in the industrialized world has replaced weathering as the dominant means of oxidizing buried organic carbon. The amounts are very significant, and are possibly the main mechanism in raising the global CO_2 levels from 315 ppm to 360 ppm over the past 40 years. The cause and effect diagram allows us to predict what should happen. First, heightened CO_2 levels will raise the global temperature and precipitation levels (in other words, it will affect climate), higher CO_2 levels will cause a reduction in ice volume, assist plant growth and possibly alter global ocean circulation patterns. This is the scenario predicted for the not too distant future.

But wait … the carbon cycle also provides an answer to this crisis. Increased levels of CO_2 warm the climate, increasing the rates of chemical weathering of silicates, drawing down CO_2 and, in addition, higher CO_2 levels encourage land plants which enhance chemical weathering. These negative feedbacks will tend to dampen climate change, thus reducing the effects of burning fossil fuel, so can we ignore the evidence of climate change and continue pumping organic carbon-derived CO_2 into the atmosphere? Surely all that will happen is that plants will grow faster and mountains will erode faster, drawing the extra CO_2 into solution and cooling the planet. Where is the problem?

The problem is time, or rather the rates of geological processes. If the level of CO_2 in the atmosphere rises, the climate becomes warmer but the feedback loops of the long-term climate cycle do not act instantaneously. It may take many years for the distribution of land plants to react to the warmer environment. Chemical weathering may increase as a result of warmer conditions, but CO_2 will only be permanently removed from the atmosphere when it is converted to solid carbonates by fauna in the oceans. In the meantime, the central equatorial belt will become drier, reducing the extent of equatorial rainforests and civilizations. So, by using the cause and effect feedback diagram, we can predict the changes caused by rising CO_2 levels upon the long-term carbon cycle, but we cannot easily predict the speed of such a response.

6.3 A quantitative model of the workings of the long-term carbon cycle

After reading the descriptions of the various carbon cycles above, it might seem that the whole system would be too complex to model, even with the huge computing power available today. However, it is possible to describe many of the processes approximately, using relatively simple mathematics (in comparison with Global Climate Models, for example). In 1991, Bob Berner published an ambitious model called GEOCARB, which related atmospheric CO_2 variations throughout the last 500 Ma to the long-term carbon cycle. The model has been discussed at length and updated (GEOCARB III was published in 2001 by Berner and Kothavala), but the fundamental assumptions remain the same. The model illustrates 'what we know at the present time' about the variation of CO_2 in the past, encapsulated in mathematical form. It does not represent the final answer nor does it pretend to say 'this is how it was'. This is why the results of the model have changed slightly with each version. It is not a failure of the model, for advances in science are made by preparing models such as this and then by trying to disprove them in order to build better ones.

We have seen above how the carbon cycle has to be viewed in sub-cycles of cause and effect, but we also know that the fluxes of carbon involved in some processes are greater than in others. GEOCARB attempts to place this in a quantitative framework by making sensible assumptions about natural processes. The model involves formulations of the reservoirs and fluxes as discussed above but also formulates the feedbacks; in other words it tries to estimate the strength of each feedback mechanism at any time during the past 500 Ma. It is not necessary to follow the mathematics of the model, but we will consider the most important feedbacks before looking at the implications of the model results for the Cretaceous.

6.3.1 The feedback between weathering and atmospheric CO_2

How can past weathering rates be determined? Answering this question is obviously crucial because weathering leads to the drawdown of CO_2, and it provided Berner with a difficult problem: how can past weathering rates be estimated? Two possible methods of estimating weathering rates present themselves: (i) estimates of past sediment volume and (ii) strontium (Sr)-isotope variation in seawater which is known to vary with weathering rates (Section 1.3.1).

The sediment volume technique uses a set of estimates of the volume of sedimentary rocks lain down during each geological period, prepared by carefully measuring on maps the areas of sedimentary rocks. The amounts of sedimentary rock remaining decrease with increasing age as they are more likely to have been eroded. However, the change is gradual and it is possible to estimate how much sedimentary rock has been lost since deposition and thus how much weathering there was during each period. The drawback of this technique is that the error for the estimates is quite large. The other technique uses the ratio of Sr produced by the radioactive decay of Rb (^{87}Sr) to non-radiogenic strontium (^{86}Sr). This ratio can be used to determine the balance between Sr in the oceans derived from the oceanic crust and that derived from the continental crust. The variation of the oceanic Sr-isotope ratio is related to variations in the supply of dissolved Sr reaching the oceans in rivers. The current ^{87}Sr/^{86}Sr value of seawater is 0.7093 which is a combination of Sr from basalts (~0.7035) and Sr derived from continental erosion which yields a relatively high value (~0.7119). The Sr-isotope value of seawater is recorded in marine carbonates and has been measured in rocks over a wide range of ages.

The output from the GEOCARB model is a graph of atmospheric CO_2 against time, generally over the last 550 Ma, although we will focus on the last 250 Ma from Figure 6.8 onwards. Figure 6.6 shows how the two different formulations of the weathering rates affect the variations of atmospheric CO_2 estimates of the model. As you can see there is close agreement for much of the time except for 550–480 Ma (Late Cambrian and Early Ordovician), and 240–100 Ma approximately (Triassic to Early Jurassic and, unfortunately, much of the Cretaceous). The amounts of CO_2 predicted to be in the atmosphere in the mid-Cretaceous range from six times to more than ten times the present-day value. Berner preferred the Sr-isotope record because it provided a better fit to theoretical curves, but he incorporated the scatter between the two curves into the model to demonstrate the degree of uncertainty involved. This uncertainty is the range of values within which we are confident the true value will lie.

Figure 6.6 GEOCARB model results showing variations in atmospheric CO_2 for the last 500 Ma of Earth's history, as a ratio with the present-day level. The model takes account of variations in the rates of weathering (remember we saw above that weathering can draw down atmospheric CO_2) by estimating the variations of weathering with time. The red line shows the best estimate based upon the variations of Sr isotopes in seawater, and the brown line shows an earlier attempt using estimates of the volumes of sediments of known ages. (Berner and Kothavala, 2001.)

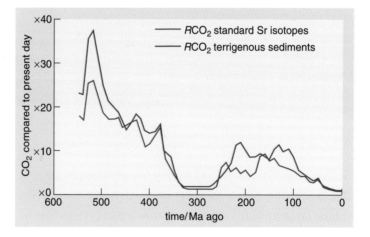

○ Does the model prediction confirm what we know about global temperatures in the Cretaceous climate system?

● Yes, the model indicates much higher atmospheric CO_2 levels during the Cretaceous (142–65 Ma) which led to the establishment of a 'greenhouse world'.

There is a further complication in that weathering does not directly respond to changes in CO_2 in the atmosphere, but responds to the level of CO_2 in soils, which is much higher than in the atmosphere because of biological activity. As we saw earlier, much of the organic carbon in the terrestrial carbon cycle is, in fact, in soils. Following the cause and effect loop in Figure 6.5c, higher atmospheric CO_2 causes higher global temperatures which, in turn, means more continental runoff (a more vigorous climate system — more rain, faster mineral dissolution and hence formation of HCO_3^-) because the chemical processes are also faster at higher temperatures. The total HCO_3^- flux is accelerated by both increased runoff and higher temperatures, and this combination is a very strong brake on increases in atmospheric CO_2. In the GEOCARB model, Berner used a function which in effect mimicked the variation that weathering had upon the atmospheric CO_2 through time, using the equation:

$$f(CO_2) = RCO_2{}^{0.22}$$

where $f(CO_2)$ is the feedback function expressing weathering rate as a function of atmospheric CO_2, and RCO_2 is the ratio of CO_2 in the atmosphere at the time compared to the present-day atmospheric concentration.

○ What would be the effect of atmospheric CO_2 levels five times the present-day level upon weathering rates according to this formulation?

● All weathering reactions would be 1.42 times as fast as the present-day ($5^{0.22} = 1.42$). Soil breakdown, and erosion rates would be affected, along with concentrations of carbonate and organic carbon in rivers running into the sea. Note that weathering rates do not increase proportionally with CO_2, so large increases in CO_2 lead to small but significant increases in weathering rates.

6.3.2 The feedback between land area variation and atmospheric CO_2

Heightened sea-levels during the Cretaceous meant that more of the land was submerged. In fact, there is more land currently exposed than at any time since the Cambrian Period.

○ How does the area of land affect the model?

● If there is less land, there will be less area for weathering, decelerating the rate of CO_2 removal from the atmosphere. The Cretaceous oceans covered between 10% and 45% of the present land surface (see Figure 3.3).

6.3.3 The feedback between river runoff and atmospheric CO_2

One of the most important input parameters for the GEOCARB model is the results of global circulation models (GCMs). These huge mathematical models have been run to monitor the effects of change in continental disposition upon the climate. The GCM model most recently used by Berner, was produced by the National Center for Atmospheric Research (based in Boulder, Colorado, USA) and is called CCM-3 (CCM stands for Community Climate Model). Berner also used the outputs of this model to account for variations in the effects of temperature upon river runoff. Although it may not seem so at first, river runoff is an important factor in controlling climate, because it controls the rate at which carbonate ions reach the sea and thus, in turn, controls the feedback rate of the silicate–carbonate cycle.

Finally, Berner used data on continental orography (the vertical disposition of land) from another model developed by Bette Otto Bliesner of the University of Texas at Arlington, which compared present-day continental runoff with ancient palaeogeography.

○ How does the position of the continents affect runoff and, in turn, climate?

● As the continents drift across the surface of the Earth, they may sometimes lie in zones of high rainfall, and if a majority of continents were in high rain areas, global runoff would be higher. However, if the continents were concentrated around the Equator, they would receive less rain and thus the global runoff would be low.

Prior to the Cretaceous, the majority of land was concentrated in one supercontinent, Pangaea. The vast size of this continent tended to restrict rainfall in its interior, resulting in low river runoff. From about 200 Ma ago, as the continents progressively separated, the continental fragments moved to higher latitudes and runoff increased. The model used by Berner showed that runoff rose rapidly from 0.8 times present-day level around 150 Ma ago to 1.3 times at 90 Ma ago where it peaked and declined slowly to around 1.2 times by the end of the Cretaceous.

The resulting model is markedly different from previous versions of the GEOCARB model because of the improvements in the GCM which have been achieved by using a higher resolution model (and a lot more computer time). Figure 6.7 shows just how much difference this improvement has made and it is particularly important in the Cretaceous Period. The new model predicts CO_2 levels almost double the earlier model. This gives us a further insight into the Cretaceous Period, in that if rivers were supplying more water, the continents saw more precipitation than seen in the present-day. In other words, it rained a lot during the Cretaceous!

Figure 6.7 GEOCARB model results showing the best estimate (red line) and model results (blue line) using older climate models. Note the older climate models do not predict such high CO_2 levels in the early Palaeozoic (550–400 Ma ago) or the Triassic, Jurassic and Cretaceous (250–65 Ma ago). *(Berner and Kothavala, 2001.)*

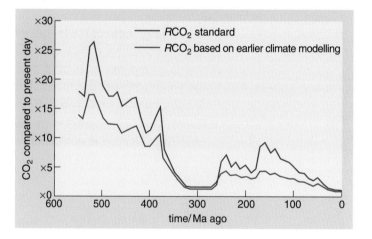

6.3.4 The feedback between mean land elevation and atmospheric CO_2

In mountainous areas, physical weathering rates are higher, because fresh rock is constantly exposed to the elements. This process also supplies rock fragments to finer-grained soils downstream where chemical weathering is greatest. Berner used the same Sr-isotope curve that was used as a proxy for weathering to derive the effect of mountain uplift, by combining it with areas of land and sizes of drainage basins. Models incorporating no variations in mountain uplift, using only present-day values, yield much lower CO_2 values during the Cretaceous (Figure 6.8). Yet, we know that the rates of mountain building have varied a great deal in the past and correlate well with the Sr-isotope curve. As we saw in Section 2.3 and mentioned in the discussion of Cretaceous river runoff (Section 6.3.3), there were few continental collisions in progress during the Cretaceous, so there were proportionally fewer mountain ranges.

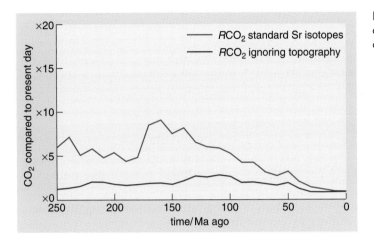

Figure 6.8 Results of the standard GEOCARB model (red line) and results without taking account of variations in mountain relief over the last 250 Ma (purple line). *(Berner and Kothavala, 2001.)*

6.3.5 The feedback of biological activity upon weathering rates

Land plants have a profound effect upon weathering and, in particular, the rise in angiosperms in the Early Cretaceous (Box 4.1) probably increased global chemical weathering rates because they recycled nutrients less efficiently than the earlier gymnosperms (conifers) and pteridophytes (ferns and horsetails), leaving soils enriched in organic acids (Berner, 1998). Berner assumed that the angiosperms started to affect weathering rates around 130 Ma ago, just after the beginning of the Cretaceous, and increased in importance until they were dominant by 80 Ma ago. He also recognized that the final ratio of gymnosperms to angiosperms would be important and tested several different values before choosing a ratio of 0.875 for his standard value (Figure 6.9). Unfortunately, existing measurements were too scattered to be helpful. Estimates of the ratio of gymnosperms to angiosperms from other studies range from 0.75–1.25, and this range of ratios was incorporated into the uncertainty of the model. Increased land plant activity can be seen to increase weathering rates for both silicates and organic carbon, and to increase rates of organic carbon burial in Figure 6.5c. Clearly, this is an important effect across a range of climate processes.

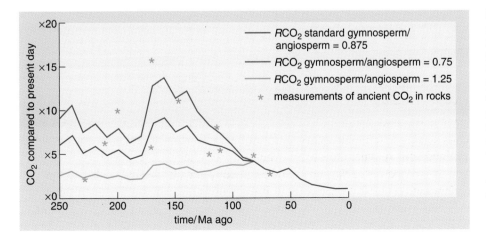

Figure 6.9 Results of the GEOCARB model showing the effect of varying the ratio of gymnosperms to angiosperms over the last 250 Ma. Berner used a ratio of 0.875, an average number he obtained from palaeontological estimates. However, varying the ratio from 0.75–1.25, the full range of estimates, makes a significant difference to the predicted CO_2 level of the atmosphere. The lines appear to come together at around 80 Ma ago, but this reflects Berner's modelling which assumed the final values, and extrapolated back through time. *(Berner and Kothavala, 2001.)*

6.3.6 The effect of the variable global outgassing rate

CO_2 degassing processes were important in Berner's model and, although they are not feedbacks (Figure 6.5c) as they are not subject to other climate processes, we will discuss them briefly here and return to them in Chapter 7. Berner tied the model CO_2 degassing rates to varying sea-floor spreading rates, although he recognized that this method was only possible during times for which there is still sea-floor in existence today. The degassing rates are particularly important for the Cretaceous Period because there was a rapid increase in ocean floor spreading rates and volcanic eruptions in the mid-Cretaceous which contributed to CO_2 levels and thus had a strong driving effect upon climate change (Figure 6.10). Remember that CO_2 outgassing from volcanoes and metamorphism is not linked via feedback loops to other carbon cycle processes (Figure 6.5c). However, more volcanoes outgassing CO_2 does not necessarily mean more CO_2 in the atmosphere and much higher temperatures over the long term. What it does mean is that the whole carbon cycle is more active as the extra CO_2 feeds into the system and leads to more weathering, and more burial of organic carbon and carbonates.

Figure 6.10 Results for the GEOCARB model showing the effects of varying CO_2 degassing rates over the last 250 Ma. The standard model is shown as a red line and the model run with no variation in CO_2 outgassing rates as a black line. The difference between the two is particularly important for the Cretaceous Period. *(Berner and Kothavala, 2001.)*

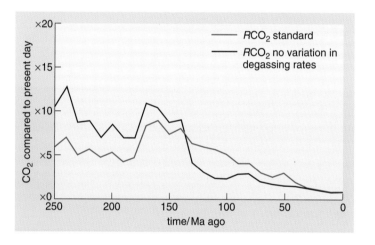

Strangely, the other important effect on degassing rates involves evolution. Large volumes of carbonate-secreting plankton emerged around 150 Ma ago. Whereas in earlier times all carbonates were deposited in shallow continental shelf seas, by the Cretaceous, carbonates were being deposited in deeper water on oceanic crust (see Box 6.1). Unlike continental crust, oceanic crust may be subducted and when it is, the thin sedimentary veneer is heated as the oceanic plate descends and carbonates break down to release CO_2. Berner incorporated this effect into his model as an increased rate of carbonate recycling into the atmosphere over the last 150 Ma. However, it may be that Berner overestimated the importance of this effect because thick deposits of deep-sea calcareous ooze do not appear until latest Cretaceous times.

One last parameter which Berner had to incorporate was the slow increase in heat from the Sun, which is increasing in temperature as it evolves. Although it did not vary much over the relatively short period of the Cretaceous, its changes are very noticeable in comparing Precambrian solar energies with Cretaceous values.

○ Compare the CO_2 levels of the early Palaeozoic in Figure 6.6 (say CO_2 levels between 500–400 Ma) with those in the mid-Cretaceous. Would you expect temperatures to have been significantly higher during the early Palaeozoic?

● Although CO_2 levels were much higher during the early Palaeozoic (Figure 6.11), surface temperatures were actually lower than those during the Cretaceous and the poles were probably covered in ice. The main reason why such high CO_2 levels did not lead to very hot conditions is that solar intensities were significantly lower. In fact, the heightened CO_2 levels and consequent greenhouse effect kept the climate reasonably equable for life in those times.

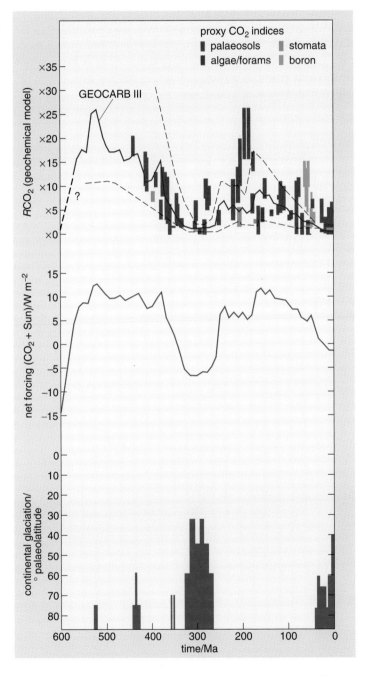

Figure 6.11 Comparison of the predictions of the GEOCARB model for CO_2 levels over the last 550 Ma with measurements of CO_2 from various sources. Geologists have used several different methods to measure the CO_2 content of the ancient atmosphere, including palaeosols (ancient soil horizons), algae and forams (silicate-secreting organisms), and stomatal densities (distance between stomata on fossil leaves). The times of major glacial periods are also indicated, together with the latitude they reached from 0° (a measure of glacial extent and intensity). *(Crowley and Berner, 2001.)*

We have spent a good deal of time discussing the problems and assumptions of the GEOCARB model and you might be forgiven for thinking it would give almost any answer, depending upon the values chosen for the input parameters. However, Berner was able to test the performance of the model against estimates of ancient atmospheric CO_2 from geological sources. Variations in the atmospheric CO_2 levels over the last 200 000 years have been measured in deep drill ice cores (see Figure 6.2), but no ice has been preserved to record more ancient atmospheric levels. Geologists have however found several ingenious ways to measure ancient atmospheric CO_2 levels: palaeosols (fossil soils) preserve ancient atmosphere in mineral coatings and cements; algae and forams sometimes preserve ancient atmosphere in their skeletons; and stomatal densities (the distance between leaf stomata) vary with varying atmospheric CO_2 levels (Figure 6.11). Global temperature changes over the last 100 Ma have been calibrated using oxygen isotopes, producing a remarkably good fit to the estimate from the GEOCARB model (Figure 6.11). In addition, glacial deposits have been discovered from several major glaciations, and the times correspond closely with times of low atmospheric CO_2 predicted by the model (Figure 6.11). The fact that the model predictions seem valid gives us some confidence that the input parameters are reasonable, a conclusion which throws up several important points about the Cretaceous:

- The CO_2 levels in the Cretaceous were high, and if the model is accurate, they fell from around six times the present value at the beginning of the Cretaceous to around three times the present value at the end of the Cretaceous (Figure 6.11). However, there is considerable uncertainty and values may have been as high as ten times the present value in the early Cretaceous (dashed lines in Figure 6.11 indicate limits of uncertainty).

- One of the most important advances made to the GEOCARB model has been the improvement in our understanding of global climate change using global climate models, and, in particular, the changes in river runoff from continents. This shows that rainfall and river runoff were much greater during the Cretaceous, thus enhancing weathering. However, because the continents were more commonly breaking up, and not colliding, during the Cretaceous, there were relatively fewer mountains compared with the present day (Section 2.3).

- Evolution was also very important to climate change during the Cretaceous and the changing proportion of gymnosperms and angiosperms made a significant difference to model CO_2 values for the Early Cretaceous. In other words, terrestrial plants were having an important effect upon global CO_2 values. Furthermore, the evolution of carbonate-secreting plankton to live in open oceans in the Cretaceous led to greater subduction of carbonates and a subsequent release of CO_2, though this effect may not have been an important factor until the Late Cretaceous

- The global degassing rates had a significant effect upon the model of CO_2 in the Early Cretaceous, but they were not a dominant factor in the high CO_2 values. In fact, the model shows they make very little difference to the latter part of the Cretaceous. However, the release of CO_2 as a result of volcanic activity is not part of the feedback system of the long-term carbon cycle. Thus, volcanic input is not immediately balanced by an opposing process, but makes the whole system more vigorous. It does not necessarily lead to more CO_2 in the atmosphere but it does cause faster weathering and more carbonate and organic carbon burial.

- High temperatures led to rapid weathering which accelerated the supply of nutrients to the continental shelves. In the Cretaceous, the high sea-levels resulting from paucity of ice at the poles, led to an explosion of shallow carbonate platform building.

6.4 The sulfur cycle

The short-term cycling of sulfur in the atmosphere has significant effects on climate over time-scales of a few years, particularly during large volcanic eruptions, and this is discussed in detail in Chapter 7 covering volcanoes and climate. Here we will discuss only the long-term geochemical cycling of sulfur.

Unlike the carbon cycle, the sulfur cycle is not dominated by huge rock reservoirs. The main rock reservoirs storing sulfur are evaporites ($CaSO_4$) containing some 5300×10^{15} kgS, and sedimentary pyrite (FeS) which represents around 5700×10^{15} kgS. However, the oceans contain 1282×10^{15} kgS as sulfate. Therefore, whereas the carbon reservoirs in rock are over 1000 times larger than the oceanic reservoir, the sulfur reservoir in rock is only ten times that of the oceans. The main implication of this closer balance between rock and ocean reservoirs is that the release of small proportions of sulfur from the rock reservoirs by weathering or outgassing does not overwhelm the oceanic system. In order to understand ancient sulfur cycles, we must look to the pre-industrial sulfur cycle shown in Figure 6.12.

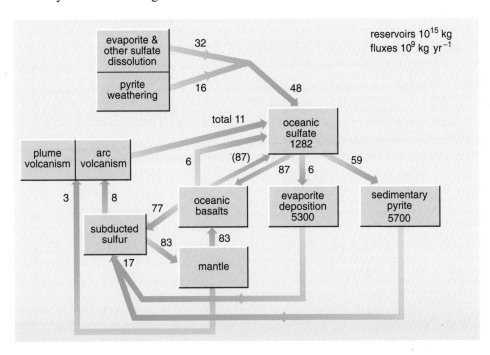

Figure 6.12 The pre-industrial sulfur cycle showing the sizes of terrestrial reservoirs and fluxes between the reservoirs. *(Arthur, 2000.)*

If the sulfur cycle is balanced, rock sulfates and pyrite are weathered and recycled in a similar manner to carbonates and organic carbon-rich rocks in the carbon cycle. However, there is no similar strong feedback in this loop for the sulfur cycle. The rates of sulfate and sulfide weathering are controlled by temperature, precipitation, the proportion of land area and proportion of mountains. All of these factors are feedbacks in the carbon cycle and the sulfur cycle is therefore linked to some extent with the carbon cycle. Note, however, that weathering of organic carbon and pyrite are related to the abundance of O_2 in the atmosphere. Knowing the relative sizes and fluxes of carbon and sulfur with time, it should be possible to estimate the burial and weathering fluxes for carbon and sulfur and thus estimate the long-term variation of O_2 in the atmosphere. The importance of O_2 is obvious from a purely selfish point of view since we depend upon it for life, but it is also one of the crucial parameters in determining the rates and pathways of evolution. For example, large-bodied arthropods which breathe through air vessels traversing their hard exoskeleton could only evolve when oxygen levels were high. There is good reason to suppose that levels of oxygen in the atmosphere have not varied by more than a factor of two over the last 550 Ma. However, models based upon the combination of carbon and sulfur cycles always seemed to indicate catastrophic O_2 variations with sudden excursions to both extremely high values (which would lead to spontaneous burning of biomass) and extremely low values (with consequent mass extinction of flora and fauna). This has been an enigma since the earliest attempts to reconcile the interaction of the carbon–sulfur–oxygen cycles and only recently has the dependence of these cycles upon another cycle, the phosphorus cycle, been understood. We will discuss the coupling of the different cycles again in Section 6.5.

There is another area of uncertainty in the sulfur cycle, namely the seawater–oceanic basalt interaction. As hot lava is erupted along ocean ridges, seawater flows through cracks in the warm basalt to form huge hydrothermal system. Sulfate is extracted from seawater during this process and precipitated as anhydrite at temperatures of 150–200 °C. At the same time, sulfide emanates from central hydrothermal vents or 'black smokers' although there may be no interaction between the two forms of sulfur. Estimates of the mass of sulfate entering the oceanic crust from seawater (around 25.7×10^9 kgS yr^{-1}) and the flux of sulfide at vents close to ocean ridges (around 3.2×10^9 kgS yr^{-1}) are very different and the two forms of sulfur also have very different isotopic signatures. If the sulfur cycle is in balance, much of the sulfate must be redissolved prior to subduction of the oceanic crust, otherwise this process would have drained all the sulfate out of the seawater in a few tens of millions of years. There is no evidence of this having happened but it makes it very difficult to estimate the flux of new sulfur introduced into the cycle from the mantle. Most models consider the input to be small or even treat the mid-ocean ridge system as having no effect on the global sulfur cycle. Recent estimates of the sulfur output from the world's volcanoes suggests a total flux of around 9.6×10^9 kgS yr^{-1}, which is an insignificant quantity in comparison to the sulfate reservoir of the oceans but a significant input over long periods of geological time. Historically, the largest eruption, Laki (which erupted on Iceland in 1783–4), emitted around 15×10^9 kgS and the giant Roza flow of the Columbia River flood basalts (erupted around 15 Ma ago) released a total of about 6×10^{12} kgS, but this is only around 0.0005% of the sulfur already in the ocean as sulfate. The aerosols from volcanic eruptions fall from the air as H_2SO_4 and quickly dissolve in the ocean so that even transient surface water effects during the eruption may be small. The climatic effects of volcanic sulfur are discussed in Chapter 7.

6.5 Coupling between the geochemical cycles of carbon, phosphorus, iron and sulfur

For the carbon cycle, we saw how important it was to understand the causes and effects of feedback mechanisms. For example, when an orogeny builds a large new mountain range, this causes global weathering rates to increase, and draws CO_2 from the atmosphere which, in turn, reduces global temperatures (Figure 6.5c). However, cooler temperatures decrease weathering rates which in turn slow the rate of CO_2 drawn from the atmosphere, slowing any temperature decrease effect of the mountain range. Thus, the feedback loop acts to maintain balance.

Consider the problem of boom and bust in coupled models for carbon, and sulfur which was highlighted in the previous sulfur cycle Section. The carbon and sulfur cycles seem to be relatively well understood, but when combined to understand the variation of oxygen partial pressure in the atmosphere the model produces very unsatisfactory results. It seems to predict sudden excursions, which had they happened in reality, would have led to massive biomass burning and mass extinctions by asphyxiation, neither of which is recorded in the geological record. In fact, there is another component to the system which we have so far missed from all the models. The coupling between carbon, sulfur and oxygen cycles is stabilized by the element phosphorus (Petsch and Berner, 1998). Phosphorus is an essential nutrient particularly in oceanic ecosystems and, in fact, is the limiting nutrient in many systems, though many argue that land-based systems are nitrogen limited. In other words, most of the nutrients necessary for life may be present in excess of the amounts needed, but the concentration of phosphorus is only just sufficient. In this case, the concentration of phosphorus will dictate whether or not life can thrive.

One pathway where phosphorus exerts control involves the oxygenation of ocean bottom waters. Phosphorus forms part of the organic matter (it is an essential component of DNA and ATP (adenosine triphosphate — an important control on metabolism)) and inorganic matter (skeletons and shells) which sink to the bottom of the ocean. Some phosphorus-rich material is recycled into the ocean and some becomes buried, thus sealing it from further use as a nutrient by life close to the ocean surface. Measurements of modern and ancient sediments have shown that when bottom waters are rich in oxygen, phosphorus is quickly absorbed onto the surface of abundant iron oxyhydroxide particles or is actively sequestered by bacteria. In this situation, the bottom water PO_4^{3-} (phosphate ion) concentrations are kept low and little is returned to the surface by upwelling. However, if the bottom waters are anoxic (oxygen supply is restricted), the iron oxyhydroxides are very much less abundant so more phosphorus remains in the water. It can be returned to the surface during upwelling to act as a nutrient promoting life in the surface layer. How can this affect atmospheric oxygen? Consider the feedback loop in Figure 6.5c, including atmospheric oxygen, iron-phosphate burial, nutrient aqueous phosphorus and organic carbon burial. If the oxygen content of the atmosphere increases, it should lead to more oxygen-rich ocean water and thus greater burial of hydrous ferric oxides and phosphorus in marine sediments. Less phosphorus in the surface waters limits production and thus induces lower burial rates for organic carbon. Lower organic carbon burial rates means, in turn, lower production of O_2. This sub-cycle thus acts to slow the organic carbon sub-cycle discussed earlier, the organic carbon cycle which seemed to be able to create a runaway effect.

Phosphorus also acts to slow the apparent runaway by another route, via land plants. High O_2 levels lead to more burning and wild fires. However, fewer land plants should lead to lower chemical weathering rates. Phosphate in rocks is the primary source of phosphorus and less chemical weathering thus means less supply to the ocean. Less nutrient oceanic phosphorus will lead to less organic carbon burial and thus a lower production of O_2, lessening the probability of a runaway organic carbon sub-cycle (Figure 6.5b). Thus, both cycles involving phosphorus dampen the feedback effect of the organic carbon cycle.

6.6 Summary

- This Chapter examined some of the biogeochemical cycles involved in controlling climate change. Many of the concepts considered here concern geochemical feedbacks between climate, the biosphere (life) and the geosphere (rocks) in relation to the Cretaceous climate.

- There are three interrelated carbon cycles. The short, and intermediate, time-scale cycles are dominated by biomass and marine reservoirs, whereas the long-term carbon cycle is dominated by the geosphere and, in particular, the reservoirs of carbon locked up in carbonate (limestone, chalk, etc.) and organic carbon (coal, oil, gas and organic-rich mudrocks)-rich rocks. These huge reservoirs dominate the long-term carbon cycle and have an important influence on climate and climate change. The interactions of the long-term carbon cycle must be invoked to understand the difference between the modern climate and the workings of the Cretaceous world.

- The long-term carbon cycle is a complex system but we can break down individual feedback loops to understand how they work and how they interact with each other. Positive and negative feedback loops are used to describe the interactions of the various reservoirs and fluxes. This way of describing the system can be illustrated in a cause and effect diagram, as a series of loops, allowing us to understand graphically the workings of the carbon cycle.

- Two of the most important feedback loops are the silicate–carbonate sub-cycle and the organic carbon sub-cycle. The silicate–carbonate sub-cycle describes the interaction of climate (temperature and precipitation) with silicate weathering and drawdown of atmospheric CO_2. The organic carbon sub-cycle involves the burial and weathering of organic carbon, which affects both atmospheric O_2 and CO_2.

- We have seen that while the silicate–carbonate sub-cycle is a negative feedback loop and tends to dampen changes to atmospheric CO_2, the organic carbon sub-cycle is a positive feedback loop and tends to accelerate change. However, the organic carbon sub-cycle is dampened by the supply of phosphorus, a limiting nutrient.

- The GEOCARB model, developed by Bob Berner and others, uses a series of simplifying assumptions to quantify the relationships between feedback parameters including: silicate weathering; land area and elevation; river runoff; biological activity; and global degassing rates. Given a set of reasonable input parameters, the model yields variations in atmospheric CO_2 in line with geological measurements.

- Unlike the long-term carbon cycle, the sulfur cycle is not dominated by huge geological reservoirs. It is closely linked to volcanic inputs and will be discussed in detail in Chapter 7.

- Although models of carbon and sulfur cycles seem to work independently of each other, when combined to model O_2 changes in the atmosphere they yield an unstable system prone to times of excess oxygen and severe oxygen depletion. In fact, the level of O_2 over geological time must have varied across a narrow range since the Devonian at least. Any larger variation would have had severe consequences, so the model must lack some other stabilizing input parameter. In fact, the biogeochemical cycle of the element phosphorus is closely linked with the carbon cycle, since phosphorus is the limiting nutrient in the marine environment. Additionally, its release from phosphorus-rich rocks and cycling within the marine environment are controlled by weathering and biological cycling. Phosphorus acts to dampen the organic carbon cycle in particular, and a combined carbon, sulfur and phosphorus model provides a far more stable O_2 level in the atmosphere, and thus a more reasonable approximation to the workings of the real world.

- Studying the long-term carbon cycle during the Cretaceous has shown that CO_2 levels in the atmosphere were raised by a combination of factors that acted to drive the system. Important factors include: the positions of the continents and river runoff; the scarcity of high mountain ranges; the evolution of angiosperms and their domination over gymnosperms; and the evolution of plankton to live in deep water. While volcanic CO_2 release was not the sole cause of heightened CO_2 levels during the Cretaceous, the heightened Cretaceous sea-floor spreading may have assisted in driving the system to become more vigorous, accelerating weathering and burial of organic carbon and carbonate-rich rocks.

6.7 References

BERNER, R. A. (1991) 'A model for atmospheric CO_2 over Phanerozoic time', *American Journal of Science,* **291**(4), 339–376.

BERNER, R. A. (1998) 'The carbon cycle and CO_2 over Phanerozoic time: the role of land plants', *Philosophical Transactions of the Royal Society of London Series B-Biological Science,* **353**(1365), 75–81.

BERNER, R. A. (1999) 'A new look at the long-term carbon cycle', *GSA Today,* **9**(11), 1–6.

BERNER, R. A. AND KOTHAVALA, Z. (2001) 'GEOCARB III: A revised model of atmospheric CO_2 over Phanerozoic time', *American Journal of Science,* **301**(2), 182–204.

PETSCH, S. T. AND BERNER, R. A. (1998) 'Coupling the geochemical cycles of C, P, Fe, and S: The effect on atmospheric O_2 and the isotopic records of carbon and sulfur', *American Journal of Science,* **298**(3), 246–262.

7 Volcanic inputs

Simon P. Kelley

Volcanoes only enter most people's consciousness through the TV news, for instance when Mount Etna erupts in Sicily and news footage shows lava flows encroaching on local villages and towns. If you do not live close to a volcano and are not subjected to its local hazards, it is easy to dismiss them as remote and unimportant. However, volcanoes play a very significant role in the Earth's climate and climate change, releasing gases including H_2O, HCl (hydrochloric acid), HF (hydrofluoric acid), CO_2 and SO_2 into the atmosphere. Several of the gases, particularly SO_2 which forms H_2SO_4 aerosols (a suspension of tiny liquid particles in air), affect the transmission of sunlight. Additionally, as we saw in Chapter 6, variations in the level of atmospheric CO_2 are strongly correlated with climate change via feedbacks through the biosphere and geosphere. So, where do volcanoes fit into the picture, and are they implicated in climate change?

7.1   Volcanoes and climate

Volcanoes throw rocks, dust, and gas into the atmosphere and, although the explosive eruption plumes or rivers of flowing lava make the best TV pictures, it is the invisible gases released into the atmosphere during volcanic eruptions which affect the weather over the following months and years. An important point to understand here is the difference between what we call 'weather', a short-term local variation in precipitation and temperature, and 'climate', a long-term variation of global weather patterns. Single volcanic eruptions affect the local weather, but only the largest eruptions cause global climate change. Even when a volcanic eruption does induce climate change, it is short term on a geological time-scale and conditions rapidly return to the norm for the time.

Are most gases released from volcanoes during big eruptions or do they continuously leak out? In fact, significant amounts of gas are emitted in both ways. Volcanoes emit most gases during major eruptions as part of explosive eruptions and also slower effusive eruptions like those on Mount Etna (Figure 7.1), but very significant amounts of gaseous emission occur almost constantly on some volcanoes such as Masaya in Nicaragua (Figure 7.2). Such emissions can be a major hazard to life in the immediate vicinity, but the effects are largely confined to the flanks of the volcano.

7.1.1   Volcanic gases

Although small eruptions may not be accompanied by large amounts of lava or ash, they are also a significant volcanic hazard and illustrate just how much gas accompanies volcanic eruptions. For example, on 20 February 1979, the Sinila water-filled crater of the Dieng volcanic complex in Indonesia erupted, emitting blocks of rock, mud and gas, followed by a hot mudslide (lahar) which flowed 3.5 km downslope. A few hours later, a new crater opened 300 m west of Sinila close to the village of Koputjukan, and villagers fleeing along the road to Batur were asphyxiated by a gravity flow of gases (probably CO_2 and H_2S which are heavier than air) from the new crater. Others who probably witnessed the deaths retreated to a local school, but were also killed by other gas flows emanating from many small vents around the new crater. When the area was safe a few hours later, rescuers also found many wild animals that had been killed by the gas flows.

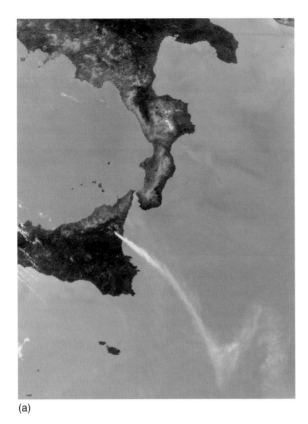

(a)

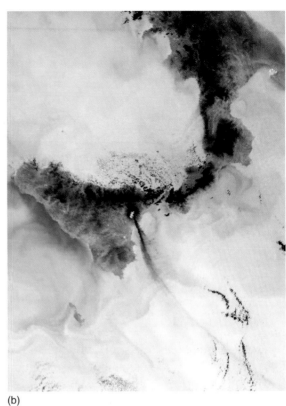

(b)

Figure 7.1 ASTR satellite images of the plume from the recent eruption of Mount Etna (July 2001). (a) The plume in daytime, using the infra-red and 11 μm channel. The plume of smoke is very evident in the image due to the temperature difference between the smoke and the sea. Interestingly, the smoke is cooler than the sea, much like the vapour trails that follow a jet aircraft. (b) This image was taken at night and is made by combining two channels from the satellite's detectors, the infrared 3.7 μm and the 11 μm channel. The 3.7 μm channel distinctly shows the extremely high temperature of the lava. The images are around 500 km across. (Copyright © ESA.)

In contrast to the eruptive release of gases seen at Dieng, the Masaya volcano constantly emits a plume of gas between eruptions. Many volcanoes including Etna emit significant amounts of gas in this way, but at Masaya the gas is emitted close to the ground rather than from a high volcanic cone. This has severe effects on the local environment. Figure 7.2a shows the low-level plume from Masaya which emits up to 2500 tonnes of SO_2 per day causing severe damage to vegetation downwind. In recent years the SO_2 has devastated the local coffee, palm and soya crops. In addition, concrete, metal fences, telephone wires and all exposed metal in the area quickly becomes corroded (Figure 7.2b).

The effects of the Masaya SO_2 plume described above are very localized. Large-scale damage and climate effects only arise when volcanoes release large quantities of gas into the atmosphere. For example, during its eruption in 1991, Mount Pinatubo released 3×10^{10} kg of gaseous aerosols high into the stratosphere (see Figure 7.4 for the structure of the atmosphere) in only a few hours, and of course far larger quantities into the troposphere. Compare this amount with the annual release from Mauna Loa in Hawaii (1.4×10^9 kg yr^{-1}). Clearly, the scale of climatic effect of an eruption depends upon its style and ferocity, but global effects can also depend upon where on the globe it occurs.

7.1.2 The atmosphere and volcanic aerosols

First, we need to consider the structure of the atmosphere and how it can be perturbed by volcanic eruptions. The Earth's atmosphere has no sharply defined outer edge; it simply gradually becomes thinner and thinner until it merges into outer space. Over 80% of the atmosphere is held by gravity within 20 km of the Earth's surface, and only the very lightest gases, such as hydrogen and helium, can escape its pull. The atmosphere limits heating at the Earth's surface by absorbing and reflecting solar radiation, but it also insulates the surface against

(a)

(b)

Figure 7.2 (a) A continuous gas plume rising from Masaya volcano, Nicaragua *(Glynn Williams-Jones, University of Hawaii.)*. (b) A metal gate damaged by prolonged exposure to acid gases emanating from Masaya volcano. *(Pierre Delmelle.)*

extremes of temperature by limiting outgoing re-radiated heat (Figure 7.3). In the absence of climatic disturbances resulting from a volcanic eruption, around 8% of incoming solar radiation is reflected by the atmosphere, about 17% is reflected by clouds and another 6% is reflected by land and sea surfaces. In addition to this, around 23% of the incoming solar radiation is absorbed at various levels in the atmosphere, meaning that only about 46% of the incoming radiation is available to warm the Earth's surface. Individual volcanic eruptions alter these parameters by introducing aerosols, dust and gas which reflect and absorb solar radiation, thus decreasing the amount reaching the Earth's surface. In addition, CO_2 released from volcanoes interacts with the atmosphere by increasing the absorption of heat, as we saw in Section 6.3.

Figure 7.3 Schematic diagram showing the overall radiation budget of the atmosphere, and the rough proportion of radiation absorbed at each level (figures are percentages). Values for outgoing radiation have been measured by satellite-borne radiometers; whereas the re-radiated radiation (back radiation) has a longer wavelength than the incoming radiation. Radiation reflected by the Earth's surface, clouds or molecules in the atmosphere has the same wavelength after reflection as before. Other values are more difficult to measure precisely, and you may find slightly different values given elsewhere.

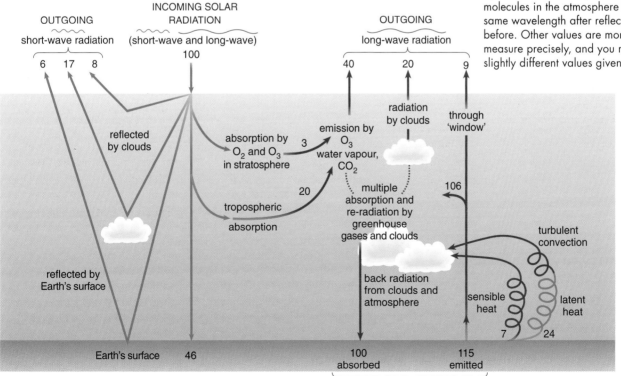

The lowest region of the atmosphere, known as the troposphere (Figure 7.4), is also the warmest region, and is warm enough to support life. The temperature decreases with height in the troposphere, reaching a minimum at the tropopause, the boundary between the troposphere and stratosphere. The tropopause varies in height between the Equator, where it lies at around 18 km at a temperature of −80 °C, and the poles, where it lies only 8 km above the surface, at a maximum temperature of around −50 °C. The height of the tropopause also varies with the seasons as the air temperature changes. Above the tropopause lies the stratosphere, where temperatures increase as a result of increased concentrations of ozone which absorb high energy ultraviolet rays from the Sun. Temperatures continue to increase to the top of the stratosphere (stratopause) at a height of about 50 km. Above the stratosphere, in the mesosphere, temperatures decrease again to a minimum at the mesopause with a temperature of around −100 °C at a height of about 85 km (Figure 7.4).

Figure 7.4 Vertical temperature 'structure' of the atmosphere defined by temperature variations with altitude. At each successive level, the temperature gradient is reversed. The temperature decreases with height in the troposphere, so that conditions are conducive to convection. In contrast, temperature increases with height in the stratosphere, and little convection occurs.

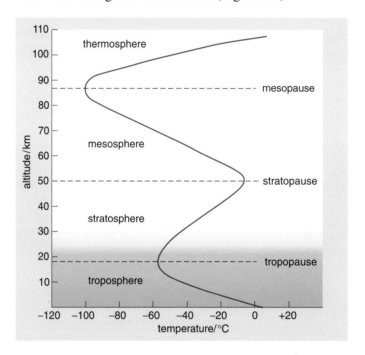

The troposphere is the wettest, dirtiest region of the atmosphere. Infra-red heating of the Earth's surface locally heats the air, which rises through colder air above it, causing turbulence. This process entrains countless particles and surface-derived gases, but since these are to be rained out within a few weeks to months, any weather effects are short-lived. All volcanic dust and gases ejected into the troposphere are returned to the surface in this way over a few months. 'Raining out' in the troposphere is the main reason why the worst effects of volcanic eruptions are very localized and only the largest eruptions which eject gases into the stratosphere have truly global effects upon climate.

The stratosphere is far more stable than the troposphere, because warm air rises above cold air and the stratosphere is hotter towards the top. Although small amounts of air from the troposphere enter the stratosphere via storms in the tropics, most of the water vapour freezes and falls back into the troposphere, leaving the stratosphere much drier. The lack of water vapour in the stratosphere is an important factor in controlling the climatic effects of large volcanic eruptions. SO_2 released during large volcanic events weakly absorbs solar radiation, but is converted to an H_2SO_4 aerosol by the addition of H_2O. Droplets of H_2SO_4 with a diameter of 0.1–1 µm, formed after major eruptions, strongly absorb solar radiation resulting in a cooling effect. Air in the stratosphere moves much more slowly than the troposphere, so when air from the stratosphere re-enters the troposphere at mid-latitudes, it has been in the stratosphere for around two years. At the poles, air transferred by rapidly moving winds during the winter can be three to five years old. Consequently, volcanic eruptions which eject dust and gases into the stratosphere can cause changes to the climate which last several years, whereas gases released into the troposphere are rapidly rained out and the effects are more localized.

The atmosphere is composed mainly of N_2 (78%) and O_2 (21%) with a small amount of the inert gas argon (0.93%), but these gases are not significant absorbers of solar radiation. Other gases make up less than 0.1% of the total, but CO_2 is noteworthy, because even at such an apparently low concentration (0.032%), it is the most important gas contributing to the greenhouse effect and is strongly linked to climate change. In contrast to the atmosphere, the major gases released by volcanoes are H_2O (> 80% by volume) and CO_2 (10%) with smaller and variable components of SO_2, H_2S, CO, H_2, N_2, HF, HCl and HBr.

Despite being the most important gas by far, the amount of water vapour released in volcanic eruptions ($\sim 10^{12}$ kg) is far less than the mass of water in the atmosphere (around 13×10^{12} kg), which is enough to completely cover the Earth to a depth of 2.5 cm. Other gases released by explosive eruptions such as CO_2, CO, H_2 and N_2, are already present in the stratosphere in higher quantities than those released during even the largest eruptions. Highly soluble species such as HF, HCl and HBr, are rained out of volcanic plumes as they rise and rarely reach the stratosphere. Moreover, volcanic ash particles in eruption plumes provide nuclei upon which HF, HCl, HBr and water vapour condense, forming water droplets or ice particles which fall back to the surface. SO_2 is considerably less soluble than HF, HCl and HBr and reaches the stratosphere in large quantities. Once in the stratosphere, SO_2 reacts with hydroxyl radicals (OH) from water vapour to form products which quickly break down to sulfuric acid (H_2SO_4). Satellite observations of the plume from the Mount Pinatubo eruption in 1991 showed that this process was completed within a month of the eruption.

From this brief description, you will see that the structure of the atmosphere, in particular the height of the tropopause relative to the height reached by the volcano plume, is an important factor regulating the effect any volcanic eruption may have upon the global climate. Gases, particularly SO_2, introduced into the stratosphere remain there for several years, extending the effects of an eruption upon the climate. Although short-term effects are minor, the cumulative effect of CO_2 release probably played a vital role in the Cretaceous climate, as we saw in Section 6.3. A final consideration of the effects of volcanoes during the Cretaceous is that in that greenhouse world, the climate system was much warmer than the present day. In such a hot world, the tropopause may also have been at a greater height, so that only the very largest eruptions penetrated the stratosphere and caused long-term effects.

We know that some single volcanic eruptions have been sufficiently large to affect the climate: numerous early historical records of volcanic eruptions provide anecdotal evidence of climate changes lasting for up to a decade. For example, in 44 BC, the Romans recorded a red-stained Moon, red-coloured Venus and a red comet, and they also suffered extraordinarily cold weather which ruined the harvest. They took these events to be signs of the gods' displeasure at the murder of Julius Caesar on the ides of March that year. However, there was a more earthly explanation for the environmental problems of Rome. Although the Romans knew that Mount Etna in Sicily had erupted, throwing dust and clouds of ash into the atmosphere, they did not realize that the gas plume and fine-grained ash cloud might scatter blue light (~0.4 μm wavelength) as it passed over Rome, causing the red (~0.7 μm wavelength) coloration in the sky. A similar effect was observed after the 1883 eruption of Krakatau (Figure 7.5). The Romans also did not realize that gaseous aerosols would cause cooling by absorbing and reflecting solar radiation long after the ash cloud had disappeared. It has also been noted that the following year was the first of six years of famine in China, in which harvests continuously failed as a result of summer frosts.

○ Can these reports from Rome and China be linked to the single eruption of Etna?

● The failed harvest in Rome, with some certainty, can be linked to the Etna eruption since the Romans observed the plume directly. However, the Chinese reports offer only the possibility of a connection. This is the main problem in looking to ancient volcanic eruptions for quantitative evidence of climate effects. We have to recognize that parts of China may have been suffering from an unconnected weather change.

Another example of a volcanic eruption affecting global climate was in the year 1453 AD, when the full Moon was eclipsed by ash for four hours on 22 May as Constantinople was being attacked by the Ottoman Turks. A few days later, the city was pelted by an exceptionally violent hailstorm, and subjected to a dense fog. Chinese records of the same year talk of tens of thousands of people and animals freezing to death, and the Yellow Sea freezing over 20 km from the coast. Records also speak of a southern region of China with a normally equable climate experiencing 40 days of continuous snowfall. European records also show extensive crop damage for nine years following 1453 and tree rings record widespread frost damage in that year. Research into these ancient climate data found an oral tradition among the islanders of what is now the Republic of Vanuatu in the South Pacific, that the island of Kuwae savagely erupted around 500 years ago. There is evidence that vegetation on the island was burned in a blast dated to between 1420 and 1475 and this seems to establish a link between this eruption and the climatic effects described above. This hypothesis is confirmed by Antarctic ice cores which record volcanic eruptions for several years following 1453, although these events were only weakly recorded in the Greenland ice cores, indicating a Southern Hemisphere source, possibly Kuwae. As in the previous example of the Chinese records of 44 BC, the evidence from further afield is more difficult to quantify. We have to unravel the chronology and distinguish between what is coincidence and what is evidence of the volcanic eruptions causing climatic change.

Perhaps one of the most famous ancient historic volcanic eruptions was that of Thera (now Santorini) in the 17th century BC which is widely held to have destroyed the Minoan civilization on Crete. However, this case also illustrates the problems of incorporating older data into a model to quantify the climatic effects of volcanoes. First, it is difficult to be precise about the date of the eruption which has been dated by tree rings, and indicates adverse climatic conditions in

Figure 7.5 During the autumn of 1883, after the eruption of Krakatau, spectacular sunsets around the world attracted the attention of artists. This is one of a series of six paintings by William Anscom of the Sun setting on a November evening seen from Chelsea, London. It was published in the contemporary Royal Society report on the great eruption.

1628 BC. It has also been dated by a thin layer of ice with high acid concentrations (from volcanic SO_2) in Greenland ice cores at 1645 BC. Clearly, it could not have been at both dates but there is some uncertainty in both ages. Furthermore, the view that falling ash and tsunami (tidal waves) suddenly overwhelmed the Minoan civilization does not fit with the recently unearthed evidence. Research on Santorini has established that the eruption was one of a series of massive eruptions which happened throughout the Holocene and that at the time, Santorini was probably a near-perfect natural harbour. An older collapsed volcano caldera had created a complete island ring with a sea-filled centre, probably containing a small island, much like the island of Santorini today. The island was a natural centre for Minoan trade, yet the lack of any bodies preserved in the local town of Akrotiri, which was buried by pyroclastic flows, shows that the eruption did not come without warning, unlike the ash cloud which engulfed Pompeii in 79 AD. Moreover, although the island of Crete was certainly hit by tsunami, there is no record of a substantial ash layer on Crete and, in fact, the Minoan civilization did not finally decline until c. 1450 BC when the mainland Greeks invaded and subjugated the remaining Minoans. In this case, the clear evidence of the longer-term geological history of the volcano, the archaeological evidence from Santorini, and tree ring and ice core data, indicate that the eruption was big but not sufficient to throw a thick layer of dust as far as mainland Crete. What probably finally caused the decline of the Minoan empire was the loss of Thera as a centre for trade. In other ancient historic cases, anecdotal evidence is often very enticing but great caution has to be exercised in extrapolating quantitative effects upon climate.

In order to quantify accurately the atmospheric effects of volcanism, we must look to recent eruptions when precise climate records are available, and when modern high-precision recording techniques have been applied. This will give us clues as to the mechanisms responsible for climatic effects, and will allow us to quantify the magnitude of eruption-induced climate change. In fact, gas release from volcanoes makes a continuous contribution to the climate (Box 7.1), but major eruptions have the potential to release amounts of gas that are several orders of magnitude greater.

Box 7.1 Quantifying the amount of gas volcanoes release

There are two aspects to volcanic effects upon the climate: short-term disruptions of the climate (over the time-scale of the eruption), and long-term effects resulting from volcanic gases involving strong feedback mechanisms with the climate and the biosphere.

Before we can consider gases released from volcanoes during the Cretaceous, we must establish a baseline of emissions from volcanoes today. This is not as simple as measuring all of the Earth's volcanoes and calculating a cumulative total. It is not possible to monitor all of the world's volcanoes constantly, and so we have to make estimates based on a small number of them.

Measurements of CO_2 emissions are particularly difficult, but SO_2 can be measured remotely. This is done by measuring the sunlight or a laser beam passing through the volcano plume using COSPEC (correlation spectrometer) or via satellites recording light over a range of wavelengths such as that used to record the images of Etna (Figure 7.1). In some cases, direct measurements of SO_2 can be made and the CO_2 estimated using measured carbon/sulfur ratios in volcanic gases. The problem with using very small numbers of volcanoes to estimate the global total is well illustrated by the degassing estimates for Mount Etna in Sicily. Etna outgasses around 4.4×10^{10} kg yr^{-1} CO_2, and contributes around two-thirds of all volcanic CO_2 currently entering the atmosphere.

○ Is it likely that in the past the global total has been dominated by one volcano?

● Yes. In fact, it is probably quite normal for world output to be dominated by a few large volcanoes, although Etna is unusual. Volcanologists have shown that by knowing the outgassing rates of the 20 largest volcanoes in the world, they can predict the world total quite accurately.

Combining the release from subaerial, mid-ocean ridge and plume-related volcanism, almost 2×10^{11} kg yr^{-1} CO_2 is released. Although this amount of CO_2 sounds huge, it represents only around 2.5% of the amount released every year by fossil fuel burning and farming activities. We will return to this figure later because it helps us to understand how greenhouse conditions may have varied during the Cretaceous and how the Earth's climate may change in the future as we release more CO_2 into the atmosphere.

Estimates for the SO_2 released from subaerial volcanoes is more precise and a comprehensive data set for 72 volcanoes indicates a global outgassing rate of 2.6×10^{10} kg yr^{-1}. Submarine volcanism is not quite as simple to quantify because most sulfur released at mid-ocean ridges is recycled within the still warm basaltic oceanic crust in hydrothermal systems close to the mid-ocean ridge. It is precipitated as sulfides and sulfates within the oceanic crust and on the ocean floor at hydrothermal vents (black smokers). Although hard to estimate, this oceanic sulfur contribution may be as high as 30×10^{10} kg yr^{-1}.

Two main styles of volcanic eruption, explosive and effusive, have caused measurable climatic effects in recent times. Explosive volcanism, e.g. Mount St. Helens in 1980 or Mount Pinatubo in 1991, is the more spectacular and immediate of the two volcanic styles. Explosive volcanic plumes commonly penetrate into the stratosphere and have the potential to cause widespread climate disruption. On the other hand, effusive volcanism, e.g. present-day Mount Etna on Sicily or Mauna Loa on Hawaii, is more commonly associated with lava flows or lava fountains which rise a few hundred metres into the troposphere and consequently only affect the local environment. However, one recent eruption, the Laki eruption in Iceland in 1783, stands out for its climatic effects which were felt over most of Western Europe. In the longer term, effusive volcanism can be equally important because it typifies the eruptions at mid-ocean ridges, which dominate gas release over geological time.

7.1.3 The climatic effects of explosive volcanism

The eruption of Mount St. Helens in America in 1980 and Mount Pinatubo in 1991 provided modern-day examples of what explosive volcanoes can do. Both eruptions ejected huge amounts of rock, dust and gas high up into the atmosphere. Silicate dust particles (produced when dissolved gases are released by the explosion of volcanic glass during the eruption) are entrained in the volcanic plume to become the main solid aerosol. The main effect of a volcanic plume is to block out sunlight locally under the plume. Gases such as H_2O, CO_2 and SO_2 are also released in large quantities along with smaller amounts of HCl and HF. However, it is mainly gaseous SO_2, combined with water to form aerosols (in this case, a suspension of acid droplets in air), which has significant effects upon global climate.

During an explosive volcanic eruption, a plume of hot volcanic ash and gas is ejected (Figure 7.6) through the troposphere and into the lower stratosphere. Silicate dust in the plume plays a minor role in climate change, because is has a short lifetime in the atmosphere. This was demonstrated in the Mount St. Helens eruption, when even the finest dust particles rapidly aggregated into larger particles and fell to the ground, hours after the eruption, within a few hundred kilometres of the source (Figure 7.7). However, gaseous aerosols produced in such eruptions produce a range of effects over the full height of the volcanic plume as shown in Figure 7.6.

Figure 7.6 The interaction of a volcanic eruption with the atmosphere. The eruption releases sulfur dioxide (SO_2), which is eventually converted to sulfuric acid (through a reaction with atmospheric oxygen and water), producing a mist of droplets, or aerosol. This aerosol reflects the incoming solar radiation, so cooling the troposphere; it also absorbs some solar energy, warming the stratosphere. If the sulfuric acid aerosol is sufficiently large, it may absorb and re-irradiate thermal energy warming the lower atmosphere. The eruption also produces large amounts of CO_2, which mixes with atmospheric water to form a much weaker acid that falls locally as acid rain. Smaller amounts of hydrogen chloride are also released into the atmosphere, and mix with water to form hydrochloric acid and again form a component of acid rain.

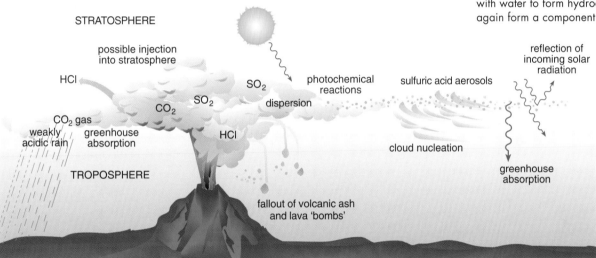

Figure 7.7 An isopach map (lines of equal thickness) of ash fallout from the Mount St. Helens plume after the 18 May 1980 eruption. Numbers on the isopachs are thickness of ash in metres. (Houghton et al., 2000.)

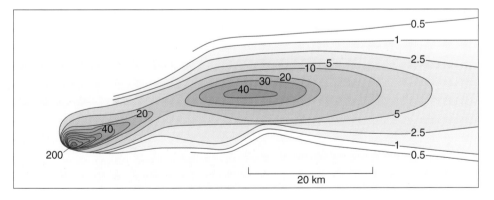

The eruption of Mount Pinatubo on 14–15 June 1991 provided a graphic example of what explosive volcanoes can do, by producing 120×10^{11} kg of magma and pumping over 30 Mt (1 Mt = 10^9 kg) of gaseous aerosols into the stratosphere. Aside from providing graphic TV images of ash fall on the local US Airforce base, the Pinatubo eruption was observed by climatologists working on the Earth Radiation Budget Experiment. As a result of satellite coverage during the eruption, for the first time scientists were able to measure global temperature decreases, and to monitor the slow dispersion of stratospheric SO_2 over the three years following the eruption (Figure 7.8).

Figure 7.8 Images from the Stratospheric Aerosol and Gas Experiment (SAGE II) monitoring the distribution of stratospheric aerosols following the 1991 eruption of Mount Pinatubo: (a) 15 April–25 May 1991; (b) 14 June–26 July 1991; (c) 13 February–26 March 1993. Initially confined to the tropics, the aerosols increased the 1 μm optical depth (the fraction of light scattered by particles in a column of atmosphere) by two orders of magnitude. Over six months, the aerosol spread to higher latitudes, globally changing optical depths. The stratospheric aerosol continued to be dominated by steadily decreasing Pinatubo input for three years. (*McCormick et al., 1995.*)

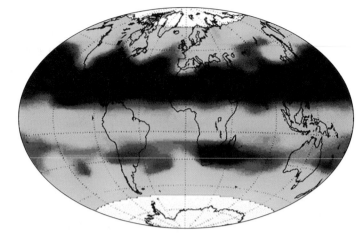

1020 nm optical depth

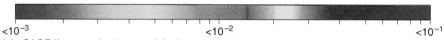

$<10^{-3}$ $<10^{-2}$ $<10^{-1}$

(a) SAGE II stratospheric aerosol April-15-91 to May-25-91

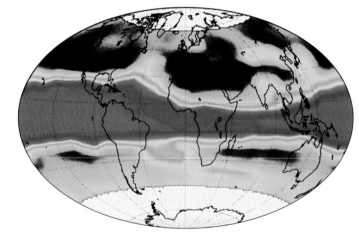

1020 nm optical depth

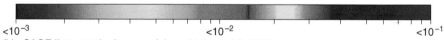

$<10^{-3}$ $<10^{-2}$ $<10^{-1}$

(b) SAGE II stratospheric aerosol June-14-91 to July-26-91

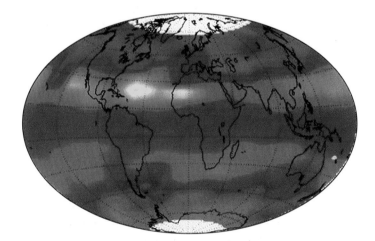

1020 nm optical depth

$<10^{-3}$ $<10^{-2}$ $<10^{-1}$

(c) SAGE II stratospheric aerosol February-13-93 to March-26-93

The global temperature decrease as a result of the Mount Pinatubo eruption was 0.5 °C over 1–2 years following the eruption, and although this does not sound like very much, it is a standard against which we can compare other eruptions and conditions in the past. Compare this figure with the extremes of temperature experienced by the Earth. The global mean temperature during the peak of the last ice age was 5 °C lower than that of today, and the global mean temperature during the Cretaceous Period was up to 15 °C higher than that of today.

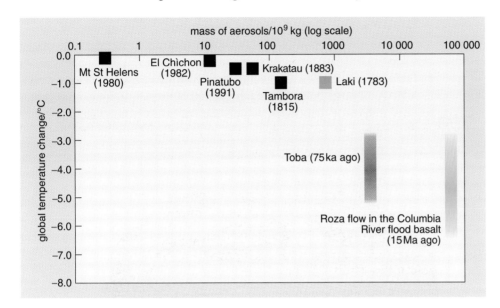

Figure 7.9 A graph of the mass of aerosols injected into the stratosphere, plotted against the measured climate change for historic eruptions. The estimated effects in the cases of the prehistoric Toba eruption and the Roza flow of the Columbia River flood basalt are marked by larger symbols, indicating the uncertainty in global temperature change estimates. Explosive eruptions are marked as black or grey symbols. Many effusive eruptions are too small to have any global effects but the larger ones are marked by orange symbols. Note that the mass of aerosols is presented as a logarithmic scale.

Figure 7.9 shows the relative mass of gaseous aerosols from historical eruptions and the estimated effect on global mean temperatures based upon the radiative forcing produced by the release of SO_2 (a measure of the potential reduction in solar input). Here you can see that Tambora, the largest historical eruption, probably caused a global mean temperature decrease of around 1 °C. The Tambora eruption was followed by the 'year without a summer' in Europe when many harvests failed. The eruption of Santorini around 1628–1645 BC, with a volume of perhaps twice that of Tambora, might have been expected to produce a far greater cooling effect. However, because of the non-linear relationship between cooling and the mass of aerosol (note that the aerosol mass scale on Figure 7.9 is logarithmic), Santorini probably only produced a cooling effect slightly more than Tambora. In ancient prehistory 75 ka ago (recent times by geological standards), the eruption of the Toba volcano in Indonesia produced some 7.5×10^{15} kg of magma and may have ejected as much as 3.3×10^{12} kg of gaseous aerosols into the stratosphere.

However, the relationship between aerosol mass and temperature change based upon changes in radiative forcing (Figure 7.9) cannot be used to estimate the effects of the Toba eruption. The theoretical radiative forcing (a measure of the reduction in solar radiation) resulting from Toba would have been more than the total solar radiation budget. Clearly, the amount of solar radiation cannot be less than zero, and the reason why aerosol from the Toba eruption did not completely block out the sunlight is the lack of water in the stratosphere. Remember that global cooling is caused by aerosols of H_2SO_4, and not gaseous SO_2. Each SO_2 molecule released during an eruption requires three water molecules to hydrate it, and aggregates to form minute 0.1–1 μm H_2SO_4 droplets which absorb solar radiation, thus warming the stratosphere but cooling the Earth's surface. Measurements show the stratosphere contains only around 8×10^{11} kg H_2O above a height of 20 km and since the Toba eruption is estimated to have ejected around 2.1×10^{12} kg SO_2 into the stratosphere, it would have quickly exhausted the supply of water.

One view of this effect is that the low abundance of H_2O in the stratosphere limits the effects of SO_2 input, meaning that even the largest eruptions in the geological past could not have caused climatic effects much worse than those we have seen in the recent past. Another view is that the lack of H_2O in the stratosphere merely prolongs the cooling effect. As the H_2SO_4 aerosol forms, it aggregates into larger droplets and falls to the ground, thus limiting the length of time the effect can last. However, if the stratosphere becomes completely exhausted of water, not all SO_2 is converted to H_2SO_4 and remains in the stratosphere to react with new H_2O, slowly added from the troposphere. Water exhaustion may prolong the cooling effect over many years, rather than the one to three years seen in the Pinatubo eruption, and might have even more severe climatic implications as the flora and fauna of the Earth's biomes start to respond, altering the ecology of large areas of the continents. However, there is no recent evidence for this, and older records are simply not sufficiently accurate to be able to discern such an effect. We will return to this possible effect in the following Section on effusive eruptions.

7.1.4 The climatic effects of effusive volcanism

Present-day examples of effusive volcanism are slow, but continuous, eruptions. For example, Mount Etna in Sicily is erupting at the time of writing, sending lava down its flanks (the previous eruption lasted for two years). Mauna Loa in Hawaii has supported a molten lava lake for many years and continuously erupted from 1988 to 1990. The atmospheric effects of these effusive eruptions are limited to eye irritation and respiratory problems in the local population from the release of gases such as HCl and SO_2. The only recent effusive eruption to which we can look for quantified climatic effects was the eruption at the Laki ridge in Iceland in 1783–4. Laki was a 27- km-long fissure eruption that covered much of Western Europe in a dry sulfurous fog. The fog was observed and recorded by Benjamin Franklin, while in Paris as the diplomatic representative of the newly formed USA. It lasted for five to six months and was detected all over the Northern Hemisphere, indicating that it continued to spread within the troposphere rather than being rapidly rained out. Like all effusive volcanism, the worst hazards from the solid eruptive products were limited to the valleys filled by lava during the months of the eruption, while gaseous HCl, HF and SO_2 poisoned the surrounding land (Figure 7.10).

Although not globally important, such gases are important to the local environment. In particular, HF binds to particles in the eruption column and, during the Laki eruption, acid rain rich in HF fell on grazing land where local cattle died two to three days after they ate the contaminated grass. The consequent famine in Iceland resulted in the deaths of 75% of livestock and 25% of the human population. At the height of the eruption, magma flow rates as high as $10\,000\,m^3\,s^{-1}$ drove fire fountains up to a height of 1–1.5 km producing thermal plumes that took gases and fine-grained ash to heights of 10–14 km. The local weather systems were probably responsible for the eventual dispersion of the plume over Europe. Although it was an effusive eruption, Laki probably did have global climatic implications because SO_2 and fine-grained dust reached the stratosphere. Remember that the tropopause (the boundary between the troposphere and stratosphere) varies in height between around 18 km at the Equator and around 8 km at the poles. Over Iceland in June (the peak of the initial eruption) the tropopause is at a height of around 9 km and the Laki eruption probably pushed >30 Mt of dust and gases into the stratosphere, a

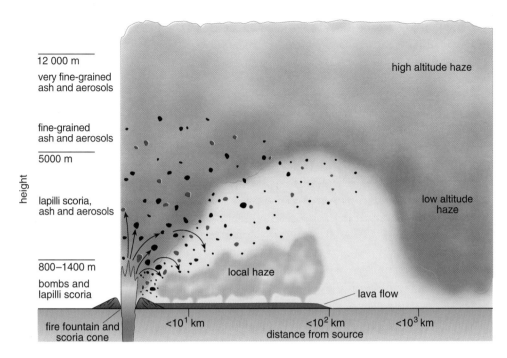

Figure 7.10 Diagrammatic representation of the plume from the eruption of Laki in Iceland in 1783–4, showing the distances over which different aspects of the eruption were effective. The areas close to the volcano were affected by lavas, gases, bombs and ash ejected up to a height of 5 km into the atmosphere. Parts of Europe at a distance of >1000 km from the eruption experienced a dense low altitude haze or fog. *(Rampino and Self, 2000.)*

greater amount than the 1991 Mount Pinatubo eruption. Laki also ejected around 12.5×10^{10} kg SO_2 into the atmosphere, a very large amount in comparison with the current average SO_2 release from volcanoes (around 1.8×10^{10} kg SO_2 per annum) and from the major industrial countries (around 4.8×10^{10} kg SO_2). However, the Northern Hemisphere location of the volcano meant that the effects of Laki were never felt in the Southern Hemisphere.

Basaltic magmas which dominate effusive volcanism generally contain a higher abundance of sulfur (around 0.0013 kg sulfur were actually released into the atmosphere by Laki for every kilogram of magma erupted), but rarely produce eruptions of sufficient ferocity to penetrate into the stratosphere. This is an important point because much of the Cretaceous volcanism, including continental flood basalts such as the Deccan Traps in India, was dominated by effusive eruptions. Estimating their global effects depends critically upon whether the plumes reached the stratosphere. The Laki eruption may have been unusual in recent history, but many aspects of its eruption style and the resulting climatic effects have been well characterized, enabling us to attempt to scale up the effects to Cretaceous volcanism. However, in order to comprehend the scale of eruptions that might have affected the climate during the Cretaceous, we need to consider an eruption on a different scale. There is strong evidence that increased rates of sea-floor spreading and flood basalt magmatism during the Cretaceous contributed to the unusually high CO_2 levels in the atmosphere. Moreover, climatic effects of the famous Deccan volcanism in India at the end of the Cretaceous have been linked by some to the end-Cretaceous mass extinction. Although the Laki eruption may be a taster for this style of volcanism, combined with the extensively studied Columbia River flood basalts, it provides a better understanding of the scale of eruptions which probably occurred during the Cretaceous.

The Laki eruption in 1783–4 lasted around eight months and caused severe environmental effects throughout Iceland and Europe. However, would a larger eruption of similar style cause even greater environmental effects? Although the effect of water exhaustion in the stratosphere is difficult to quantify for the past, it seems likely that climatic effects would be prolonged and this might compound the effects by curtailing growing seasons over several years and disrupting the whole food chain. Since there have been no flood basalt eruptions in recent times, we have no direct evidence for eruptions over such long time-scales, and this has fuelled a long-standing debate concerning the style of eruption of flood basalts, which has very important implications for their environmental effects. Rapidly erupted basalt flows would have been very dramatic, but they would also have short-lived effects upon the atmosphere. The difference between Cretaceous volcanism, such as the Parana basalts in South America which heralded the opening of the South Atlantic Ocean in the early Cretaceous, and volcanic eruptions we see today, is not composition; it is quite simply that larger volumes of basalt were erupted. Although individual Cretaceous flood basalt eruptions have not been quantified, the younger Roza flow of the Columbia River flood basalts is one of the best studied of all ancient lava flows. It erupted over 1200 km^3 of lava around 14.7 Ma ago, compared with the Laki eruption which erupted just 14 km^3. The Roza flow also reached over 300 km from its vent, whereas the Laki eruption only flowed around 50 km down-valley from its source.

Therein lies the dilemma of these huge flood basalt fields: how do the flood basalt flows manage to travel so far before they cool and solidify? If flood basalt flows with similar compositions to modern flows erupted at a similar temperature, they would have moved at similar speeds to modern flows, and should travel similar distances before solidifying. However, if the flood basalt flows erupted at a higher temperature or flowed more rapidly, there must be a mechanism whereby an eruption lasting a few months could have remained liquid and flowed to travel hundreds of kilometres. It is important to be able to estimate lava flow lifetimes since this parameter strongly affects the climatic consequences. Although short-lived eruptions would have been more dramatic, the effects would have been short-lived, and long-term change would have been less important. Large explosive eruptions such as Pinatubo or Toba cause effects lasting a few years, but longer-lived eruptions may have caused more severe climatic effects because they were sustained eruptions continuously pumping gas into the stratosphere.

Let us consider the geological evidence for different lava flow lifetimes and flow rates. There are two basic lava flow types (which incidentally correspond to the rapid and slow flow types above), named from their first description in Hawaii: these are 'aa' or rocky flows and 'pahoehoe' or lobed flows (Figure 7.11). In conventional terms, aa flows are rapidly flowing with blocky surfaces and tend to form in channels. Pahoehoe flows are smooth-surfaced lava flows that move forward slowly and look like the wax lobes which form around the base of a candle burning in a draught. However, in recent years careful studies on Hawaii have discovered that the formation of pahoehoe flows is more complex, and may be a more important flow mechanism than was previously recognized. The new work shows that pahoehoe flows form by a process of extrusion and inflation, and some pahoehoe flows, notably on the volcano Pu'u O'o on Hawaii, have continuously flowed for over a decade (Self *et al.*, 1998).

Figure 7.11   Dark, rough-surfaced aa lava flows (left) that have been covered by younger smooth-surfaced, shiny pahoehoe lava flows (right). Both were erupted in 1973 during the Mauna Ulu eruption on the east rift zone of Kilauea volcano, Hawaii. *(Dave Rothery, Open University.)*

Pahoehoe flows form as lobes characterized by a smooth continuous flexible skin of cooler lava. In the outermost layers, stretched vesicles (bubbles) are preserved. A few millimetres into the lobe, tiny crystals are preserved and bubbles are circular since they are able to return to their ideal shape before the lava solidifies. Ropes, wrinkles and other small-scale features form during flow as the cooler skin deforms (Figure 7.11). There are, in fact, several self-descriptive sub-types of pahoehoe flow, including 'shelly', 'spiny', 'slabby', 'toothpaste' and 'shark-skin', which are transitional between pahoehoe and aa flows. Each lobe forms by initial breakout from the flow and the formation of a crust. Following this, the lobe inflates like a hot lava-filled balloon, to many times its original thickness.

Typical advancing lobes in Hawaii are around 20–50 cm across, 20–30 cm wide and 0.5–5 m long. They can consist of up to 50% bubbles making them stiff foams rather than liquids. Thinner lobes can be formed by less gas-rich lavas. The larger pahoehoe flow lobes can coalesce during inflation, producing wide sheet flows. Lava flows constructed by pahoehoe flow inflation involve a series of internal processes (Figure 7.12).

As each new lobe is formed at the front of the flow (Figure 7.12a), it first extrudes and then expands like a balloon filling with water (Figure 7.12b,c). The early-formed crust remains plastic and deforms as the lobe advances and, as the crust forms, bubbles in the lava are trapped, forming vesicles. The outer solid crust cracks, and overall cooling is often enhanced by the formation of deep clefts in the flow. The solid crust under the flow grows more slowly than the upper crust since heat escapes faster to the air than to the underlying crust or earlier flow (Figure 7.12b–d). As new lobes break out from the flow, the sudden pressure reduction causes new bubbles to form in the liquid and these collect at

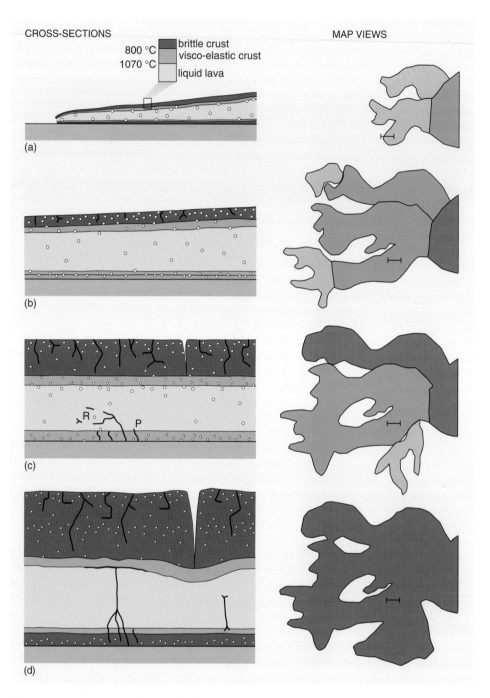

Figure 7.12 Cartoons showing the development of a pahoehoe flow sheet by extrusion and inflation. The cartoons can apply equally to inflated pahoehoe flow sheet lobes that range from 1–100 m in thickness and <100 m to >10 km in lateral dimensions. On the left are cross-sections at a fixed location shown by the short line which appears in the right-hand side, first at the edge of the flow on the right and then at the same point in each diagram as the flow develops. The cross-sections show the flow first extruding and then progressively inflating and cooling. Temperatures within the flow are shown by colours in the sections on the right, corresponding to liquid lava at temperatures of over 1070 °C, to solid lava, which has cooled below 800 °C. Although vesicles form throughout the flow (open circles in the cross-sections), the rising pipe vesicles (P) in part (c) form only in the lower crystallization front. Buoyant silicic segregations or residuum (R) rise from the lower crystallizing magma and mix into the flowing magma. Shading of flows on the right darkens with age, showing their progressive formation. *(Self et al., 1998.)*

the base of the crust as it advances into the still-liquid portion of the flow. As the flow slows, crystal formation and differentiation at the solid/liquid interface leads to silica-rich residuum rising from the lower interface and being entrained in the flow (Figure 7.12c). Finally, the flow stagnates and the remaining bubbles rise to the top of the still-liquid portion of the flow within a few days to a few weeks. The silica-rich vesicular residuum also rises through the flow, forming vesicular vertical and horizontal sheets at the base of the upper crust (Figure 7.12d).

Many of the smaller pahoehoe flows on Hawaii form thin 'hummocky' flows which are discontinuously inflated as slow-flowing parts of the lava flow solidify, leaving the lava flowing through 'tubes'. Larger sheet flows probably contain a continuous body of liquid throughout. Large flows associated with high effusion rates will tend to be sheet flows while slower flows will tend to be 'hummocky'. Although the Laki eruption had a very significant proportion of aa flows, due in part to it having filled river valleys, ~ 70% of the flows have inflated pahoehoe form. Pahoehoe-style lava forms have also been recognized in other flows on Iceland, Hawaii, and in 15 000–200 000-year-old flows in Australia.

Having recognized pahoehoe flow features in what are geologically recent flows, we can return to the Columbia River Flood Basalts in the western USA. The province contains around 300 separate flows identified by their different geochemical signatures, over 50% of which were erupted in only 1 Ma. Pahoehoe flow surfaces have been recognized in the Columbia River Flood Basalts ranging from hummocky-style flows reminiscent of those on Hawaii, to huge sheet lobes tens of metres thick and with a lateral extent of the order of kilometres (Thordarsson and Self, 1998). Internal features of the flows are also very similar to examples in Hawaii, indicating that inflation took place during their eruption. The 14.7-Ma-old, 1200 km^3, Roza flow within the Columbia River Flood Basalt has been studied in detail, showing that it consists of five sub-flows dominated by inflated pahoehoe lobes. The period of active inflation has been estimated from the size of vesicles in the upper crust, and it seems that each of the five flows was active for between just over six months and two years. By assuming that the five lobes inflated in sequence, a total flow time of the order of a decade has been deduced for this flow. Dividing the total volume of the Roza flow by ten years yields a flow rate of 4000 km^3 per year, similar to the peak eruption rate for the Laki eruption of 1783. Samples of the Roza flow at various distances from the vent show that around 77% of the volatiles were degassed in fire fountains and 23% escaped from the lava flow as it moved downslope. Again, like the Laki eruption (Figure 7.10), the Roza flow would have formed a two-layer cloud releasing over 800×10^{10} kg SO_2 into the atmosphere, which dwarfs the release from Laki of around 12×10^{10} kg SO_2.

○ Can we now estimate the global environmental effects of the 14.7-Ma-old basalt eruption which formed the Roza flow?

● It is not a simple task, for the climatic effects depend critically upon the height of the plume relative to the local tropopause, 14.7 Ma ago. This is something which we cannot measure and only the SO_2 gases which reach the stratosphere cause long-term climatic effects. We can be sure that the local effects would have been extreme during the eruption but without corroborating evidence from the rocks, we cannot yet determine the quantitative global effect of such a flow.

The extrapolation of the climatic effects of very ancient eruptions becomes even more difficult for flood basalts such as the Deccan which occurred around 65 Ma ago, and is less well exposed and less well studied than the Roza flow. However, what we have learnt from the Laki eruption and the Roza flow helps us to understand what may have been possible during the eruption of Cretaceous volcanism.

7.2 Volcanoes and the Cretaceous climate

Volcanoes are a major feature of the Earth today, but there was far more volcanic activity during the mid-Cretaceous. In this Section, we will explore when and where the volcanism was concentrated, and consider to what extent enhanced volcanism in the mid-Cretaceous heated the planet by releasing gases which enhanced the greenhouse effect. Then, we will try to consider just how much of an effect this might have had on the climate at the time.

The evidence for intensified volcanic activity during the Cretaceous comes mainly from the oceans because the greatest volume of additional volcanism occurred at ocean ridges and within oceanic plates. On the continents, large volcanic fields called flood basalts erupted as the continents rifted apart. Although they were volumetrically less important, the fact that they erupted into the atmosphere makes them far more likely to have caused short-term climate change. We will consider the evidence for faster oceanic spreading, and the rapid formation of both oceanic and continental volcanic fields. We will also look at the evidence for climate change caused by increased volcanism during the Cretaceous, bearing in mind the problems associated with interpreting these data, which were discussed in the previous Section. Finally, we will consider the Deccan continental flood basalts in India, the eruption of which has been linked to the end-Cretaceous mass extinction. Was the Deccan just another flood basalt in the Late Cretaceous, or did it have more profound effects?

7.2.1 The story of the Atlantic Ocean and Pacific Ocean floors

In Section 1.1.5, we saw that whereas the Atlantic is characterized by deep abyssal plains covered in pelagic mud, the abyssal plains of the Pacific Ocean are strewn with underwater plateaux and the remnants of volcanoes planed flat by wave action, and topped by thick carbonate platforms. So, why are the two oceans so different? To answer this question we need to explore the development of the Atlantic and Pacific Oceans.

Until quite recently, little was known of the structure of the ocean floor. In fact, there had been little systematic exploration of the ocean floor since the *Challenger* expedition in 1872–76 which returned thousands of dredged samples. Only after the Second World War, as maps of the world's ocean floors were released to the scientific community, did it become apparent that the Pacific Ocean hosted not only volcanic chains above water, like the Hawaiian volcanic chain, but also a great many extinct submarine volcanoes. The ocean floor studies initiated from around 1950 onwards were an extremely important part of the development of the theory of plate tectonics. One of the main failings of the older theory of continental drift, which Alfred Wegener first developed in 1915, was the lack of any reasonable explanation of how the continents moved around the globe through apparently solid oceanic crust.

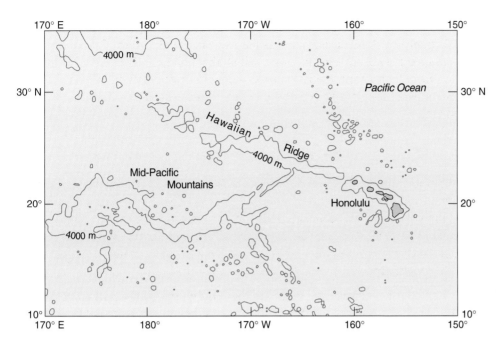

Figure 7.13 Bathymetric map of the mid-Pacific showing the Mid-Pacific Mountains and the Hawaiian chain. Note that the Mid-Pacific Mountains cover a larger area than the Hawaiian ridge, but the lack of major surface expression has meant that, until recently, little was known of the construction of this huge plateau. See Figure 1.15 for location in the Pacific. *(Ocean Drilling Program, Texas A & M University.)*

In 1950, the Scripps Oceanographic Institute and US Navy mounted a joint expedition to determine the structure of Bikini atoll in the Pacific Ocean, and to investigate some of the guyots west of Hawaii. What they found was not a few isolated submarine volcanoes, but a vast submarine mountain range extending over 2000 km to the west, which they named the Mid-Pacific Mountains (Figure 7.13).

Since the 1950s, the Ocean Drilling Programme (ODP) has uncovered several other large igneous provinces (LIPs) in the oceans, and other important features of the ocean floor. One of the particular achievements of ocean floor exploration has been the mapping of linear magnetic anomalies which parallel ancient mid-ocean ridges. These bands mark zones of similar-aged oceanic crust formed at mid-ocean ridges which moved like a 'conveyor belt' away from the ridge. As scientists began to understand the process of oceanic crust formation, they recognized that the oceanic crust is generally younger than the continental crust, much of which formed initially over 2000 Ma ago, and oceanic crust is continually recycled in subduction zones. It became apparent that nearly all oceanic crust is eventually subducted and the hunt was on to find areas of oceanic crust that had survived the longest; in other words, the oldest oceanic crust.

Finding the oldest areas of continental crust is a very difficult process because we can only identify areas of old crust in ancient continental areas by dating them using radiometric techniques, since there is no global pattern to the outcrops. This process should be much easier in oceanic crust because the distance of any piece of oceanic crust from the centre of the ocean, divided by the spreading rate, roughly equals the age of the oceanic crust. This assumption works well in the Atlantic Ocean where the ocean margins roughly parallel the mid-ocean ridge (Figure 7.14) and the ages of basalts dredged from the ocean floor match the predicted values. However, in the Pacific Ocean, the spreading centre is not in the centre of the ocean. By the late 1960s, geologists had worked out that the oldest oceanic crust should be an area the size of the continental USA in the western Pacific. More importantly, the oceanic crust in that area should date back to the Jurassic Period, between 142–200 Ma ago. Several expeditions dredged or drilled rocks in the region, but the material recovered was always younger sediments or basalts formed by volcanic eruptions during the mid-Cretaceous, not during the older Jurassic Period. The Mid-Pacific Mountains were an integral part of this problem from their first exploration in 1950, yielding Cretaceous basalts and fossils from various guyots (Figure 1.18).

This mystery was not finally solved for almost 30 years until December 1989, when the international ODP drilled into western Pacific mud containing Jurassic microfossils. Two days later, the drill reached volcanic basement which proved to be of middle Jurassic age. This breakthrough had been achieved by drilling into the ocean floor in an area away from all volcanic ridges and guyots. Not only did this confirm earlier suggestions that old oceanic crust did exist, but it also showed that previous sampling by dredging and drilling had never been able to recover material from deeper levels. In fact, we now know that there are large areas of Jurassic oceanic crust in the western Pacific but they are almost completely covered by later Cretaceous volcanism. This is an observation which intrigued Roger Larson, the geophysicist on board the ODP ship in December 1989. Why should the ocean floor formed during the Jurassic become covered by volcanic eruptions in the mid-Cretaceous, whereas more recent oceanic crust is dominated by basalts formed at mid-ocean ridges?

The discovery of Jurassic oceanic crust, where geophysicists had predicted it should exist in the western Pacific, confirmed a simple model for oceanic crust formation and allowed compilations of palaeomagnetic data (gathered during many expeditions) to be combined to map the age of the ocean floor (Figure 7.14). Although not all of the oceanic crust has been characterized, particularly in complex areas such as east of Australia (see Figure 7.14), we can use the results to explore the formation rate of oceanic crust and differences between the Atlantic Ocean and Pacific Ocean.

The variation in oceanic crust ages (Figure 7.14) shows the contrast between the world's two biggest oceans: the Atlantic and the Pacific. Geologists have made two simple, and yet crucial, observations on the difference between the Atlantic and Pacific Ocean floors from this map: the first concerns the fate of oceanic crust, and the second concerns the pattern of volcanoes. Together they tell an amazing story about the Cretaceous Period.

The topography and age variation of the Atlantic Ocean floor is symmetrical about the mid-ocean ridges which fall roughly in the middle. The oldest oceanic crust is adjacent to the continental margins with ages ranging from early Jurassic (~180 Ma old) at the edge of the Central Atlantic, to Early Cretaceous (140–120 Ma old) in the South Atlantic, and early Palaeogene in the northernmost Atlantic (between Europe and Greenland). The surface of the older ocean floor on the Atlantic abyssal plain is relatively flat, but interrupted by a few volcanic ridges such as the Walvis ridge and Rio Grande Rise that trend perpendicular to the ocean ridge. In addition, occasional individual volcanoes and rare volcano chains draped in pelagic sediments break up otherwise topographically even abyssal plains.

The Pacific Ocean ridge falls to the east of the ocean centre in the Southern Hemisphere. In the north, the ocean ridge is so far from the centre of the ocean that it approaches North America, eventually running onto the continent through the San Andreas fault system. Unlike the Atlantic, crustal ages in the Pacific Ocean, are not symmetrical about the centre of the ocean and a huge area of the western Pacific forms the oldest oceanic crust in the world (Figure 7.14).

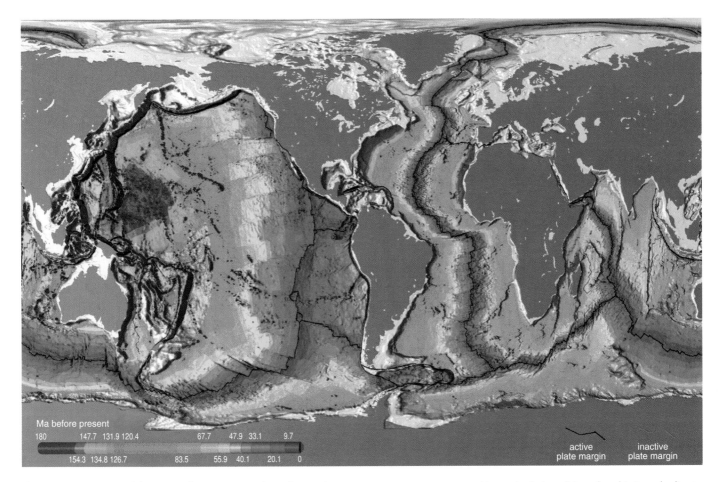

Figure 7.14 Map of the ages of oceanic crust based on palaeomagnetic measurements and isotopic dating. (Note that this is a duplicate of Figure 1.15 and has been reproduced here for ease of reading.) The coloured age zones have been overlain on a map of ocean floor bathymetry which highlights the many ancient submarine volcanoes in the Pacific. The colours representing different ages grade from red and yellow, indicating young Cainozoic crust, to green representing Cretaceous oceanic crust, to blue representing the oldest (Jurassic) crust. (National Geophysical Data Center, USA.)

The crust of the Pacific Ocean represents a much longer-lived system Atlantic, which only started to open between the Early Jurassic and Early Cretaceous as the supercontinent Pangaea broke up (Section 2.2.1). Note that the narrower colour bands in the Atlantic Ocean show that it is spreading relatively slowly in comparison with the Pacific Ocean where wider bands indicate faster spreading. Another striking difference is that there are no subduction zones at the margins of the Atlantic Ocean (save for two small zones in the Caribbean and at the very southern end of South America), preserving even the oldest oceanic crust at the passive margins. The Pacific Ocean is actually much older and was subducting under the earlier Pangaea supercontinent for hundreds of millions of years before the formation of the oldest preserved oceanic crust. Much of the Pacific Ocean crust, formed during the Jurassic and Cretaceous, has been subducted, either beneath North and South America and the Antarctic Peninsula, or beneath the 'ring of fire' lining the western Pacific rim (Section 2.2.4). Additionally, unlike the Atlantic, the oldest ocean floor in the western Pacific is not dotted by occasional volcanoes and ridges but, as we have seen, it is mountainous (albeit underwater mountains) and covered by thousands of eroded volcanoes known as guyots. Some of these are isolated but many form broad plateaux or linear chains, like the well-known Hawaiian chain (Figures 7.13, 7.15).

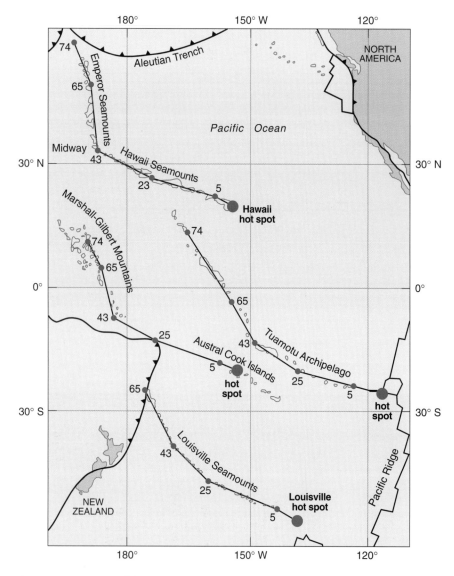

Figure 7.15 Map of the eastern Pacific showing chains of volcanic islands which formed as the oceanic crust moved over mantle plumes during the last 75 Ma. All of the island chains exhibit linear age progression from north-west to south-east. Note the characteristic bend in all the trends at around 43 Ma, indicating that they were formed by relatively static plumes rising under moving oceanic crust. *(Courtillot, 1999.)*

The island chains in Figure 7.15 like many others, such as the Mid-Pacific Mountains (Figure 7.13), initiated in the Late Cretaceous but continue to the present day, and the narrow tracks they trace contrast with the broad oceanic plateaux such as the Mid-Pacific Mountains (Figure 7.13), formed during the Cretaceous. We will consider the formation of these plateaux in more detail below.

7.2.2 Oceanic spreading rates over the last 150 million years

The linear magnetic anomalies observed in the oceans have been tied to the geological time-scale by careful dating of lava samples recovered from cruises in the Pacific and other oceans and by comparison with sections exposed on the continents. Geologists have been able to use this information to map areas of similar oceanic crust age and began to quantify the rates of oceanic crust creation through time. We know the formation rate of present-day oceanic crust because we can measure the spreading rates and we know how thick the oceanic crust is, but being able to measure areas of crust formed in the past using the palaeomagnetic anomalies has allowed geologists to estimate the changing spreading rate. Perhaps the best-known set of estimates were undertaken by Roger Larson (from the University of Rhode Island) and published in 1991 (Figure 7.16b). He estimated the volumes of oceanic crust formed for the whole of the last 150 Ma years in all the world's oceans by calculating the volume of basalt created in 10 Ma periods assuming an oceanic crustal thickness of 6.5 km. This is a simple process for recent crust but becomes progressively more difficult in older crust.

Figure 7.16 Compilation showing the International Ocean Drilling Project view of changes in climate tracers and geological events through the Cretaceous and Cainozoic. (a) Geological time-scale and pattern of geomagnetic reversals (black is normal polarity, white is reversed polarity) against which scale all the other trends/events are plotted. Note the long normal period in the mid-Cretaceous (Section 1.3.1). (b) Crustal production rates show total global production (purple), crustal production at mid-ocean ridges (red) and crustal production in oceanic plateaux and volcanic island chains (mauve). (c) Formation times of large flood basalt fields are shown as brown bands corresponding to the total eruption period although the majority of lava in these provinces erupts in 1–3 Ma. (d) Sr-isotope variations from carbonates precipitated in seawater show little change through the Cretaceous but rise rapidly in the later Cainozoic. The rise in the oceanic Sr-isotope ratio has been linked to supply of radiogenic Sr from the eroding continental crust via rivers and has strongly influenced climate models (see Section 6.3.1). (e) Generalized global average sea-level changes from Haq *et al.* (1987) emphasizing the very high levels reached during the mid-Cretaceous, over 200 m above present-day level (variations during the recent glaciation are not shown, Section 3.2). (f) Global mean temperature level illustrates the very marked contrast between the high temperatures and ice-free conditions of the Cretaceous and glaciation in the late Cainozoic, though it does not show the transient glaciations of the early Cretaceous (Section 5.3). Upper part of curve was drawn using oxygen-isotope data from cores in the Atlantic; lower part drawn from several lines of evidence and plotted against temperature only. (g) Times during which organic-rich mudstones were lain down are shown by wide grey bands, marking periods of ocean stagnation and anoxia (Section 5.2). *(Courtesy of M. F. Coffin, University of Texas at Austin.)*

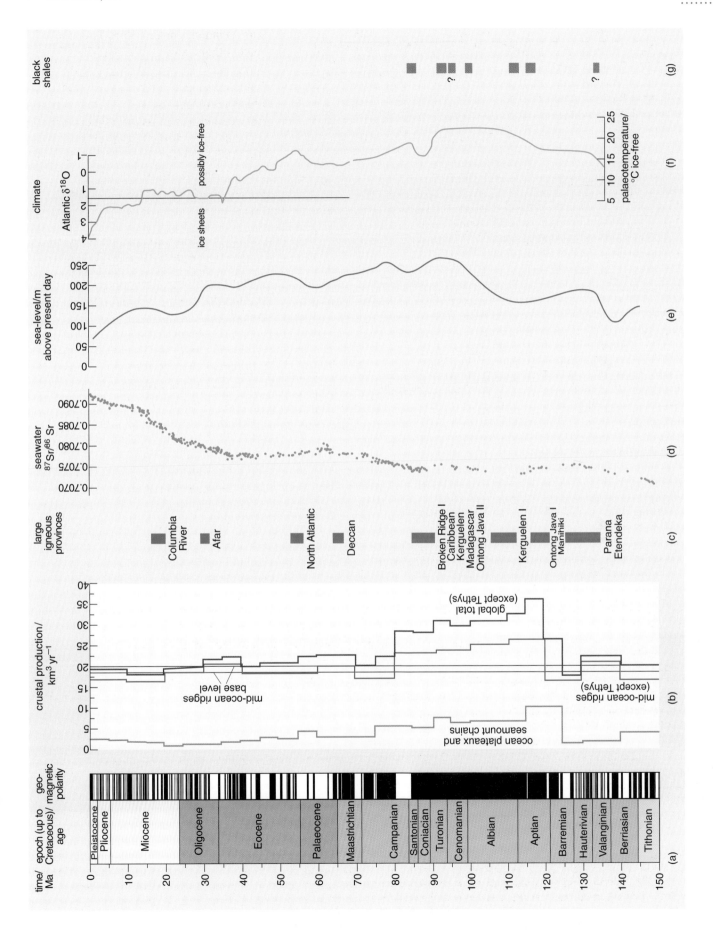

○ What problems can you see that the loss of older oceanic crust might cause in the Pacific Ocean?

● In many of the world's oceans, particularly those with slower spreading rates, the Cretaceous oceanic crust still lies at the continental margins and at the bottom of the abyssal plains. However, in faster spreading areas, particularly the eastern Pacific, much of the oceanic crust created during the Cretaceous has been subducted and the spreading rate information has been lost.

The situation is even more extreme in the eastern Pacific because two ocean spreading ridges which existed during the Cretaceous, known as the Farallon–Phoenix and Farallon–Kula Ridges, have been consumed at subduction zones. No one ever observed these ridges directly, since they were subducted before humans appeared on Earth, but they have been inferred from very good evidence that two triple junctions must have existed in order to explain the disposition of the present-day plates. Their disappearance means that all estimates of the spreading rates are less certain, but there is one very effective way of estimating the amount of lost oceanic crust. Wherever it has been measured, mid-ocean ridge spreading is nearly symmetrical about the spreading centre, and thus if only one side of the spreading centre remains intact, we can estimate the amount lost by assuming it was equal to that still present on the other side. This was the technique used by Larson and illustrated in Figure 7.17.

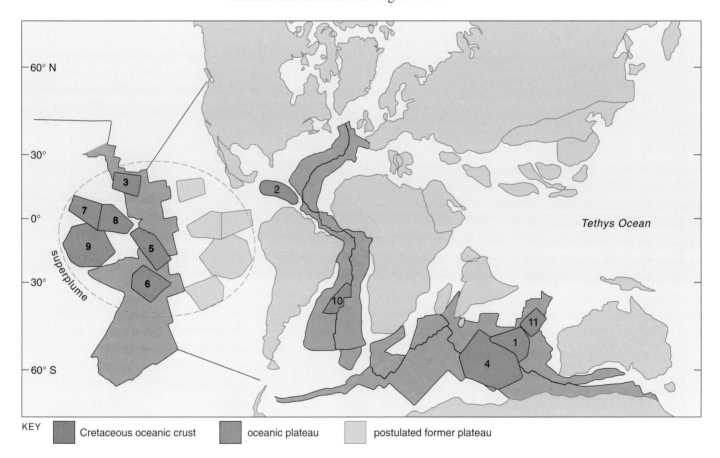

Figure 7.17 Reconstruction of the continents and oceans 83 Ma ago from Larson (1991), showing the oceanic crust created during the mid-Cretaceous in a dark shade and oceanic plateaux in red. The numbered plateaux are (1) Broken Ridge; (2) Caribbean Plateau; (3) Hess Rise; (4) Kerguelen Plateau; (5) Line Islands; (6) Manihiki Plateau; (7) Marcus Wake Seamounts; (8) Mid-Pacific Mountains; (9) Ontong–Java Plateau; (10) Rio Grande Rise; (11) Wallaby Rise.

The resulting ocean production rates can be displayed as a histogram through time (see Figure 7.16b) and this shows that oceanic crust has been produced consistently at around $18-20 \times 10^6 \, km^3$ per Ma, over the last 70–80 Ma. However, about 120–125 Ma ago, oceanic crust production suddenly doubled to over 35 million km^3 per Ma and remained high for over 40 Ma. This increase in crustal production was greatest in the Pacific Ocean, which was covered by erupting volcanic islands and plateaux, and experienced rapid ocean floor spreading rates. The term Larson coined for this phenomenon was 'superplume', which is explained in a bit more detail in Box 7.2. Crustal production rates in the Indian and Atlantic Oceans did not show the same variation and in fact have increased gradually over the last 150 Ma, whereas Pacific Ocean production has reduced in recent times. The maximum spreading rates for individual ridges during the Cretaceous pulse were as high as $17 \, cm \, yr^{-1}$, about the same as the fastest spreading rates seen today on the Pacific–Nazca ridge. Thus, individual ridges did not spread abnormally fast but there was an abnormally large length of fast-spreading ridge.

Box 7.2   Superplumes

Most geologists are familiar with the concept of hot plumes rising through the mantle, and the role they play in forming hot-spot tracks in oceans such as the Hawaiian volcano trail in the Pacific. Laboratory models of plumes using viscous liquids show that plumes initiate at thermal instabilities at a boundary within the lower fluid and rise rapidly to the upper surface. There has been a great deal of debate concerning the boundaries which might initiate plumes within the mantle; the two most likely boundaries are the core/mantle boundary and the boundary between upper and lower mantle at around 670 km depth. The models show that as the plumes rise, the initial surge of material develops a spherical head which rises through the mantle followed by a narrow tail. As they reach the bottom of the oceanic or continental crust the head deforms to a wide flat disk, which may reach over 2000 km in diameter, providing a wide area of hot material which can lead to the eruption of huge volcanic fields over wide areas of oceanic crust (like the Ontong–Java Plateau) or continental crust (like the Parana–Etendeka flood basalts or Deccan flood basalts). Huge volumes of magma might be produced by the impact of a plume head but this quickly dies away as the plume is entrained in mantle flow, leaving the smaller plume tail to cause continued volcanism. The tails of many currently active plumes such as the Réunion Island chain in the Indian Ocean can be traced back to a large flood basalt eruption such as the Deccan flood basalt. Flood basalt eruptions on continental crust are also closely linked to times of continental rifting events; for example the Parana–Etendeka field erupted across the opening rift between Africa and South America between 138 and 130 Ma ago, with the major pulse at 133–132 Ma.

During the mid-Cretaceous, a large area of the Pacific Ocean experienced increased ocean spreading rates and saw a huge increase in volcanism, including the short-lived eruptions of the Ontong–Java Plateau and Manihiki Plateau and the slower formation of the Mid-Pacific Mountains, the Hess Rise, the Line Islands and the Marcus Wake Seamounts. These events all took place in the Pacific Ocean in a zone which may have been as much as 5000 km across. The combination of high magma production rates at ocean ridges and the eruption of volcano chains and broad volcanic plateaux exceed anything which might have been produced by a single mantle plume, and the term 'superplume' was coined for the phenomenon. The huge area outlined in Figure 7.17 shows the extent of the superplume eruptions as they would have appeared at the end of the period of rapid ocean floor spreading, around 83 Ma years ago. The cause of this paroxysm of the mantle is not clear and much debated, though all the models seem to require a large cold block like a zone of cold mantle (possibly accumulations of old subducted oceanic crust), or a supercontinent such as Pangaea insulating the lower mantle. Heat builds up under the barrier and eventually a superplume results as the mantle overturns and a large zone of hot mantle rises to the surface. Although we cannot yet say for certain why a superplume formed during the Cretaceous, the concept of superplumes has been applied to other large events at other times on the Earth. Periods of heightened igneous activity, crustal growth, and atmospheric change have been linked to superplume events, around 2700 and 1900 and possibly 1200 Ma ago, and in addition a Cretaceous-scale event has been hypothesized to have occurred around 300–350 Ma ago.

Finally, a new theory is emerging about the evolution of volcanoes on Mars. Although Mars is only half the size of Earth and has a very thick lithosphere which prevents plate tectonics, some volcanism continued to recent times. Large areas of the Martian surface are extremely old and pitted with many impact craters like the lunar surface (Figure 7.18), but other areas show much younger surfaces with few craters and even perhaps signs of fluid flow and lakes. Recent mapping of the surface by the Mars Orbiter Laser Altimeter (MOLA) has greatly refined our understanding of the surface features showing that a large area, known as the Tharsis bulge (pale brown on Figure 7.18), is higher than the rest of the surface and in addition it is much younger (because it exhibits few craters). The Tharsis bulge also encompasses most of the Martian volcanoes, including the massive Olympus Mons (the largest of the white-capped mountains north of the bulge), and is also the area showing evidence for fluid movements such as river traces and deltas from catastrophic fluid release.

Victor Baker and Shigenori Maruyama (2001) recently proposed that several features of the Martian surface are linked by a single mechanism, the development of a superplume which impacted under the base of the Martian lithosphere causing the Tharsis area to bulge. This might also explain the location of many of the largest Martian volcanoes in this area, and perhaps the release of fluids accompanying the plume eruptions or remobilization of sub-surface fluids by heat from the eruptions gave rise to the river features which seem to run out from the bulge. If life existed on the surface of Mars, it might have been during relatively brief periods when fluids flowed on the Martian surface, caused by the mantle superplume.

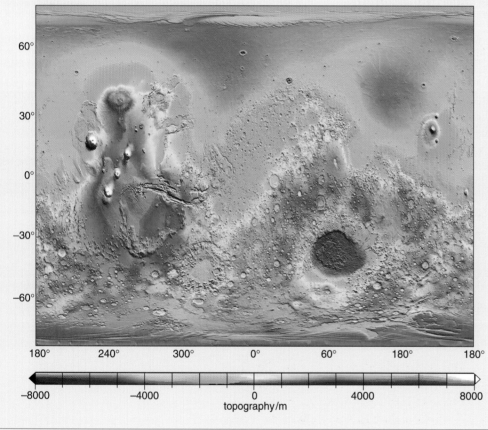

Figure 7.18 Contour map of the Martian surface from the recent high resolution Mars Orbiter Laser Altimeter (MOLA) survey. Note that much of the less-cratered surface lies in the eastern quarter around the volcanic centres including one of the largest volcanoes in the Solar System, Olympus Mons, which rises 24 km above the dome and is over 550 km in diameter. Compare this with Mauna Loa on Hawaii, one of the Earth's largest volcanoes, which is 9 km high and 120 km across. (NASA.)

○ Why is this 35 km³ yr⁻¹ a minimum for the amount of oceanic crust, particularly the older oceanic crust?

● Figure 7.17 shows the position of the continents 83 Ma ago but shows no Cretaceous spreading in the large Tethys Ocean stretching from Spain to eastern Asia, which was such an important feature for Cretaceous climate. The Tethys Ocean closed in the latest Cretaceous/early Tertiary and remains today only as a few fragments of oceanic crust trapped between Asian and Indian or African continental crust. When Larson calculated the volumes of oceanic crust production, he excluded any attempt to estimate crustal production in the lost ocean of Tethys.

The corollary to the increased Pacific oceanic crust production must be increased rates of subduction, particularly around the Pacific Ocean, but is there any evidence for this? In fact there is a good deal of evidence for an increased intensity of tectonic processes on the surrounding continents indicating that the normal conditions around passive margins were reversed during the mid-Cretaceous. Areas from as far afield as Japan and Antarctica, on the margins of the Pacific Ocean, show compressive tectonic events during the mid-Cretaceous (Figure 7.19).

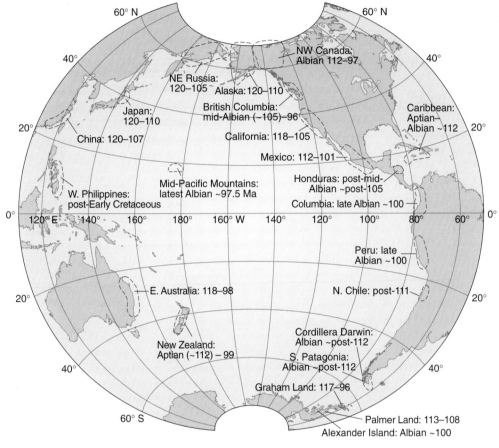

Figure 7.19 Tectonic events around the Pacific margin during the Cretaceous. Note that the areas which experienced tectonic events all abut the ancient subduction zones and the ages centre around the time of the accelerated ocean floor spreading in the Pacific Ocean. *(Vaughan, 1995.)*

7.2.3 Oceanic volcanic chains and plateaux

We have many present-day examples of oceanic island chains, but are they a good model for volcanism in the mid-Cretaceous? The Hawaiian chain in the western Pacific (Figure 7.15) for example, stretches west from the presently active island of Hawaii, joining the NW–SE-trending Emperor chain at the bend which occurs close to islands around 43 million years old. The oldest islands of the Emperor chain in the north are at least 74 Ma and the oldest may be as old as 81 Ma. The Hawaiian/Emperor chain is prominent and convincing evidence that mantle plumes can punch through cold oceanic crust forming a long line of volcanoes as the 'conveyer belt' passes over a near stationary hot mantle plume (Figure 7.15). For many years, the Hawaiian chain has also been cited as the best evidence for fixed mantle plumes which remain at the same location as oceanic crust moves over them. The bend in the Hawaiian chain was thought to represent a change in the movement direction of the Pacific Plate. However, if this were so, all the other Pacific islands of the same age should exhibit chains with exactly the same orientation. The problem is that although they are similar, attempts to fit the directions of other Pacific volcanic chains to the Hawaiian model showed that the mantle plumes which feed the volcanic chains have moved relative to each other through time. Detailed work measuring the eruption ages of the Magellan seamount trail in the western Pacific, showed that they were erupting at the same time as the Musician seamounts which lie in the Central Pacific, north of Hawaii. The trails of the two plumes moved relative to each other at rates of up to 20 km per Ma (Figure 7.20) during the time when both were erupting.

The Hawaiian chain is distinctive for another reason: it has a very linear trend, and has always erupted through old, and thus cold, oceanic crust, unlike many of the Cretaceous volcanic plateaux and volcanic chains. For example, the Mid-Pacific Mountains represent not a narrow chain but a broad plateau of ridges and volcanoes over 2000 km long which formed on thin oceanic crust close to the mid-ocean ridge in the Pacific at the time. Many of the other Cretaceous Pacific Ocean plateaux such as the Hess Rise and Marcus Wake seamounts (Figure 7.16) also probably formed close to mid-ocean ridges. Moreover, in the Atlantic, the Walvis Ridge and Rio Grande Rise formed close to the Mid-Atlantic Ridge at the site of present-day Tristan da Cunha, starting around 130 million years ago. So perhaps the Hawaiian island chain is not a good model for all mid-Cretaceous oceanic volcanism.

Figure 7.20 Eruption ages for submarine volcanic seamount chains: the Magellan seamounts (shown in Figure 7.21) and the Musician seamounts which lie north of Hawaii. (Koppers et al., 1998.)

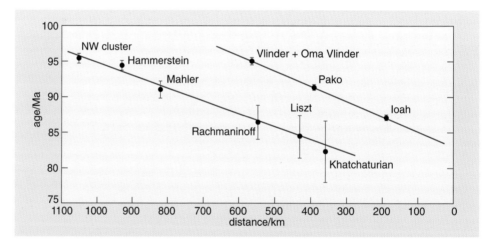

Many oceanic plateaux like the Mid-Pacific Mountains are simply broad sections of thick, normal composition oceanic crust (up to 25 km thick) formed close to or at spreading ridges. They show age progression but not as well defined as those of the seamount trails such as the Magellan or Musician seamounts. Still other plateaux such as Ontong–Java do have distinctly different basalt chemistry and were not directly associated with spreading ridges. There seem to have been two distinct types of plateau forming during the mid-Cretaceous:

1   Common volcanic island chains and plateaux like the Magellan seamount chain (Figure 7.21) and the Mid-Pacific Mountains took tens of millions of years to form at roughly similar times in the mid-Cretaceous. Some formed close to ocean ridges, others erupted through thick oceanic crust.

2   A very different type of eruption produced even larger amounts of basaltic lava over very much shorter time-scales, perhaps as little as a million years in some cases; these are the 'Large Igneous Provinces' (LIPs). The Ontong–Java Plateau (Figure 7.21), the largest LIP known on Earth, formed in the western Pacific, in one short burst of volcanic activity lasting perhaps a million years at 122 Ma, with two minor later eruptive episodes around 90 and 60 Ma (Figure 7.16).

Other oceanic LIPs formed in the Caribbean and Indian Oceans during the Cretaceous (Figure 7.17). We should note at this point that LIPs formed during the Cretaceous were not limited to the oceans, as flood basalts also erupted on continents such as Madagascar, Brazil, Africa, Antarctica and finally in India where the Deccan flood basalt erupted at the very end of the Cretaceous. Although we have no record of the eruption of oceanic flood basalts before the Jurassic, the record of oceanic and continental flood basalt volcanism over the last 200 million years shows a very marked concentration in the Cretaceous (Figure 7.16c).

Figure 7.21 The Western Pacific contains several different manifestations of mid-Cretaceous oceanic volcanism including: narrow seamount trails such as the Emperor chain, the Hawaiian chain, the Line Islands and the Magellan Seamounts; broad slowly formed plateaux such as the Mid-Pacific Mountains and the Hess Rise, and rapidly formed plateaux such as the Ontong–Java Plateau and Manihiki Plateau. (Neal et al., 1997.)

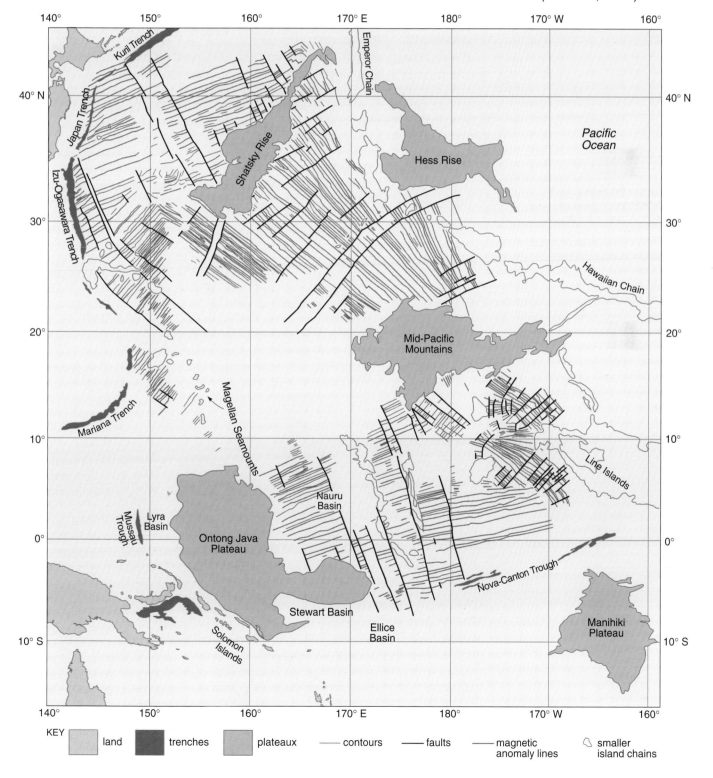

KEY land | trenches | plateaux | —— contours | —— faults | —— magnetic anomaly lines | smaller island chains

○ Many of the narrow volcanic chains continued to develop through the Cainozoic but are there any modern analogues for the broad plateaux like the Mid-Pacific Mountains, which formed close to spreading ridges in the Cretaceous Pacific?

● Most island chains are now erupting through thick cold oceanic crust. However, Iceland is a good example of a broad oceanic plateau forming at a ridge today.

An upwelling mantle plume impacted the base of the oceanic crust beneath the mid-Atlantic spreading ridge around 62 million years ago forming the North Atlantic Igneous Province (flood basalt province), bringing hotter mantle into contact with the base of the oceanic crust and causing an increase in the proportion of mantle melting beneath the ridge (Section 2.2.1). The smaller tail of the plume still rises beneath the mid-Atlantic spreading ridge today, enhancing basalt production at the ocean ridge, and forming the island of Iceland.

Iceland is part of the North Atlantic spreading ridge system and as the 'conveyer belts' move apart, the two sides of Iceland move away from the ridge. New sections of Iceland thus form in the centre of the island and progressively move out (Figure 7.22). As time goes by and the Atlantic continues to open, Iceland will evolve into an elongate plateau. The Mid-Pacific Mountains probably formed in a similar way to present-day Iceland as a plume impacted the base of the oceanic crust close to the Pacific–Farallon Ridge.

Studies of guyots like Allison (see Section 1.1.5) in the Mid-Pacific Mountains, have shown that they rise only a few hundred metres above the plateau which is already 1–2 km above the rest of the ocean floor. Although we cannot be certain of the relationship between the guyots and the plateau above which they rise, they may also be explained by events on present-day Iceland. As the outer parts of Iceland move further from the mid-Atlantic ridge, they cool and subside beneath the ocean surface, yet the west of Iceland appears to remain above the waves longer than the east (Figure 7.22). This is partly because the area of Snaefellsnes in the west is still an active area of volcanism today. Magma is still being supplied to this outlying volcanic centre from the ridge buoying up the area and it thus rises above the surrounding older basalts. As time goes on, the peninsula of Snaefellsnes may become isolated from the mainland and form a seamount, subsiding finally to become a guyot.

The Ontong–Java Plateau did not form in this slow progressive way at a ridge but formed very rapidly close to an existing ridge. Ontong–Java is the largest LIP in the world, measuring some 1500 km across, and has a crustal thickness of around 36 km. The erupted volume was in excess of $1.5 \times 10^6 \, km^3$. The plateau was drilled in a series of ODP cruises in the late 1980s, and has also been extensively studied because its southern tip has collided with a subduction zone, lifting parts of the plateau above sea-level to form the Solomon Islands (Figure 7.21). The drilling, combined with Ar–Ar dating has established that parts of the plateau separated by hundreds of km, were erupted within 1–2 Ma, 122 Ma ago, with later additions at around 90 Ma and 60 Ma. The chemistry of the basalts was not like those produced at normal ocean ridges but similar to other plume-related basalts, like those of present-day Iceland. Huge volumes of basalt magma were erupted at rates similar to those which have been proposed for continental flood basalts like Deccan, $1–2 \, km^3 \, yr^{-1}$. The very rapid eruption rates have prompted much speculation over the climatic effects of such massive eruptions, particularly the involvement of the Deccan basalts with the end-Cretaceous mass extinction, 65 Ma ago.

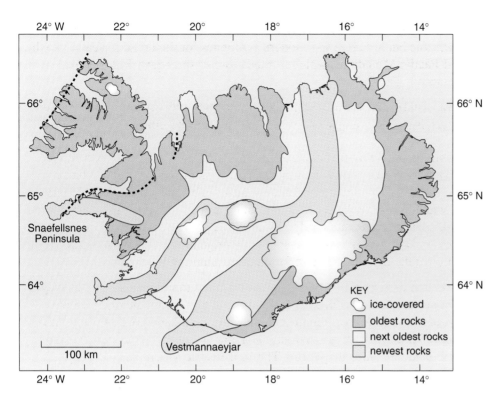

Figure 7.22 A contoured age map of Iceland showing the ages of different zones progressively moving away from the ridge. The oldest rocks of Iceland are around 16 Ma old and lie at the extremities of the island in the east and west. Rocks formed between 16 and 3 Ma ago are shown in the darker brown shade. Volcanoes and lavas formed during the last 3 Ma (but before the most recent Ice Age) fall in a central zone shown in the paler shading. Almost all volcanism since the last Ice Age lies in a narrow inner zone (coloured green) contained within the 3 Ma zone and defines the two active volcanic axes. The exception to this is Snaefellsnes Peninsula where eruptions started again in the last 10 000 years. *(Hardarson et al., 1997.)*

○ Is there any simple relationship between the eruption of the Ontong–Java Plateau and oceanic anoxic events or mass extinctions?

● The simple observation is that there is **no** discernible relationship between the timing of the Ontong–Java eruption and OAEs. Figure 7.16 shows that the eruption of the Ontong–Java Plateau pre-dates the Aptian–Albian OAEs and associated mass extinction by at least 4 Ma (see also Figure 5.1). In addition, dinosaurs were common 122 Ma ago, but did not suffer any obvious ill effects from this enormous eruption.

7.2.4 Oceanic plateau formation over the last 150 Ma

When Larson quantified oceanic crust formation at spreading ridges, he also attempted to quantify the formation of oceanic plateaux. As we saw above, there are two types of plateau: rapidly formed flood basalts such as the Ontong–Java Plateau, and slowly formed island chains like the Hawaiian chain and Mid-Pacific Mountains. Both types of plateau were prevalent during the mid-Cretaceous pulse in ocean spreading rates in the Pacific. In 1991, Larson estimated the amount of oceanic crust involved in plateaux and, like the spreading ridge data, found that he had to estimate the volume of oceanic plateaux in the eastern Pacific because like the normal oceanic crust, any plateaux formed in that area have subsequently been subducted. Larson overcame this problem by assuming that all the western Pacific plateaux had mirror images in the eastern Pacific (Figure 7.17). However, there was a problem with this approach to estimating the plateau volumes. ODP expeditions to the large oceanic plateaux since Larson originally used this technique have demonstrated the two styles of plateau formation. We now know that rapidly formed plateaux such as Ontong–Java and Manihiki did not form at ocean ridges and thus never had mirror images across the ridge. By 1995, Larson's estimates of the oceanic plateau formation took the rapid formation model into account, and yet still the formation rates show the same general trend, a hint at the robustness of this model. Figure 7.16c shows the variations in oceanic plateau formation over the last 150 Ma.

Like ocean spreading at ridges, the highest rate of oceanic plateau formation during the last 150 Ma was in the mid-Cretaceous, and was also highest in the Pacific Ocean. Global eruption rates have normally averaged $1–2 \times 10^6$ km^3 per Ma but, in the mid-Cretaceous, they reached almost 1.0×10^7 km^3 per Ma. This was still less than half the rate of oceanic crust production at spreading ridges during the mid-Cretaceous, but they significantly enhanced the total world crustal production rate which reached around 3.5×10^7 km^3 per Ma. Nevertheless, there is another factor about the rapidly formed oceanic plateaux which marks them out as important to the Cretaceous climate. Considering the production of flood basalts in 10- or even 5-Ma boxes dilutes their relative importance. Oceanic plateaux formed in the same way as the Ontong–Java Plateau, were created in a very short time period, perhaps as little as 1 Ma.

○ How would this affect a graph of world volcanic production such as Figure 7.16?

● If it was erupted in as little as 1 Ma, the Ontong–Java Plateau would have effectively doubled world oceanic crust production. We will return to this observation in a later discussion of the effects of Cretaceous volcanism on the climate.

7.2.5 Cretaceous continental flood basalts

If the rapid eruption of oceanic plateau basalts might have caused significant climatic effects, then the continental flood basalts were the truly menacing volcanic events, and are more likely to have caused short-term climatic fluctuations. The Cretaceous continental flood basalts were of comparable size to the oceanic flood basalts and their environmental effects would have been more severe than the oceanic events simply because they erupted directly into the atmosphere, whereas the oceanic flood basalts erupted partly under the ocean. Gases such as SO$_2$, released from sub-oceanic eruptions, would have become rapidly dissolved in the water, reducing any atmospheric effects.

Unlike oceanic flood basalts, continental flood basalts are not subducted and thus preserve a more complete record. Large continental flood basalts are known to have occurred thousands of millions of years ago and well-preserved 250- Ma-old flood basalts are found in Siberia. Nearly all the largest continental flood basalts of the last 200 Ma have been associated with continental rifting and break-up of the southern supercontinent, Gondwana (Figure 7.23a). This started with the Central Atlantic Province around 200 Ma ago, then 183 Ma ago a huge province called the Karoo–Ferrar Province erupted in Southern Africa, Australia and Antarctica. In the Cretaceous Period, the Parana–Etendeka Province reached its peak eruption rate as Africa and South America rifted apart (Figure 7.23b) around 133–132 Ma ago. As Australia, India and Africa rifted apart, the Kerguelen Plateau Large Igneous Province (LIP) started to form, along with the Rajmahal Province in India (Figure 7.23c). The Marie Byrd Land province erupted peaking around 107 Ma ago as New Zealand rifted from Antarctica and Australia (Figure 7.23d). The Madagascar flood basalts formed as India and Madagascar rifted apart around 88 Ma ago. Finally, though it is not shown in Figure 7.23, the Deccan volcanic field erupted as an area of crust, which now forms the Seychelles bank, rifted from northern India. Any attempt to model the climatic effects of the Deccan Flood Basalt Province must also take account of the many earlier flood basalts and oceanic LIPs in the Cretaceous Period.

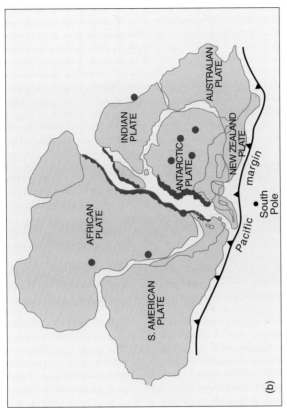

(a)

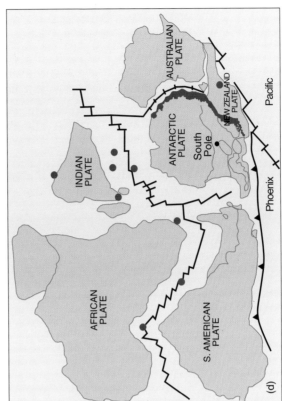

(b)

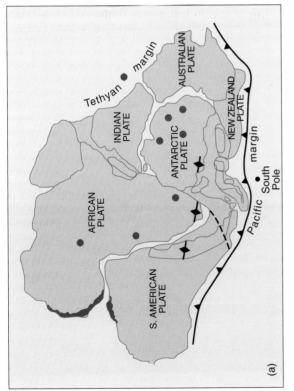

(c)

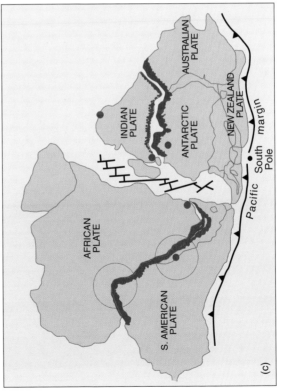

(d)

Figure 7.23 Continental reconstructions for (a) 200 Ma, (b) 160 Ma, (c) 130 Ma and (d) 100 Ma. Hypothetical positions of the hot spots (mantle plumes) at each time are shown as small filled red circles and as large open 2000-km-diameter circles as they produced flood basalt eruptions. 2000 km is the largest diameter such plume heads are likely to reach as they impact the base of the oceanic crust. Rifting margins are shown in red. *(Storey, 1995.)*

7.2.6 The feedback relationship between climate and Cretaceous volcanism

The crucial message from studies of the long-term carbon cycle is that it is feedbacks that drive and change the climate system over long time-scales. We can estimate the volumes of volcanism in the mid-Cretaceous but the increased CO_2 output is not directly proportional to the CO_2 level in the atmosphere at the time. Remember from Chapter 6 that increased levels of CO_2 in the atmosphere also increase weathering rates which in turn remove CO_2, so the atmospheric levels which determine the climate are not a simple function of the volcanic output. So how can we establish the true importance of volcanism to the Cretaceous climate? The best way is to test how susceptible the climate is to changes in CO_2 using a simple model.

Consider a very simplified version of the GEOCARB model without all the feedbacks looked at in Chapter 6. In this version of the model, we take no consideration of land area, mountains, or the effect of land plants or burial of organic carbon. On land, atmospheric CO_2 is drawn down by silicate weathering, and in the oceans CO_2 is sequestered by carbonate precipitation. CO_2 supply to the atmosphere by degassing is in balance with the weathering and carbonate precipitation processes in our model. Let us start with two atmospheric CO_2 concentrations: modern pre-industrial values of around 280 ppm and values ten times greater, 2800 ppm, which we might expect for a greenhouse world like the Cretaceous (Figure 7.24). This model has volcanoes giving off CO_2 but it is in balance with the other processes.

First, let's try reducing volcanic emissions by 25%: not simply switching off all the CO_2 degassing from volcanoes and other sources, but just reducing their emissions by a quarter. Weathering continues at the same rate, because there is no feedback with the atmospheric CO_2 level. Nothing will happen for a little while, but then slowly at first, silicate weathering will start to have a significant effect upon the atmospheric CO_2 level (Figure 7.24). In a modern pre-industrialized world, CO_2 levels will fall significantly after 20 000–30 000 years and Ice Age conditions ensue not long after. In the greenhouse world, atmospheric CO_2 levels are reduced to the pre-industrial levels after around 1.2 Ma and it too falls into an Ice Age after less than 2 Ma. This effect occurs when the CO_2 emission is reduced by only 25%, in other words the feedbacks in the real world must act to stabilize variations very rapidly. This reduced CO_2 effect is as devastating as huge volcanic eruptions and would be certain to cause mass extinction on time-scales of far less than a million years. It is a devastating extinction mechanism created simply by artificially altering the balance of feedbacks in the carbon cycle.

What if, instead of reducing volcanic CO_2 emissions, we increase them by 25%? This is what would have happened as the ocean ridge spreading rates increased as the superplume event was initiated in the mid-Cretaceous. In fact it may have been more extreme than this, but for the purposes of the model, the outcomes are the same. Remember that weathering rates will not increase because there are no feedbacks in this system. In the pre-industrial system, atmospheric CO_2 levels rise linearly (although represented as a curve on Figure 7.24 because the vertical scale is logarithmic) and reach Cretaceous greenhouse levels after just over 1 Ma. Atmospheric CO_2 levels in the greenhouse world also increase, although the effect is smaller as there is already a great deal of CO_2 in the atmosphere. Maintained for 40 Ma, this situation would extinguish most life and leave the Earth with an atmosphere of almost 10% CO_2. Clearly this did not happen in the Cretaceous but it illustrates the importance of the feedbacks.

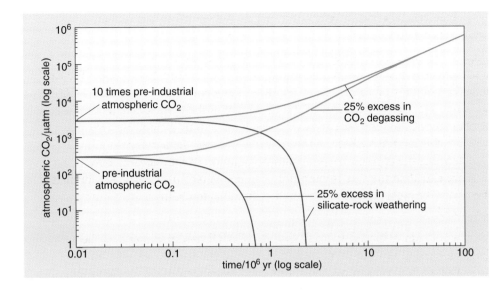

Figure 7.24 The effects on atmospheric CO_2 of a 25% imbalance in degassing and silicate-rock weathering in the absence of feedback mechanisms. Two trends are shown: one starts from 280 μatm (ppm by volume; modern pre-industrial levels) and the other starts from 2800 μatm (approximate Cretaceous levels). The model also assumes present-day uptake of CO_2 by weathering reactions (about 8×10^{18} mol C per Ma); equilibrium removal CO_2 from the ocean into carbonates; no feedback mechanism to remove the extra CO_2 and a balanced organic carbon sub-cycle. *(Berner and Caldeira, 1997.)*

Prolonged volcanic CO_2 input during the mid-Cretaceous virtually doubled the volcanic input into the carbon cycle. However that did not lead to the runaway greenhouse effect we saw in the simple non-feedback model above. In fact if the GEOCARB model is correct, levels of atmospheric CO_2 during the Cretaceous were lower than those earlier in the Phanerozoic, a hypothesis backed up by measurements of atmospheric CO_2 levels (Figure 6.11). If we assume that the atmospheric CO_2 levels were around 2800 ppm (the equivalent of around eight times present-day levels) for the period of the mid-Cretaceous from 120 Ma to 80 Ma, only around 0.14% of the CO_2 release from volcanoes would be required to raise atmospheric CO_2 to that level, based on a doubling of the present-day volcanic CO_2 outgassing. What the additional volcanic CO_2 did was to accelerate feedback reactions which effectively increased weathering rates, added nutrients (along with dissolved CO_2) to the drowned continental shelves and also fertilized biomass production on land. If we want to look for the lost volcanic CO_2 of the mid-Cretaceous, it can be found in abundant Cretaceous coal and oil deposits (Chapter 4) and in huge Cretaceous carbonate deposits (Chapter 5).

7.3 The Deccan Flood Basalt and climate change at the end of the Cretaceous

Like other Cretaceous LIPs, the Deccan Traps (Traps is an old name for layered basalt deposits) covered huge areas with thick successions of basalt flows. In the case of the Deccan, there seems to have been a progression of the volcanic centre from NW to SE during the eruptions which fits the direction in which the Réunion mantle plume would have been moving at the time. Careful work on ancient drainage patterns has shown that the area was uplifted by the plume and that volcanism produced a series of very broad shield-like volcanic edifices.

Early work on the age of the Deccan Traps placed them roughly at the boundary between the Cretaceous and Tertiary but the age range suggested prolonged volcanism lasting from around 80 Ma to 30 Ma, i.e. around 50 Ma. Vincent Courtillot and a team of French and Indian colleagues had analysed rocks from Tibet to study their palaeomagnetism, and determine the rate at which India and Asia had been travelling to their collision at around 50 Ma which led to the formation of the Himalaya. In 1984, Courtillot and colleagues started to analyse the Deccan Traps with the intention of using the variations in palaeomagnetic pole direction they found in basalt layers of different ages to monitor the movement of India prior to and during the collision.

However, rather than the many palaeomagnetic reversals they expected to find in the Deccan lavas, Courtillot and colleagues found that most of the specimens they had collected had the same polarity, in this case the opposite direction to the present day (reversed polarity). By reanalysing all previous data on Deccan, they determined that the whole lava pile exhibited only two reversals, and over much of its area only one reversal. Although palaeomagnetism can be used to measure time by the thicknesses of magnetic reversals and compare these with the magnetic reversals on the ocean floor, it is not an absolute dating technique so some of the samples were dated using K–Ar dating. However, the samples had been altered by many millions of years of exposure to tropical weathering and it was suggested that this alteration might be the reason why the ages were apparently scattered. The samples were re-dated using the Ar–Ar technique which can date very small samples only a few milligrams in weight (the size of a few grains of sand). This analysis showed that the Deccan Traps erupted between 67 Ma and 63 Ma, a duration of only 5 Ma rather than the 50 Ma derived from earlier attempts. More recent work has broadly confirmed the short time-scale of eruptions and more importantly has shown that the basalts initiated around 68 Ma, reached their peak eruption rate within 500 000 years of the end of the Cretaceous and died out around 63 Ma. The total volume of the Deccan Traps is difficult to estimate precisely because a great deal may have been eroded, but estimates range from 5×10^5 km^3 to 2.6×10^6 km^3, similar to those for other Cretaceous LIPs. We will assume an erupted volume of 1×10^6 km^3 for the following discussion but the exact amounts do not affect the final conclusions.

As we saw in the earlier part of this Chapter, the strongest volcanic effects upon climate are felt as a result of the releases of the gases including HCl, HF, SO_2 and CO_2 (Figure 7.6). SO_2 causes short-term cooling as it hydrates to form H_2SO_4 aerosol droplets in the stratosphere. CO_2 gas can contribute to weakly acid rain but the amounts released during individual eruptions are too small to cause significant atmospheric effects. The effects of volcanoes which we consider important today, such as dust and cooling which lasts a few years, are not significant on geological time-scales. Even large flood basalt flows such as the Roza flow in the Columbia River flood basalt did not release sufficient dust and debris to cause significant climatic effects. In the longer term, continued elevated levels of flood basalt volcanism might contribute significant amounts of CO_2, causing a warming effect.

7.3.1 The effects of SO$_2$ released by Deccan volcanism

We can probably assume that SO_2 released from oceanic eruptions during the mid-Cretaceous pulse in ocean spreading rates would have had little long-term effect upon climate. The eruptions were dominated by submarine eruptions and recycling of sulfur within the spreading ridge system would have prevented much reaching the atmosphere. However, as we saw in Section 7.1.4, larger effusive eruptions on the continents, like the 1783–4 eruption of Laki in Iceland, can have global effects. The difficulty comes in scaling this up to something the size of the Deccan Traps. If we make the simplifying assumption that the main part of the Deccan, around 1×10^6 km^3, was erupted within half a million years, that corresponds to an eruption rate of 2 km^3 per annum. This is far slower than the eruption rate of the Laki eruption of 1783–4, which erupted around $14 \, km^3$ in less than one year (and of the order of $12.5 \times 10^{10} \, kg \, SO_2$). Remember however that the Deccan did not erupt continuously but possibly erupted as enormous pahoehoe sheet flows, each lasting a decade or more. The eruptions would have been separated by long periods of inactivity in much the same way as eruptions on Iceland. There are very roughly 500 flows in the main part of the Deccan which indicates only around one major flow every millennium. The volumes of such flows can be calculated by dividing the total volume by the number of flows and the average flow may have been $2000 \, km^3$, larger than the Roza flow of the Columbia River flood basalts which reached around $1300 \, km^3$.

If a $2000 \, km^3$ pahoehoe-style sheet flow took a decade to erupt, it would have an annual eruption rate of $20 \, km^3 \, yr^{-1}$, an order of magnitude greater than Laki. If it had similar sulfur contents to the lava erupted by Laki, it may have pumped some $1.8 \times 10^{12} \, kg \, SO_2$ into the atmosphere per annum, over 100 times the present-day volcanic release rate (Box 7.1). This would have been devastating to the local environment but would it have had global effects? The climatic effects would have depended upon two crucial factors:

1 The height of the tropopause and height of the volcanic plume during the eruption. Had this been a present-day eruption and Deccan had been situated close to the tropics, as indeed it was 65 million years ago, the tropopause would be around 18 km high. The thermal plume above the Laki eruption probably reached 10–14 km, and although this penetrated the stratosphere close to the pole, it would be well below the tropopause and would not have caused global climate effects. Although the Deccan eruptions may have seen higher fire fountains and plumes, the hotter later Cretaceous world would also be expected to see a higher tropopause. We simply cannot determine whether the Deccan volcanism would have reached the stratosphere.

2 Had the SO_2 reached the stratosphere, the effects might still have been limited by water exhaustion. As we saw in Section 7.1.3, when large eruptions pump more SO_2 into the stratosphere it becomes hydrated to H_2SO_4 which causes global cooling by absorbing solar radiation. However, this process can only completely hydrate the volcanic SO_2 if there is sufficient H_2O in the stratosphere. Water exhaustion would limit the initial effects of the volcanic eruption but might prolong the overall duration of any global climatic effects to decades.

Although it looks as if SO_2 emissions may have had the potential to disrupt climate during the major Deccan eruptions, this would only have happened if the SO_2 in the plume reached the stratosphere. Moreover, in the following years, long periods of inactivity (almost 1000 years) mean there was ample time for the global climate system to recover between eruptions. In order to quantify the effects of SO_2 release from Deccan upon the climate, we will have to determine whether the volcanic plumes reached the stratosphere and how much SO_2 was released. The hypothesis that Deccan eruptions might lead to rapid cooling, global catastrophe and mass extinction is not disproven. However, in the absence of evidence for the quantity of SO_2 released into the stratosphere, the height of the tropopause in ancient times, and the effects of water exhaustion upon the lifetime of H_2SO_4 aerosols in the stratosphere, the hypothesis remains unproven.

7.3.2 The effects of CO_2 released by Deccan volcanism

SO_2 emissions have the potential to disrupt climate because the amounts released are large in relation to the amounts normally in the atmosphere, but CO_2 emission is very different. The amounts of CO_2 released during effusive eruptions are of the same order as other sources, even for large Deccan eruptions. The Laki eruption in 1783–4 released around 3×10^{10} kg CO_2, but that was probably only one-tenth of the normal annual release from other volcanic sources such as mid-ocean ridges and subaerial volcanoes (see Box 7.1). Deccan flows were releasing perhaps an order of magnitude more per annum (around 3×10^{11} kg CO_2) but that is still only around 1.5% of the current annual output from the industrialized world (about 2×10^{13} kg yr^{-1} CO_2). Remember that each of the Deccan flows may only have continued for a decade, so industrial and agricultural CO_2 release is currently outdoing the Deccan by almost two orders of magnitude and we have been doing it for longer than a decade.

The total CO_2 released by Deccan may have been in the range 2.6×10^{15} kg to 9×10^{15} kg and models including feedbacks for the carbon cycle indicate that Deccan may have increased the CO_2 level of the atmosphere by as much as 75 ppm leading to global warming of the order of 1–2 °C, too little to have caused global climate change and a major extinction at the Cretaceous/Tertiary boundary.

○ Having dismissed the effects of CO_2 emission from the Deccan, does this affect the conclusions we drew that CO_2 emissions from volcanism during the superplume event in the mid-Cretaceous contributed to global climate change?

● Certainly not. All the calculations above were premised upon sudden but short-lived increases in CO_2 emission. Remember that during the mid-Cretaceous, global spreading rates increased from around 18–20 km^3 yr^{-1} to 35 km^3 yr^{-1}, an increase which was sustained for around 40 million years.

Although the short-term effects of CO_2 released by Deccan volcanism seem unlikely to have been a major cause of climate disruption at the end of the Cretaceous, we should mention an alternative hypothesis for the apparent correlation between flood basalt volcanism and mass extinction events. In Chapter 6, we noted that methane clathrates (methane gas trapped in ice) are a major reservoir for carbon in the oceans. The sudden release of methane from clathrates in ocean floor sediments could occur if the oceans warmed significantly or ocean circulation patterns were disrupted so that bottom waters warmed. This hypothesis has been advanced on the basis of carbon-isotope changes in carbonates and organic sediments for the large Karoo–Ferrar Province which erupted during the Toarcian Age (Early Jurassic) 183 million years ago, and the North Atlantic flood basalts which erupted 60–55 million years ago in the early Palaeogene. However, although the climate was cooling, the seas were still warm at the end of the Cretaceous and there may not therefore have been a large reservoir of methane in ocean floor clathrates.

7.4 Summary

- Present-day volcanic emissions include dust, ash and several gases including H_2O, SO_2, CO_2, CO, H_2, HCl and HF which absorb solar radiation to varying extents, causing cooling of the underlying atmosphere. Although volcanic plumes can have extreme effects locally, dust and many of the gases such as HCl and HF, are rained out of the atmosphere within a few days. Two gases can have longer atmospheric lifetimes, CO_2 and SO_2, and both have the potential to cause dramatic changes in climate.

- The structure of the atmosphere plays a crucial role in determining the global effects of even the largest volcanic eruptions. Volcanic plumes in the troposphere are rapidly 'rained out' and although local effects may be severe, they do not affect the global climate. Volcanic plumes which pump significant volumes of SO_2 into the stratosphere have significantly greater effects since the aerosols can reduce solar input over several years.

- The historical volcanic eruptions of Pinatubo (1991) and Laki (1783–4) can be used to characterize explosive and effusive volcanism in the past. Although there are many anecdotal accounts of more extreme eruptions in early human history, extrapolation to quantitative climatic effects is very difficult because records are sparse and uncorroborated.

- Explosive volcanism like that of Pinatubo in 1991 rapidly pumps large amounts of SO_2 into the stratosphere, causing a reduction in radiative forcing and thus global cooling, but there is a limit to this effect. Larger eruptions may exhaust water in the stratosphere, limiting the formation of H_2SO_4 aerosols. This may reduce the immediate effects but prolong the lifetime of the stratospheric SO_2 effect. Water exhaustion may be the reason why cooling after the huge Toba eruption 75 000 years ago was not more pronounced.

- Effusive volcanism like the Laki eruption on Iceland in 1783–4 also pumps SO_2 into the stratosphere but such effects may be limited to eruptions at high latitude where the base of the stratosphere is lower. However, effusive eruptions generally release more SO_2 than explosive eruptions, last longer, and thus prolong the effects.

- The structure of the Pacific and Atlantic Ocean floors is very different. While the Atlantic has spread symmetrically about its centre, the Pacific spreading ridge is offset to the east by subduction beneath North and South America. The oldest Jurassic-age Pacific Ocean crust lies in the western Pacific but it is covered in Cretaceous volcanic plateaux, and volcano chains, the result of a superplume event in the mid-Cretaceous.

- The mid-Cretaceous superplume resulted in rapid spreading rates in the Pacific, initiating around 120 Ma ago and lasting for around 40 Ma. Two types of ocean plateau formed in the Pacific synchronously with the rapid ocean floor spreading. The Ontong–Java Plateau formed at 122 Ma, in around one million years, in a very similar manner to continental flood basalts, probably as a result of the impact of a mantle plume at the base of the ocean lithosphere. The Mid-Pacific Mountains formed over a period of at least 10 million years as the oceanic crust spread above a plume much like modern-day Iceland.

- Several large igneous provinces erupted both in the oceans and on land during the Cretaceous. Those on land are known as continental flood basalts, and they are generally spatially and temporally associated with continental break-up. For example, the Parana–Etendeka basalts erupted as South America and Africa rifted apart around 132 Ma. The most famous continental flood basalt, the Deccan Traps in India, erupted at the end of the Cretaceous, between 68 and 63 Ma ago with its peak eruption rates within 500 000 years of the end of the Cretaceous.

- It seems clear that the superplume volcanism did play an important role in the climate of the mid-Cretaceous, but the case for Deccan causing sudden climatic effects at the end of the Cretaceous is much less clear. If the Deccan flood basalt eruptions did cause significant climatic effects, it would have been cooling caused by the production of SO_2 which penetrated the stratosphere. CO_2 production by Deccan eruptions was less than current industrial production, and would only have raised the global temperature by 1–2 °C.

7.5   References and further reading

References

BAKER, V. R. AND MARUYAMA, S. (2001) 'Tharsis superplume: implications on the role of water, environmental change, and life', *Geological Society of America annual meeting*, Paper **178–0**.

HAQ, B. U., HARDENBOL, J. AND VAIL, P. R. (1987) 'Chronology of fluctuating sea levels since the Triassic', *Science*, **235**(4793), 1156–1167.

HARDARSON, B. S., FITTON, J. G., ELLAM, R. M. AND PRINGLE, M. S. (1997) 'Rift relocation — a geochemical and geochronological investigation of a palaeo-rift in northwest Iceland', *Earth and Planetary Science Letters*, **153**, 181–196.

KOPPERS, A. P. A., STAUDIGEL, H., WIJBRANS, J. R. AND PRINGLE, M. S. (1998) 'The Magellan seamount trail: implications for Cretaceous hotspot volcanism and absolute Pacific plate motion', *Earth and Planetary Science Letters*, **163**, 53–68.

LARSON, R. L. (1991) 'Geological consequences of superplumes', *Geology* **19**(10), 963–966.

NEAL, C. R., MAHONEY, J. J., KROENKE, L. W., DUNCAN, R. A. AND PETTERSON, M. G. (1997) 'The Ontong Java Plateau', in *Large Igneous Provinces: Continental, Oceanic, and Planetary Flood Volcanism*, Vol. 100, American Geophysical Union.

SELF, S., KESZTHELYI, L. AND THORDARSSON, T. (1998) 'The importance of pahoehoe', *Annual Review of Earth and Planetary Sciences*, **26**, 81–110.

STOREY, B. C. (1995) 'The role of mantle plumes in continental breakup: case histories from Gondwanaland', *Nature*, **377**, 301–308.

THORDARSSON, T. AND SELF, S. (1993) 'The Laki (Skaftár Fires) and Grímsvötn eruptions in 1783-85', *Bulletin of Volcanology*, **55**, 233–263.

THORDARSSON, T. AND SELF, S. (1998) 'The Roza Member, Columbia River Basalt Group: A gigantic pahoehoe lava flow field formed by endogenous processes?', *Journal of Geophysical Research — Solid Earth*, **103**(B11), 27411–27445.

VAUGHAN, A. P. M. (1995) 'Circum-Pacific mid-Cretaceous deformation and uplift: A superplume related event?' *Geology*, **23**, 491–494.

Further Reading

COURTILLOT, V. (1999) *Evolutionary Catastrophes — The Science of Mass Extinction*, Cambridge University Press.

HALLAM, A. AND WIGNALL, P. B. (1997) *Mass Extinctions and their Aftermath*, Oxford University Press.

SIGURDSSON, H. (2000) *Encyclopedia of Volcanoes*, Academic Press.

8   The operation of the major geological carbon sinks

Bob Spicer and Peter W. Skelton

In Chapters 6 and 7, you considered the geological sources for the high flux of carbon through the Cretaceous world. We now turn to the ultimate geological sinks — buried organic and carbonate carbon — and the circumstances that sustained the flux of carbon into them throughout the period. As before, we will deal first with the terrestrial record (Section 8.1) and then move on to the marine record (Section 8.2).

8.1   Climate and terrestrial carbon sinks

In Chapter 1, geological evidence from the terrestrial sediments of the Arctic and the carbonate factories of the seas highlighted just how different the Cretaceous greenhouse world was to our present, relatively cold, world. Nevertheless, rich and informative as these sources of evidence may have been, they only provided isolated snapshots of the Cretaceous world. Subsequently, we have looked at more comprehensive data from both marine and terrestrial realms but we are now going to examine the bigger picture, by bringing geological information together from across the globe.

A useful review of Cretaceous geological data was published in 1995 by a Russian research group headed by Nicolai Chumakov of the Geological Institute of the Russian Academy of Sciences. In their 'Warm Biosphere' programme, the group compiled palaeontological, sedimentological and mineralogical data to provide as detailed a pattern as possible of biotic and palaeoclimatic patterns throughout the Cretaceous.

8.1.1   Terrestrial organic carbon sequestration — compiling the geological data

Because it represents one of the warmest intervals during the Cretaceous, we start by looking at the Cenomanian time slice. Figure 8.1 shows the Arctic region in dark green, described as part of what the Russian workers called the 'Northern High-latitude Temperate humid (NHT)' zone. There are abundant coals (black rectangles), and symbols representing moderately thermophilic (warmth-loving) vegetation and insects. There are no quantified temperatures, but the overall pattern is consistent with what we have seen so far in Chapter 4. Southward is the Northern Mid-latitude Warm humid (NMW) belt shown in light green. The evidence for this being distinct from the NHT is in the form of thermophilic vegetation and insect remains, coals, and an abundance of bauxites (which form through chemical weathering under warm and moist climates). There are also dinosaur remains. However, there is still some debate over the physiology of these animals, the extent to which some may or may not have had external insulating coverings such as feathers or even a form of fur, and their ability to migrate between different climatic regimes. Their use as palaeoclimatic indicators is thus limited.

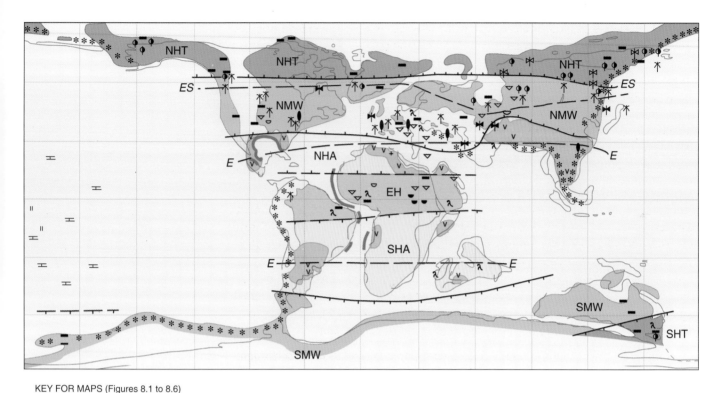

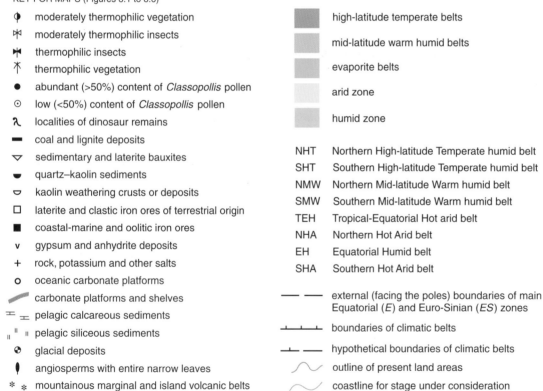

KEY FOR MAPS (Figures 8.1 to 8.6)

☥	moderately thermophilic vegetation
⋈	moderately thermophilic insects
⋈	thermophilic insects
⋏	thermophilic vegetation
●	abundant (>50%) content of *Classopollis* pollen
⊙	low (<50%) content of *Classopollis* pollen
λ	localities of dinosaur remains
▬	coal and lignite deposits
▽	sedimentary and laterite bauxites
◖	quartz–kaolin sediments
▽	kaolin weathering crusts or deposits
☐	laterite and clastic iron ores of terrestrial origin
■	coastal-marine and oolitic iron ores
v	gypsum and anhydrite deposits
+	rock, potassium and other salts
o	oceanic carbonate platforms
◢	carbonate platforms and shelves
⹀	pelagic calcareous sediments
‖	pelagic siliceous sediments
❂	glacial deposits
❚	angiosperms with entire narrow leaves
* *	mountainous marginal and island volcanic belts

▨	high-latitude temperate belts
▨	mid-latitude warm humid belts
▨	evaporite belts
▨	arid zone
▨	humid zone

NHT	Northern High-latitude Temperate humid belt
SHT	Southern High-latitude Temperate humid belt
NMW	Northern Mid-latitude Warm humid belt
SMW	Southern Mid-latitude Warm humid belt
TEH	Tropical-Equatorial Hot arid belt
NHA	Northern Hot Arid belt
EH	Equatorial Humid belt
SHA	Southern Hot Arid belt

— —	external (facing the poles) boundaries of main Equatorial (*E*) and Euro-Sinian (*ES*) zones
⊢⊢⊢	boundaries of climatic belts
⊢⊢	hypothetical boundaries of climatic belts
⌇	outline of present land areas
⌇	coastline for stage under consideration

Figure 8.1 Map of the Cretaceous in the Cenomanian according to a Russian compilation of geological data for that time interval. *(Chumakov et al., 1995.)*

Equatorward of the NMW, the evidence for aridity increases. This evidence is in the form of gypsum, anhydrite and other evaporite deposits. We also find carbonate shelves at the southern end of the seaway that covered the western interior of North America. This Northern Hot Arid (NHA) zone passes southward into a more humid equatorial region (EH) marked by more coals, bauxites, laterites (which again form under hot humid conditions) and dinosaurs. Palaeobotanical evidence for rainforests is, however, lacking except in what is now Colombia and South-East Asia.

Here, the Cretaceous humidity was very high due to wet air blowing over land from the nearby ocean. Elsewhere in the tropics though, there is no good geological evidence for constantly wet conditions. No coals are marked on the map for this region. The evidence for high organic terrestrial productivity in Colombia is marked by extensive mid-Cretaceous organic-rich shallow-marine shales where the organic material was clearly derived from land vegetation because of the characteristic signatures of terrestrial vegetation preserved in the organic geochemistry. The rest of the tropical vegetation had more the appearance of modern-day open savanna, but instead of grass (which had not yet evolved) there were open communities of ferns and cycads.

The pattern of climatic belts is repeated in the Southern Hemisphere. Offshore drilling around Antarctica indicates high organic productivity on land while pollen and other palaeobotanical evidence suggests a productive conifer-dominated forest in the Southern High-latitude Temperate humid zone (SHT). In particular, fossil floras from the Clarence Valley (north end of the South Island of New Zealand) suggest a Southern Hemisphere analogue of the forests we saw in Alaska and Russia. CLAMP analysis of the Clarence Valley woody dicots gives a mean annual temperature of 10 °C which, like the Peruc example (Section 4.5.2), is similar to nearby sea-surface temperatures obtained from oxygen-isotope data.

Figures 8.2 to 8.6 show maps for other time slices in the Cretaceous. The Berriasian map (Figure 8.2) shows an unopened southern Atlantic Ocean (Section 2.2.1) and therefore one large continental mass across the equatorial region and at low southern latitudes. Although the included data are thin in the South America/Africa landmass, the region is drawn as largely hot and arid or sub-humid and there are no reported coals. However, note that there appears to have been a thriving dinosaur population along the eastern margin of South America. Given that these animals would require water, this does at least indicate that the region was not an extreme desert and that freshwater may have been present as streams and lakes (though possibly ephemeral). The streams are likely to have been sourced on the mountainous flanks of the rift valley that was subsequently to become the southern Atlantic Ocean.

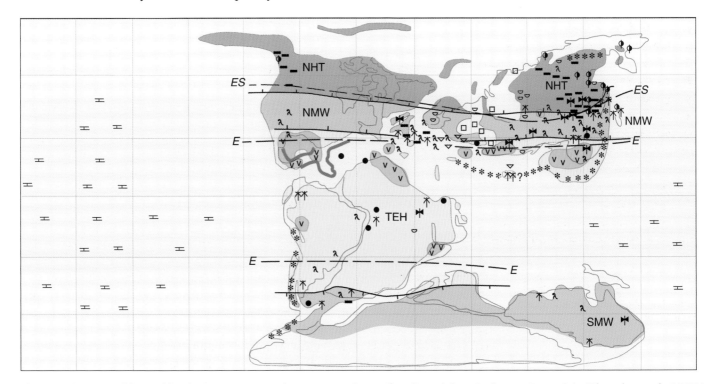

Figure 8.2   Map of the world in the Berriasian according to a compilation of geological data. For key see Figure 8.1. *(Chumakov et al., 1995.)*

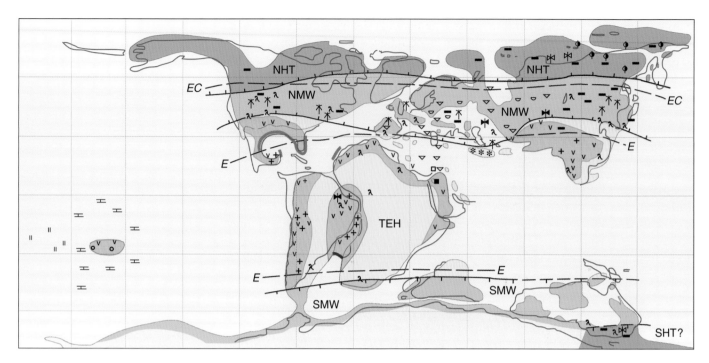

Figure 8.3 Global map for the Aptian showing geological data indicating climate. For key see Figure 8.1. *(Chumakov et al., 1995.)*

The clear message from the data for the Berriasian and Aptian time slices is that there is no evidence of tropical humidity high enough to have supported rainforests. This is in stark contrast to the modern world where terrestrial productivity is greatest in the tropics. We do see evidence for a more humid tropical environment in the Albian (Figure 8.4), however, as the proto-South Atlantic rift valley became flooded by ocean waters and introduced a source of moisture into the interior of the large tropical continental mass. In Albian times, bauxites and coals formed along the Equator, suggesting a widespread increase in equatorial moisture.

In Aptian times (Figure 8.3), this rift valley is shown to have accumulated abundant evaporites as the margins rose even further and the valley floor became even more arid (see also Figure 2.11). This increased aridity is likely to have been the result of air moving across the N–S trending valley and being forced to rise over the elevated flanks of the rift valley where it cooled and lost moisture as rain. This now drier air would have warmed as it descended into the valley so decreasing the relative humidity even further. The warming of descending air is caused in part by compressional heating (as we saw in the case of the modern temperature profiles in the Brooks Range) and, because warm air is capable of holding more moisture than cold air, the air becomes drier. Note however that some dinosaurs have even been found in this area from this Aptian period of pronounced aridity.

○ What other factors in the polar regions might have led to an increase in tropical humidity?

● We have already said that in today's 'icehouse' world of cold poles, the tropics are humid. It could be that if the Cretaceous polar regions cooled (Section 5.4), there would be an increase in equatorial moisture. However, we see no equatorward movement of the latitudinal position of the high-latitude temperate zones in Figures 8.5 (Santonian) and 8.6 (Maastrichtian), so it appears at this scale, and from these time slices, that polar cooling was not a significant cause of increased tropical humidity.

This band of relatively humid tropics persists throughout the rest of the Cretaceous and appears to get broader when the area of the southern Atlantic as a moisture source increases. However, this humid zone was still dry compared to the tropics of today. The evidence for this comes not from the sediments but from plant fossils.

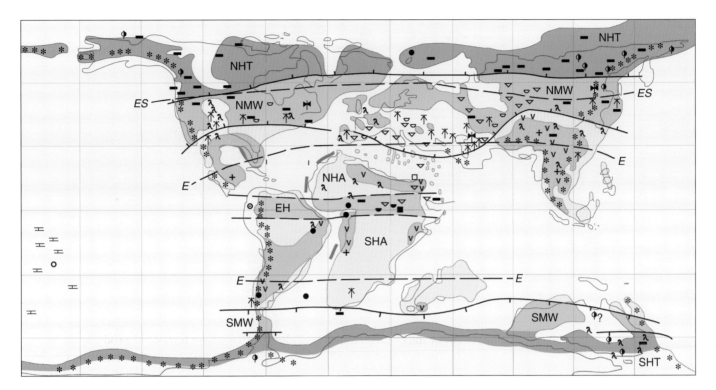

Figure 8.4   Global map for the Albian showing geological data indicating climate. For key see Figure 8.1. *(Chumakov et al., 1995.)*

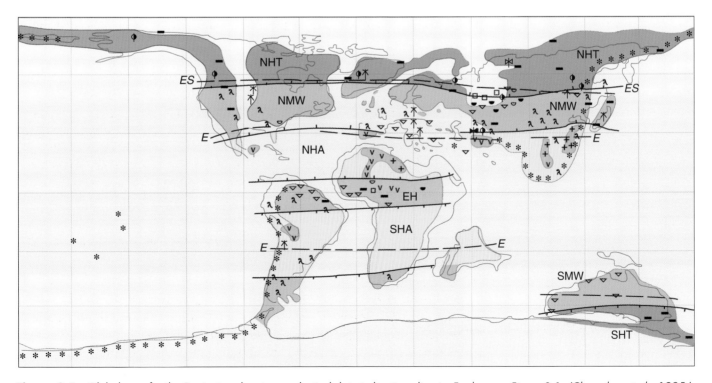

Figure 8.5   Global map for the Santonian showing geological data indicating climate. For key see Figure 8.1. *(Chumakov et al., 1995.)*

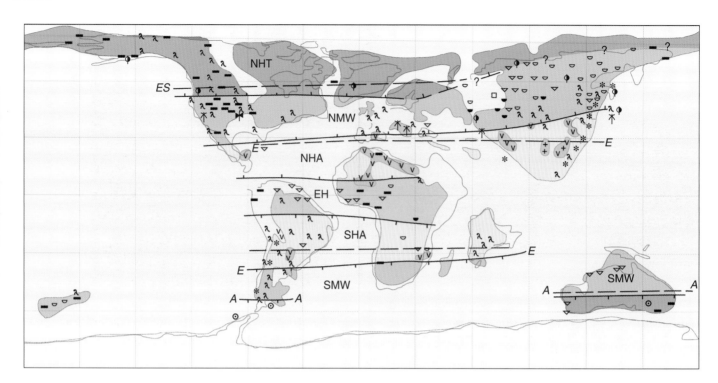

Figure 8.6 Global map for the Maastrichtian showing geological data indicating climate. For key see Figure 8.1. (Chumakov et al., 1995.)

Figure 8.7 (a) Light micrograph of an attached group of four *Classopollis* pollen grains. (Dave Jolley, University of Sheffield.) (b) SEM photograph of *Classopollis* grains. Each grain is about 50 μm in diameter. (Bob Spicer, Open University.) (c) Photograph of a branch of *Frenelopsis*. (Joan Watson, University of Manchester.)

One of the strongest phytogeographic signals that emerged from the study of Mesozoic pollen grains was the existence from the late Triassic to late Cretaceous times of a broad zone roughly between 40° N to 40° S that consistently produced large quantities of a pollen grain called *Classopollis* (Figure 8.7a, b). This grain is easy to identify because when viewed in the light microscope it has a broad equatorial band of striations (Figure 8.7a). These striations are not seen on the surface when viewed under a scanning electron microscope (SEM) (Figure 8.7b), which shows that they are an internal feature of the pollen wall. What is visible in the SEM is a sub-equatorial furrow, a triangular mark at one end of the grain and a circular pore at the other. Often these grains were shed not singly but as a group of four attached at the triangular mark. These grains are particularly common in sediments containing the foliage remains of conifers called *Frenelopsis* (Figure 8.7c)

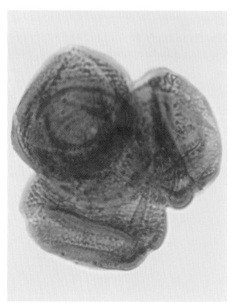

(a)

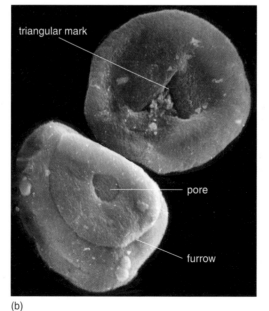

triangular mark

pore

furrow

(b)

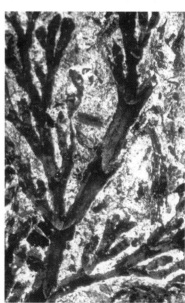

(c)

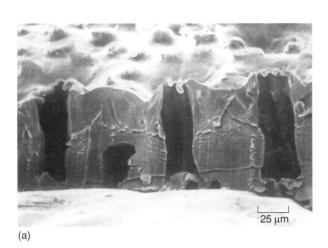

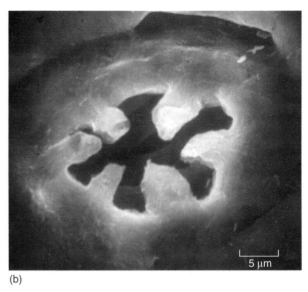

(a)

(b)

Figure 8.8 (a) Section through the cuticle of *Pseudofrenelopsis* showing depth of stomatal pits. *(Joan Watson, University of Manchester.)* (b) SEM photograph of a *Pseudofrenelopsis* stoma with the guard cells at the bottom of a pit overarched by finger-like projections. *(Bob Spicer, Open University.)*

or *Pseudofrenelopsis*, and when the grains were found in male cones of these conifers, the link to the extinct conifer family Cheirolepidiaceae was established. The foliage of the Cheirolepidiaceae is somewhat variable, and undoubtedly the family included some species that favoured a wide range of environments. However, many forms, such as those belonging to the genus *Frenelopsis*, have highly reduced leaves, and specialized stomata sunk in the bottom of deep pits within thick cuticles (Figure 8.8a, b). All these features are typical of plants living in arid situations where water conservation is at a premium. Moreover, the small triangular leaves had fringes of short hairs on which water droplets could have nucleated from moist air, such as might have been associated with sea fogs and overnight dew. This water could have been absorbed by the plant through its foliage shoots.

These xeromorphic (literally meaning 'dry form') features are seen in other plants, even ferns. A good example is *Weichselia* (Figure 8.9) which had a widespread distribution across western Europe in the early Cretaceous and occupied in part the Northern Mid-latitude Warm humid belt in Figures 8.2 and 8.3 (Berriasian to Aptian).

Weichselia was a large fern that seems to have formed extensive heaths. The community was apparently maintained in an open state by frequent wildfires. The evidence for this is that we find many *Weichselia* remains preserved as fossil charcoal, sometimes called 'fusain'. This material has the exact texture of charcoal and once formed is biologically inert and so, despite its fragility, has a high preservation potential. *Weichselia* fronds preserved in this way reveal that this plant had a very thick cuticle that even extended down into the cell layer beneath the outer epidermis. Not only that, but the leaves also had sunken stomata and the sporangia were protected by a thick outer covering that apparently protected the shed sporangia from fire. The evidence for this is that clusters of sporangia complete with the covering are often found uncharcoalified. They were obviously easily detached from the parent plant before the fire 'season' and gave rise to the next generation of ferns after the burnt landscape had been effectively cleared of vegetation. Such adaptations to, and even exploitation of, frequent fire disturbance — what we might call a 'fire ecology' — are widespread in Cretaceous communities, and fire was a powerful agent in shaping Cretaceous vegetation globally.

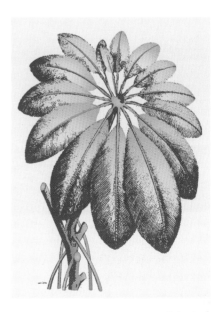

Figure 8.9 A reconstruction of the fern *Weichselia. (After Alvin, 1971.)*

○ If fire was so frequent, what implications does this have for carbon sequestering?

● It is very unlikely that significant amounts of carbon entered long-term geological storage as most of the biomass would have been fire-oxidized to CO_2 and returned to the atmosphere directly. However, the remaining charcoal is biologically inert and so cannot be subsequently oxidized by bacterial action. This carbon would enter the geological storage system but as dispersed fusain particles, not as bulk carbon deposits such as coal.

From the xeromorphism exhibited by many low and mid-latitude genera such as *Frenelopsis* and *Weichselia*, it is clear that the description 'humid' used in the Equatorial Humid and Mid-latitude Warm humid zones of the Warm Biosphere maps is only relative to the dryness of the arid zones. Clearly, water was only temporarily available, interspersed with periods of drought. Examination of the conifer tree rings produced in these climates tells us more, not only about the seasonality, but also the interannual variability of growth conditions. Fossil trees from the Lower Cretaceous of southern England show uneven sequences of ring development indicating fluctuations in growing conditions over several growth periods, and in view of the relatively low latitudes the cause is likely to be variations in water availability rather than temperature. Nevertheless, rings are formed with some regularity suggesting a more or less annual cyclicity or seasonality in wet/dry variations. Some rings, however, are imperfectly formed and frequent false rings suggest interruptions to growth during the growing season. Thus, although there is a broad annual cyclicity in moisture supply, this also varied markedly over time and perhaps there could have been several years in succession when the rains failed to materialize.

Evidence for extended intervals of moist conditions is also present. Some tree rings at some stratigraphical horizons (e.g. the Barremian of the Isle of Wight) show wider, more evenly spaced, rings perhaps suggesting periods of higher or more prolonged rainfall. Particular forms of fossil fungi found in fossil woods suggest periods of rotting in moist subaerial conditions before burial, and fungi found on the surface of some cheirolepidiaceous conifer leaves imply periods of relatively high atmospheric humidity. This is perhaps consistent with the functional interpretation of the leaf margin hairs as a nucleation mechanism for extracting moisture directly from the atmosphere.

The picture that emerges from the overall geological compilations and the features of vegetation is that the conditions under which globally significant amounts of carbon were sequestered on land, leading to the formation of peats that were transformed into coal, were climatically restricted to the high latitudes. Low latitudes were too dry except in a few spatially and temporally restricted settings where locally moisture was available borne by winds from nearby oceans, or the water table was high due to tectonic and sea-level conditions. Much of the terrestrial carbon that may have accumulated in the drier mid- and low-latitude settings was rapidly oxidized and returned to the atmosphere by decay and, more especially, by fire. The Cretaceous was a time when wildfires dominated terrestrial ecosystems and, as we have seen from the Arctic data, even occurred to some extent in settings of peat accumulation. On balance, though, in the wet and relatively warm polar regions, organic accumulation outstripped oxidation.

8.1.2 How much terrestrial organic carbon sequestering took place?

Estimates by the United States Geological Survey put the predominantly late Albian and Cenomanian coal reserves on the Arctic Slope of Alaska at 2.75×10^{12} tonnes. Similarly large deposits of Cretaceous coal are known from across north-eastern Russia while smaller accumulations occur in western Canada, Svalbard, the western interior of the US, Germany, Hungary, Spain, northern China, North Africa, Australia and New Zealand. Unfortunately, accurate assessments of the quantity of carbon preserved as coals do not exist for these areas. Even less well known is the amount of Cretaceous coal in Antarctica. The offshore evidence for high organic productivity indicates that under the present ice cap there are undoubtedly large accumulations of Cretaceous coal, particularly in East Antarctica where, in a passive margin setting, peat development would have been undiluted by sediment influx.

Coal bed thickness, lateral continuity and composition were determined largely by the factors that controlled the conditions in the mire when the peat originally formed. Such factors were the type and rate of vegetation growth, conditions affecting rate of organic decay, the level of the water table and the rate of clastic sediment input. Broadly speaking, these were a function of the biology of the organisms concerned, the climate, and the tectonic context. In the polar regions, the highly seasonal light and temperature regime resulted in abundant summer productivity, autumnal shedding of a large proportion of this productivity into the mire environment, and winter temperatures that were sufficiently low (i.e. less than 4 °C) to curtail bacterial decay. These elements would not, however, have led to the long-term sequestering of carbon were it not for the depositional setting. In areas like northern Alaska, climatic and biological factors were coupled with compactional subsidence of prograding delta floodplains and a foreland basin setting that led to high preservation potential.

Other accumulations in North America took place along the western margins of the Western Interior Seaway where basins formed as a result of thrusting and crustal loading within the Western Cordillera. Here, mires could develop on the extensive coastal floodplains associated with epicontinental seas formed during eustatic sea-level highstands. Similar highstand coastal mire settings occurred in North Africa and Europe, but in each case long-term organic preservation was facilitated by tectonic subsidence.

8.1.3 Carbon sequestering by silicate weathering

The large-scale chemical weathering of silicate minerals is often regarded as a major cause of atmospheric carbon dioxide drawdown (Section 6.2.2). This weathering is most likely to occur in mountainous regions where erosion is constantly exposing fresh minerals to the weathering process but climatic conditions also have to be conducive to the process. Most importantly, water is essential for the weathering process. It is involved in the weathering chemistry itself and it serves as a transport medium carrying the carbonate ions from the weathering site to the oceans and hence long-term storage. It follows therefore that silicate weathering can only occur where there is significant precipitation.

○ From what you have read in Chapter 2 and what you know of Cretaceous climate patterns, where is most silicate weathering likely to have taken place?

● In mountainous regions at high latitudes, because it is only at high latitudes where conditions are likely to have been wet enough.

However, it is also worth remembering that temperatures also need to be well above freezing. In cold conditions, the weathering chemistry takes place so slowly that silicate weathering is insignificant. At high latitudes in the Cretaceous, even though the polar regions were much warmer than now, temperatures associated with even modest mountains of no more than 2 km elevation are likely to have been, on average, so low that silicate weathering would have proceeded only very slowly.

At low latitudes, temperatures were conducive to the silicate weathering chemistry but conditions were, on average, rather arid. Moreover, the Cretaceous world was one in which no large mountain belts existed (Section 2.3). The highest elevations were those associated with volcanic chains alongside subduction zones, the margins of rift valleys such as that associated with the opening Atlantic, or with doming of the crust associated with mantle plume development. Even so, the elevations involved were modest compared to those produced by collisional processes (e.g. the Himalayas). Seasonal or intermittent rainfall at low latitudes would have given rise to some silicate weathering, but overall weathering rates and carbon sequestration through the silicate weathering process may have been less than those of the modern world.

8.1.4 Carbon and climate

Climate is clearly fundamental in determining the locus and rate of terrestrial carbon sequestering. Its role in mire development is clearly associated both with providing the context within which plant growth can take place and with controlling rates of decay. Decay rates are linked to climate through temperature (e.g. seasonal cold restricting bacterial decay long enough for burial to isolate the organic matter from oxygen) and through rainfall (preventing oxidation by submergence in water). Indeed, there is a comparatively narrow range of combined temperature and precipitation conditions under which mires can develop and, in the case of raised mires, high precipitation has to occur year round to prevent organic oxidation. Climate is equally important in the silicate weathering process and again it is a combination of temperature and precipitation that is the key.

The compilation of geological evidence is, as we have seen, useful in providing the general patterns of relative wet and dry belts and relative warmth and cold. In general though, geological evidence can only provide us with temporally and spatially incomplete data, because Cretaceous rocks have not been deposited and preserved uniformly over the globe. Moreover, much of the data are qualitative and not quantitative. To provide a more complete picture of the pattern and process of the Cretaceous 'Warm Earth' climate system, we need to combine the observations from the rock record with a completely different approach of mathematically modelling the Cretaceous Earth system. This we will examine in more detail in Chapter 9.

8.2 Cretaceous oceanography and marine carbon sinks

In Chapters 6 and 7, the Cretaceous Period was characterized as a time of exceptionally high long-term carbon flux: carbon was evidently flowing both into and out of the atmosphere/ocean system at a much higher rate than today, at least at the geological time-scale (Sections 6.3.6 and 7.2.6). The extent of carbon burial on the land was discussed in Section 8.1. This Section explores carbon burial in the marine realm, and the oceanographic conditions that may have facilitated it.

8.2.1 Marine carbon sinks and the conditions that sustained them

In Chapter 6 (Table 6.1), you saw that the two main geological sinks for carbon are carbonate rocks (containing 'carbonate carbon', or C_{carb}) and organic-rich deposits (containing 'organic carbon', or C_{org}), and that the ratio of their global reservoirs today is estimated to be about 4 : 1. Apart from the brief episodes of enhanced C_{org} burial during the OAEs (Section 5.2), the stratigraphical record suggests that marine C_{carb} burial likewise outstripped C_{org} burial by several times during the Cretaceous. Swiss geochemist Helmut Weissert and colleagues (1998) have estimated that the ratio of C_{carb} to C_{org} burial in marine deposits fell from 6 : 1 in the Late Jurassic to 4.7 : 1 in the Early Cretaceous, but then rose again to 7 : 1 in the Late Cretaceous.

○ Based on what you have read in earlier Chapters, what were the dominant sinks for C_{carb} in the Cretaceous seas?

● As discussed in Chapter 5, the main accumulations of C_{carb} in the Cretaceous were carbonate platforms at low latitudes and, in the late Cretaceous, the Chalk, especially at mid-latitudes.

Palaeogeographical maps of the carbonate platforms (e.g. Figure 2.16) allow estimation of their total areas. For example, French stratigrapher Jean Philip and his colleagues (1995) provided the following estimates for the Tethyan/Atlantic platforms: Early Aptian, about $7.4 \times 10^6\,km^2$; Late Cenomanian (Figure 2.16), $9.7 \times 10^6\,km^2$; and Late Maastrichtian, $5.1 \times 10^6\,km^2$ (plus part of $1.4 \times 10^6\,km^2$ of mixed siliciclastic/carbonate platforms). Using other estimates for the mean thickness of the platform successions in question, we can derive volumetric estimates of the platform carbonates for the corresponding time-slices, and hence calculate their contribution to C_{carb} burial rates.

Let us consider the platforms that developed in the Early Aptian. Different platform successions vary considerably in thickness because of variation in the provision of accommodation space for their accumulation, as discussed in Chapter 3. In fact, values of total thickness vary widely from tens to hundreds of metres. A preliminary survey of several Lower Aptian platform successions yields a mean thickness of about 130 m. Unfortunately, the duration in absolute time of the Early Aptian is not well constrained, but it is currently estimated at about 4.5 Ma, which would give a mean accumulation rate of 130 m/4.5 Ma = approximately 29 m/Ma. Hence the mean volumetric rate of accumulation of $CaCO_3$ in the Tethyan/Atlantic carbonate platforms (assuming deposition of virtually pure carbonate) can be estimated at $29 \times 7.4 \times 10^{12}\,m^3/Ma = 2.146 \times 10^{14}\,m^3/Ma$, or $2.146 \times 10^8\,m^3/year$.

Assuming a density of $2.71 \times 10^3\,kg/m^3$ for $CaCO_3$ (a slight overestimate, as such limestones may have porosities of up to about 10%), and a value of 0.12 as the mass proportion of carbon (C) in $CaCO_3$, the rate of C_{carb} burial can be calculated as $2.146 \times 2.71 \times 0.12 \times 10^{11}\,kgC/year = 0.07 \times 10^{12}\,kg\,C/year$. A similar calculation for the Late Cenomanian yields an estimate of $0.09 \times 10^{12}\,kg\,C/year$ (Skelton *et al.*, 1997, Section 7.7.2). Given the many uncertainties involved in such calculations, these values should be considered only as rough approximations, although they do at least illustrate the order of magnitude concerned.

○ Refer back to Figure 6.4 and determine how these values for C_{carb} burial rates in the Tethyan/Atlantic carbonate platforms of the Early Aptian and Late Cenomanian compare, in proportional terms, to the global flux into carbonate sinks today.

● According to Figure 6.4, the global rate of burial of C_{carb} today (in all sinks) is estimated to be about 0.24×10^{12} kg C/year. Thus, the Tethyan/Atlantic carbonate platforms buried C_{carb} at about 0.29, and 0.38, times the present global rate during the Early Aptian and the Late Cenomanian, respectively.

Hence, these platforms alone constituted a major carbon sink, and, indeed, the total contribution of carbonate platforms was even greater, as we have been considering only those that formed in or around the Tethys and Atlantic Oceans.

○ What other major shallow carbonate platforms should also be considered?

● Those of the Pacific guyots, which were mentioned in Section 1.1.5.

Our knowledge of these now deeply sunken Pacific platforms is still too scanty for us to quantify their contribution to C_{carb} burial, although, judging from the size of the oceanic promontories on which they formed (Figure 7.21), it must also have been fairly large.

Given the relatively high levels of CO_2 that have been postulated for the Cretaceous atmosphere (Section 6.3), the rapid formation of such vast amounts of $CaCO_3$ poses an apparent chemical paradox, for which we now need to consider some aspects of ocean chemistry in a little more detail.

In natural waters, including seawater, the inorganic carbon (i.e. that which is not incorporated in organic compounds) is contained in four chemical species, namely dissolved CO_2 gas (aqueous CO_2, or CO_2(aq)), carbonic acid (H_2CO_3), and bicarbonate (HCO_3^-) and carbonate (CO_3^{2-}) ions. Together, these four species make up what is referred to as total dissolved inorganic carbon, usually abbreviated to 'ΣCO_2' (the Greek letter Σ standing for 'sum of'). The balance between the different chemical species depends upon the pH of the seawater (i.e. how acid or alkaline it is, recorded as an inverse measure of the concentration of hydrogen ions, H^+) (Figure 8.10).

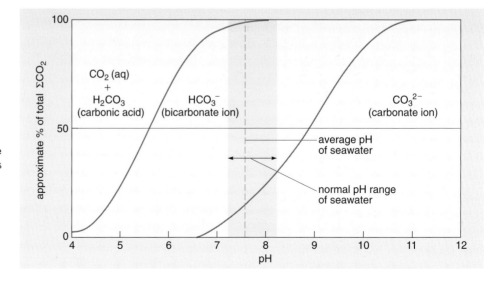

Figure 8.10 The approximate relative proportions of the main chemical species of dissolved inorganic carbon vary with pH in natural waters. Average pH of open seawater today is about 7.7 (dashed vertical line), but may range between 7.2 and 8.2. Note that the positions of the curved lines also vary with temperature, salinity and pressure.

○ Which is the dominant chemical species in seawater?

● Bicarbonate (HCO_3^-) ions, which make up between 70% (at pH 8.2) and 88% (at pH 7.2) of the ΣCO_2 over the usual pH range of seawater.

Thus, in normal seawater, the content of $CO_2(aq)$ is very small, while that of H_2CO_3 is invariably negligible (regardless of pH). Hence, to a first approximation, ΣCO_2 consists largely of bicarbonate ions with a smaller component of carbonate ions. Nevertheless, the trace of $CO_2(aq)$ in surface waters is important to our concerns, because that is the reservoir that directly equilibrates over short time-scales with atmospheric CO_2. The chain of reactions relating atmospheric CO_2 to dissolved inorganic carbon is thus as follows:

$$\text{atm } CO_2 + H_2O \rightleftharpoons CO_2(aq) + H_2O \rightleftharpoons H_2CO_3 \rightleftharpoons H^+ + HCO_3^- \rightleftharpoons 2H^+ + CO_3^{2-}$$
$$(8.1)$$

As the atmospheric reservoir of CO_2 is tiny compared with the total dissolved inorganic carbon in the oceans (Table 6.1), relatively small changes to the latter can have a big impact on the former. A small reduction of pH (i.e. to more acid conditions, with more H^+ ions), for example, will push the equilibria in Equation 8.1 towards the left, so releasing CO_2 into the atmosphere.

The apparent paradox referred to above concerns the relationship between calcification and the atmospheric content of CO_2. Calcification in seawater today is almost entirely mediated by shell-producing organisms. From Figure 8.10, we see that most of the carbonate that ends up in the solid $CaCO_3$ must come from bicarbonate ions, involving a shift to the right in the following equation (which you previously encountered in Section 6.2.2):

$$Ca^{2+} + 2HCO_3^- \rightleftharpoons CaCO_3 + CO_2(aq) + H_2O \qquad (8.2)$$

In fact, the contribution to calcification from carbonate ions in seawater is even less than you might suppose from Figure 8.10. This is because a large proportion of the carbonate ions form electrostatic bonds with magnesium ions (Mg^{2+}), which are abundant in seawater, to form ion pairs that inhibit incorporation of the carbonate in the calcite lattice.

Equation 8.2 shows that for every molecule of $CaCO_3$ formed, a molecule of CO_2 is released. Thus, although you might intuitively have supposed that calcification should consume CO_2, and so draw it down from the atmosphere, in fact — in the short term at least — it does the opposite, pushing CO_2 back into the atmosphere. This has been termed the 'coral reef hypothesis', after the observation that the rapid growth of coral reefs that accompanied interglacial eustatic rises in sea-level during the Quaternary was matched by rising levels of atmospheric CO_2 (Figure 6.2). Direct measurements of sea/air fluxes over modern coral reefs have since confirmed the hypothesis, showing them to be net releasers of CO_2 to the atmosphere.

Turning now to atmospheric CO_2: in water of given temperature, pressure and salinity, at equilibrium with the atmosphere, the concentration of $CO_2(aq)$ correlates with the volumetric proportion of CO_2 in the atmosphere (which is equivalent to its proportional contribution to total pressure, termed 'partial pressure'). Thus, as the partial pressure of atmospheric CO_2 increases, so does the concentration of $CO_2(aq)$, involving a shift towards the right in Equation 8.1 (with a decrease in pH because of the extra H^+ ions produced). This in turn leads to increased dissolution of $CaCO_3$, i.e. a shift to the left in Equation 8.2.

So it would appear that a high partial pressure of CO_2 and enhanced calcification should be incompatible, as they tend to shift the equilibrium in Equation 8.2 in opposite directions. While the calcification should push more CO_2 into the atmosphere, the high partial pressure of the latter should push it the other way, so dissolving $CaCO_3$. Yet, as this and previous Chapters have shown, the available evidence points to their co-existence in the Cretaceous. The solution to the problem must lie in the two processes having occurred in different places, under different conditions. To explore what these different conditions might have been, we must consider other aspects of the relationship between inorganic carbon and seawater.

○ From your own general experience, how would you expect the solubility of CO_2 in water to vary with increasing temperature?

● As with any gas, its solubility decreases with increasing temperature. Common experience shows that when you heat water in a pan bubbles start to appear as the dissolved air comes out of solution (long before the water actually boils to give off steam).

To express this in quantitative terms, we need to consider how the solubility coefficient of CO_2 in seawater varies with temperature. The solubility coefficient is the concentration of the dissolved gas (CO_2(aq)), in moles per litre of seawater of given salinity, when at equilibrium with the pure gas at a pressure of 1 atmosphere (Table 8.1).

Table 8.1 Selected values of the solubility coefficient of CO_2 in seawater in equilibrium with the pure gas at 1 atm pressure, for given temperatures and salinities (10^{-2} moles/litre) (from Weiss, 1974).

Temperature (°C)	Salinity					
	30	34	35	36	38	40
0	6.635	6.498	6.465	6.431	6.364	6.298
10	4.621	4.529	4.507	4.485	4.440	4.396
20	3.400	3.337	3.322	3.306	3.275	3.245
30	2.627	2.583	2.572	2.561	2.540	2.518
40	2.121	2.090	2.082	2.074	2.059	2.044

Thus, for example, if seawater of salinity 35 is heated from 30 °C to 40 °C, it will lose approximately 19% of its CO_2(aq).

○ What influence would you expect this loss of CO_2(aq) to have (a) on the equilibria shown in Equation 8.1, and (b) on the equilibrium in Equation 8.2?

● (a) By driving off CO_2(aq) into the atmosphere, heating would shift the equilibria in Equation 8.2 towards the left. (b) At the same time, however, it would move the equilibrium in Equation 8.2 to the right, enhancing the precipitation of $CaCO_3$ (that is why you get 'kettle fur', especially with hard water, which has a high carbonate content).

Table 8.1 also shows that increasing salinity likewise reduces the solubility of CO_2, a process known as 'salting out'. If, as in our previous example of heating seawater from 30 °C to 40 °C, the salinity were also raised, from 35 to 40, the loss of $CO_2(aq)$ would amount to about 20.5%. Thus, where heating is accompanied by increased salinity through evaporation, there is a double impetus to calcification. Indeed, the strong shift to the left of the equilibria in Equation 8.1 translates into an increase in pH, because of the decreased concentration of H^+ ions.

○   What influence on the proportion of carbonate, relative to bicarbonate ions, should an increase in pH have?

●   From Figure 8.10, we see that the proportion of carbonate ions increases with increasing pH.

Eventually, the concentration of CO_3^{2-} ions can cause the water to become so supersaturated with respect to $CaCO_3$ as to overcome the inhibiting effect of magnesium ion pair bonding, and clouds of minute $CaCO_3$ crystals start to form in the water. Such 'whitings', as they are called, have been observed in restricted waters in extremely hot, arid areas, such as the Bahama Banks and the Persian Gulf.

These facts suggest a possible solution to the apparent paradox of massive carbonate platform growth despite the inferred high partial pressure of CO_2 in the Cretaceous atmosphere (Figure 8.11). Given the intense heat and evaporation that the vast expanses of water overlying the shallow platforms and shelves are likely to have experienced during extreme greenhouse phases, it is probable that temperatures and salinities there rose relative to surrounding open waters. This inference, with respect to salinity at least, is confirmed in some instances by the presence of evaporites in the platform or shelf interiors (Section 5.4). By driving off the CO_2 resulting from calcification, such conditions could have shielded the carbonate factories from the contrary effect of the high levels of CO_2 in the atmosphere. Indeed, the return of CO_2 to the atmosphere arising from the calcification would have served as a positive feedback that maintained the extreme climatic conditions. By contrast, the cooler regions of the open oceans (Figure 5.11) would have borne the brunt of the high partial pressure of CO_2, with a compensatory rise in the level of the CCD (Box 6.1).

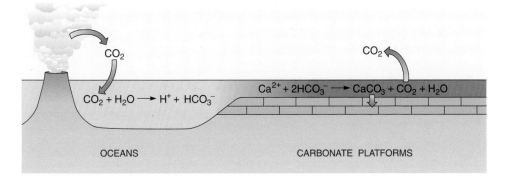

Figure 8.11   A model to explain the rapid growth of carbonate platforms despite high levels of CO_2 in the Cretaceous atmosphere, ultimately fuelled by increased volcanic emission (Section 7.2.6). CO_2 from calcification on the platform is pushed into the atmosphere because of the high temperature and salinity of the shallow water there (red), thereby both conserving the conditions favouring calcification and maintaining greenhouse warmth. Uptake of atmospheric CO_2 is essentially limited to cooler regions of the open oceans (blue), thereby lowering pH and raising the CCD.

However, a further chemical problem remains to be resolved — that of sustaining the supply of ions required for calcification. In seawater today, calcium ions (Ca^{2+}) are nearly four times as abundant as bicarbonate and carbonate ions together, while in the Late Cretaceous (as you will see in Section 8.2.3) the concentration of Ca^{2+} ions may have been even higher. Hence the problem essentially concerns the replenishment of bicarbonate and carbonate ions.

○   What was the most likely source for the bicarbonate and carbonate ions for the platforms?

●   Ocean waters, where less extreme temperatures, as well as the dissolution of calcareous planktonic remains falling beyond the CCD, would have fostered higher concentrations of ΣCO_2, and hence of bicarbonate and carbonate ions (Figure 8.10).

The question thus becomes: how were the platforms supplied with ocean water? To some extent the loss of water volume through evaporation over the platforms would have led in any case to influx from the surrounding open sea (evaporative loss from the ocean itself being compensated for by meteoric influx in other, less arid regions). However, there is one further, more drastic possibility for seasonal replenishment. Given the fairly wide latitudinal distribution of platforms (e.g. Figure 2.16), some seasonal variation in temperature seems likely, especially for those around the northern Tethys, which extended beyond 30° N palaeolatitude.

○   Refer back to Figure 5.13 and decide what might have been the effect of cooling platform water to a temperature near that of the surrounding seawater, as far as relative density is concerned.

●   If the salinity of the platform water had risen above that of the open seawater, cooling it to a similar temperature would then have made it relatively more dense.

In such circumstances, the platform water could thus have flowed off the platform, down into the neighbouring basin (as suggested in Section 5.4), to be replaced by surface waters from the open sea. At present, however, this model for seasonal overturn is merely informed speculation, which has yet to be tested.

The system described in Figure 8.11 requires the maintenance of extreme climatic warmth, implying that cooling could have had a deleterious effect on the carbonate factories of the platforms. Section 5.4 introduced the model of enhanced polar energy transport due to the latent heat of evaporation, but what would have driven the atmospheric convection cells to transport the moisture-laden air to high latitudes, if the latitudinal temperature gradient was no greater (and possibly less) than today? An ingenious twist to the water-vapour model of Hay and DeConto (1999) relies on the effect of water vapour on the density of air. Water has a lower molecular weight (approximately 18) than the main constituents of air (N_2, molecular weight 28, and O_2, molecular weight 32). Since equal volumes of gas at the same temperature and pressure contain the same number of atoms or molecules, air laden with water vapour is less dense than dry air. Conventional thermal convection from the equatorial belt could thus have been supplemented by the effect on air density of the additional water vapour content. The latitudinally extended warmth resulting from these effects would in turn have allowed the development of the carbonate platform systems well beyond the tropics. A picture thus begins to emerge of a distinctive greenhouse climate system sustained by positive feedbacks ensuing from the high carbon flux.

○   What are the two main positive feedbacks that have been mentioned so far in this Chapter?

●   First, high rates of calcification on the carbonate platforms helped to maintain the high atmospheric levels of CO_2. Secondly, high polar heat transport due to the increased atmospheric content of water vapour allowed the platforms to extend well beyond the tropics.

Moving now to mid-palaeolatitudes, what part did the Chalk play in this system? In the Late Cretaceous, especially, it constituted another sizeable C_{carb} sink, which must also have contributed to maintaining the high atmospheric levels of CO_2. However, such deposits extend far offshore in the subsurface today, including originally shallow oceanic successions (seen in DSDP and ODP cores), so it is difficult to estimate total volumes for given time intervals. You might have supposed that extensive carbonate production in oceanic waters would be inconsistent with the model suggested in Figure 8.11. Yet this material was produced in surface waters, as plankton blooms, and hence in association with high rates of organic production, through photosynthesis, as well. The latter process would have sequestered large amounts of the available CO_2(aq), and we may assume that, as with today's plankton, a sizeable proportion of the fixed carbon would have been transferred to deeper levels in the ocean in the form of debris raining down from the surface waters. Hence, conditions favourable to calcification would have been maintained at least in parts of the oceanic surface waters, as well.

As noted earlier, uplift of the CCD would have helped to boost the oceanic reservoir of bicarbonate and carbonate ions. However, an interesting contrast to today is the vast amount of calcareous ooze that was deposited on flooded continental crust (Figure 5.10), and which thus remained permanently above the CCD, incidentally also avoiding being recycled in subduction zones (Section 6.3.6).

In conclusion, notwithstanding the high partial pressures of atmospheric CO_2, the associated extreme climatic conditions may have allowed C_{carb} to be buried at a high rate through extended intervals of the Cretaceous. In the long term, therefore, C_{carb} burial did remove much of the increased volcanic output of CO_2 that was discussed in Chapter 7, despite meanwhile pushing CO_2 back into the atmosphere (helping to maintain greenhouse conditions) in the short term.

So far, we have dwelt upon the positive feedbacks that forced climatic warmth. Yet greenhouse warming did not continue indefinitely, suggesting the influence also of negative (stabilizing) feedback loops.

○   From your reading of earlier Chapters, what would have been the two main negative feedbacks to greenhouse warming?

●   Both weathering of silicate rocks (Section 6.2.2) and high-latitude burial of organic carbon on land (Section 8.1) drew down carbon from the atmosphere. Both would also have increased in extent with rising temperatures, thus serving as negative feedbacks to greenhouse warming.

Nevertheless, the supply of silicate rock for weathering would have been less than today, because of the reduction of the land surface due to eustatic sea-level rise (Section 3.2) as well as the relatively modest rates of mountain uplift throughout much of the period (Section 2.4). Perhaps these limitations to the weathering-based negative feedback also played a hand in keeping the Earth's mean climatic temperature at such a relatively high level.

Besides the burial of C_{org} on land, there were also relatively brief episodes of greatly enhanced burial of marine C_{org}, as you saw in Section 5.2. However, these were essentially associated with the OAEs, and co-incided with the crises in platform development, so will be considered in the following Section.

8.2.2 Fluctuations in the fluxes and their significance

If the carbonate platforms were important C_{carb} sinks when healthy, their global crises (Figure 5.1) correspondingly represent drastic temporary reductions in C_{carb} burial. These episodes were accompanied by significant increases in marine C_{org} burial, linked with the OAEs (Section 5.2). Figure 5.7 illustrated the positive excursion of oceanic $\delta^{13}C$ values that marked this decrease in the ratio of the fluxes to the C_{carb} and C_{org} sinks, respectively, at the close of the Cenomanian. It is not yet clear whether the increase in C_{org} burial fully matched the decline in C_{carb} burial, implying no change in the total amount of carbon entering the ocean/atmosphere system, or whether there was a change in that as well.

Similar $\delta^{13}C$ fluctuations are present in the Aptian record, for which Helmut Weissert and colleagues (1998) have documented a distinctive double spike pattern from sections in northern Italy and Switzerland. The first positive excursion lies within the middle part of the Lower Aptian, and the second, within the uppermost part of the stage (a little way below the Aptian/Albian boundary). The first of these excursions marks the start of a global decline in carbonate platform development almost on the scale of that at the close of the Cenomanian (Figure 5.1), and was likewise accompanied by a mass extinction of rudists and other platform biota. The beginning of the excursion co-incides with organic-rich black shales (known in Italy as the '*livello* Selli') that correspond to OAE 1a, so presenting an obvious parallel with the linkage between the *livello* Bonarelli (OAE 2) and the positive $\delta^{13}C$ excursion in the terminal Cenomanian (Section 5.2). The same pattern has even been observed in cores drilled from the carbonate platform successions of Pacific guyots (e.g. Figure 8.12). Moreover, the example shown in Figure 8.12 (Resolution Guyot) also records a sharp negative excursion in $\delta^{13}C$ values immediately preceding a level of laminated organic-rich shales regarded as equivalent to the *livello* Selli. Over 800 m of shallow platform facies overlie this level in Resolution Guyot, testifying to resumed platform development in the Late Aptian and Albian. However, in this core, rudists have been recorded only from below the positive $\delta^{13}C$ excursion. Rudists seen at higher levels in other guyots (e.g. Figure 1.18) are forms that radiated subsequently, as the platform biota recovered.

Weissert *et al.* (1998) proposed a model to explain the demise of Early Aptian platforms, as well as other changes, ultimately based on volcanic forcing of climate. They suggested that intensified volcanism in the Early Aptian led to extreme greenhouse warming, which in turn boosted the hydrological cycle and enhanced the weathering and erosion of continental rocks. The increased flux of nutrients to the seas and oceans could then have choked platform growth, through eutrophication, while at the same time promoting blooms of organic-walled plankton. Consistent with the latter aspect is a marked change in the plankton coinciding with the advent of the *livello* Selli. Just before deposition of the latter, there was a sharp decline of calcareous plankton, which were probably adapted to low nutrient levels. Their demise would be consistent with replacement by nutrient-induced blooms of organic-walled forms, accounting for the ensuing deposition of organic-rich sediments. A further refinement to the model, suggested by Jenkyns and Wilson (1999), is that the initial negative $\delta^{13}C$ spike (Figure 8.12) might reflect a catastrophic release of methane from deep-sea deposits of methane clathrates (Section 6.2.3), triggered by the warming. Methanogenic (methane-producing) bacteria are known to fractionate carbon isotopes very strongly in favour of ^{12}C,

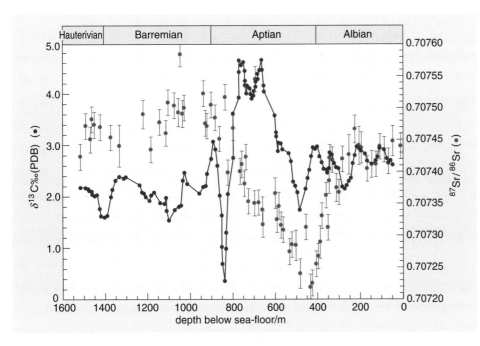

Figure 8.12   Curves of $\delta^{13}C$ values and strontium-isotope ratios ($^{87}Sr/^{86}Sr$) from the carbonate platform succession of Resolution Guyot, Mid-Pacific Mountains. Note that the horizontal axis records core depth downwards from the sea-floor (0 m), not absolute time. The level regarded as equivalent to the *livello* Selli is at just over 800 m depth. *(Jenkyns and Wilson, 1999.)*

so a sudden release of methane could have produced the sharp reduction in $\delta^{13}C$ values. As methane is a potent greenhouse gas (much more so than CO_2), such an event would have given a further boost to climate warming.

However, there are some problems with this extreme-greenhouse model for the demise of the platforms. One would have expected the effects of nutrient-induced eutrophication to have been regional, e.g. along shelf areas attached to continents and in areas of upwelling. Yet the mass extinction among the platform dwellers was global in extent, affecting even those on the isolated platforms of Central Tethys and the Pacific. Nor was the mass extinction always synchronous with the inception of organic-rich sedimentation, being later in some low palaeolatitude sites. In the subsurface of eastern Arabia (~ 12° S in the Early Aptian), for example, the likely equivalent of the *livello* Selli is a dark, clay-rich layer. Above that, platform growth resumed with a full complement of the pre-existing rudist fauna, which survived until the end of the Early Aptian, after the positive excursion of oceanic $\delta^{13}C$ values.

○ Considering the feedbacks discussed earlier, what would have been the likely climatic effect of the switch in emphasis from the C_{carb} to the C_{org} sink indicated by the positive excursion of $\delta^{13}C$ values?

● The decline in CO_2-release from calcification, together with the drawdown of atmospheric CO_2 due to burial of C_{org}, should have led to climatic cooling.

An alternative hypothesis to extreme greenhouse warming as a cause of platform demise is thus precisely the opposite — that of progressive cooling, which would have lessened the expulsion of CO_2(aq) from the platform waters, so inhibiting calcification (Section 8.2.1). This latter hypothesis would also be consistent with the evidence for cooler conditions in the Late Aptian (Section 5.3). The current oxygen-isotope evidence does not allow us to choose reliably between these alternatives, because of frequent diagenetic distortion of the palaeotemperature signal. Moreover, if, instead of straightforward warming, or cooling, there had in fact been an interval of strong climatic oscillations (analogous to those of the Quaternary), then a coherent result might prove elusive, given the problems of correlating such short-term events.

Also shown on Figure 8.12 is a curve of changing Sr-isotope ratios ($^{87}Sr/^{86}Sr$).

○ Considering the controls on the $^{87}Sr/^{86}Sr$ ratio in seawater (Section 6.3.1), what does the pattern shown for the Aptian in Figure 8.12 imply concerning any change in the relative contributions of continental weathering and oceanic volcanism?

● The $^{87}Sr/^{86}Sr$ ratio is relatively high in the lowest part of the Aptian, but, from the level considered equivalent to the *livello* Selli onwards, it declines through the rest of the stage. Hence, at the start of the Aptian, continental weathering provided the main input, but this yielded progressively to that from oceanic volcanism thereafter.

With regard to the alternative climatic hypotheses for platform demise, the implications of the strontium-isotope data are ambiguous. They could indicate increasing volcanic and hence hydrothermal activity, consistent with the hypothesis of volcanically driven warming. Alternatively, they could reflect maximum silicate weathering, associated with climatic warmth, at the start of the Aptian, followed by a decrease, due to cooling, from the time of the *livello* Selli. We are thus still left with two quite contrasting possibilities for the demise of platforms in the latter part of the Early Aptian, involving either over-heating or over-cooling. A similarly ambiguous situation exists for the Cenomanian/ Turonian transition. In both cases, however, a marked perturbation of the global carbon cycle was certainly involved, which is very likely to have impacted on climate one way or the other, doubtless with consequences also for short-term eustatic changes of sea-level. As noted earlier, an episode of climatic instability is more likely than a simple trend of change, in any case. Hence there is a need for a great deal more precise correlation between platforms in different regions, and between platform and basin successions, before we can confidently reconstruct the chronology of chemical, biotic and climatic changes, in order to try to disentangle causes and effects.

8.2.3 Other effects on ocean chemistry and life

Besides stepping up the pace of the global carbon cycle, the high rates of oceanic crust production in the Cretaceous had other profound implications for ocean chemistry.

Between about 125 Ma and 120 Ma ago (Late Barremian–Early Aptian), there was a huge increase in the rate of production of new oceanic crust (Section 7.2.2). Thereafter, it gradually declined, with some fluctuations, towards present levels.

○ What effect would you expect (a) the initial acceleration in the production of oceanic crust, and (b) the decline thereafter, to have had on concentrations of Mg^{2+} and Ca^{2+} ions in seawater?

● Hydrothermal circulation tends to sequester Mg^{2+} ions from the seawater and replace them with Ca^{2+} ions (Figure 2.5). The latter process would have increased with the initial phase of accelerated production of oceanic crust (a), and this should have led to an increase in the concentration of Ca^{2+} ions at the expense of Mg^{2+} ions in the seawater. Subsequently, with the steady decline in rate (b), the balance would have been redressed, with the Mg^{2+}/Ca^{2+} ratio starting to climb again towards present levels.

Direct evidence for such effects has recently come from studies of the magnesium content of rudist calcite by Thomas Steuber (2002). In living bivalves with a calcitic outer shell layer, such as mussels, the (small) magnesium content of the calcite shows a positive correlation with temperature. Steuber found the same kind of correlation in rudist calcite, using $\delta^{18}O$ as a negative proxy for palaeotemperature (Section 5.4), though with surprising differences between the lines of correlation for examples of different stratigraphical ages (Figure 8.13).

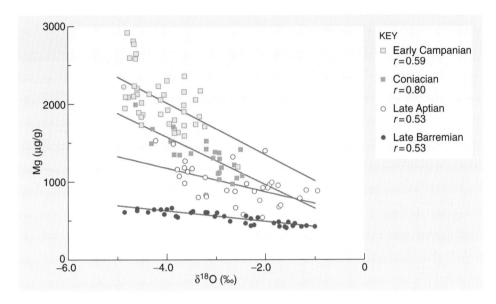

Figure 8.13   Plot of magnesium content versus $\delta^{18}O$ (relative to PDB) in the calcitic outer shell layer of rudists from different parts of the Cretaceous. *(Steuber, 2002.)*

○   If you were to try to estimate palaeotemperatures from the magnesium content values of the early Campanian rudists using the correlation line for the late Barremian samples, what would you find?

●   The Campanian values of magnesium content would correspond to extremely negative $\delta^{18}O$ values on the Barremian line, implying absurdly high palaeotemperatures (Section 5.4).

The simplest explanation for the differences in the correlation lines is that the seawater Mg^{2+}/Ca^{2+} ratio changed, with the lowest values in the oldest samples and those in the later samples beginning a long-term increase towards the present ratio (of about 5 : 1). At its minimum, the Mg^{2+}/Ca^{2+} ratio may have been as low as about 1 : 1. By reducing Mg^{2+}/CO_3^- ion-pair bonding (Section 8.2.1), a reduction in the Mg^{2+}/Ca^{2+} ratio would have had significant implications for calcification — in particular, facilitating the growth of calcite and thus giving a possible further boost to the major carbonate factories of the period. The change may also have played a role in the radiation of calcite-precipitating biota at the expense of aragonite-dominated forms during cooler episodes, when surface waters might sometimes have been supersaturated with respect to calcite but not to aragonite. Steuber (2002) also noted that rudists with thick outer shell layers of calcite became especially prevalent following the Cenomanian/Turonian mass extinction — a time when climates began to cool (Section 5.4). Such new ideas still need to be tested in detail, but they do at least illustrate the extent to which conditions in the Cretaceous world could have differed from those of today.

8.3 Summary

- Geological data allow the construction of global maps for the Cretaceous that show the first-order patterns of temperature and precipitation in qualitative terms.

- Terrestrial carbon sequestration and primary production occurred previously in the polar regions where conditions were both warm enough and wet enough for plant growth.

- Organic accumulation in the polar regions was facilitated by high summer productivity, deciduousness and winter temperatures low enough to limit decay.

- Depositional settings were also crucial for long-term carbon preservation.

- Low latitudes were seasonally arid and fire ecology often controlled the composition and structure of low-latitude Cretaceous vegetation.

- As Pangaea broke up, low-latitude humidity increased.

- With few high mountain ranges and relatively low precipitation at low (warm) latitudes, silicate weathering in the Cretaceous was likely to have been less than in the present world.

- Like today, marine carbon burial in the Cretaceous was dominated by the carbonate carbon (C_{carb}) sink, although the oceanic anoxic events (OAEs) represented short episodes of enhanced organic carbon (C_{org}) burial, while C_{carb} burial was temporarily reduced.

- Rates of C_{carb} burial in the Tethyan/Atlantic carbonate platforms can be estimated from their surface areas, revealed on palaeogeographical maps, and mean values of their thickness for given time intervals. Estimates for these platforms alone, during peak episodes of growth such as the Early Aptian and Late Cenomanian, correspond to sizeable fractions (around one-third) of the global rate, for all environments, today.

- Such high rates of C_{carb} burial appear paradoxical in view of the probably high partial pressure of CO_2 in the Cretaceous atmosphere, compared with today, because calcification releases CO_2, while increased levels of dissolved CO_2 lower the pH of the water, so opposing calcification.

- Extreme greenhouse warming may provide a solution to the apparent paradox: high temperatures and salinities in the shallow waters overlying the platforms would have driven CO_2 out of solution, so preventing it from inhibiting calcification there. The return of the CO_2 to the atmosphere represented a positive feedback to greenhouse warming. Compensatory dissolution of CO_2 would have occurred in cooler regions of the oceans, causing uplift of the CCD.

- Enhanced polar heat transport associated with the extreme greenhouse warming could have been promoted by increased equatorial evaporation, both through the energy involved in the latent heat of evaporation and the effect of water vapour on air density, hence convection. The latitudinally extended warmth allowed the development of carbonate platform systems well beyond the tropics.

- The Chalk was a further large C_{carb} sink, produced in open surface waters where high rates of organic production kept down levels of dissolved CO_2.

- High long-term rates of C_{carb} burial through the Cretaceous removed much of the increased volcanic output of CO_2, despite meanwhile pushing CO_2 back into the atmosphere, so helping to maintain greenhouse conditions in the short term. Negative feedbacks to greenhouse warming were weathering of silicate rocks and burial of C_{org}.

- Decreases of C_{carb} burial during global platform crises were accompanied by increases in C_{org} burial, representing major temporary modifications of the carbon cycle.

- Platform demise could be attributed either to intensification of greenhouse warming (with increased continental weathering and eutrophication of the waters overlying them) or, conversely, to cooling (with inhibition of calcification on the platforms) or even a combination of both. At present, isotopic data are inconclusive in this respect.

- The rapid acceleration of oceanic crust production, hence hydrothermal circulation, in the Late Barremian–Early Aptian, followed by a steady decline thereafter, appears to have caused an initial reduction in the Mg^{2+}/Ca^{2+} ratio of seawater, followed by a steady increase towards the present ratio. This would have facilitated the growth of calcite in the world's carbonate factories.

8.4 References

ALLEN, P. *et al.* (1998) 'Purbeck–Wealden (early Cretaceous) climates', *Proceedings of the Geologists Association*, **109**, 197–236.

CHUMAKOV, N. M. *et al.* (1995) 'Climatic zones in the middle of the Cretaceous Period', *Stratigraphy and Geological Correlation*, **3**, 3–14.

HAY, W. H. AND DECONTO, R. M. (1999) 'Comparison of modern and Late Cretaceous meridional energy transport and oceanology', in BARRERA, E. AND JOHNSON, C. C. (eds) *Evolution of the Cretaceous Ocean-Climate System*, Geological Society of America Special Paper **332**, 283–300.

JENKYNS, H. AND WILSON, P. A. (1999) 'Stratigraphy, paleoceanography, and evolution of Cretaceous Pacific guyots: relics from a greenhouse Earth', *American Journal of Science*, **299**, 341–392.

McCABE, P. J. AND PARRISH, J. T. (1992) 'Controls on the distribution and quality of Cretaceous coals', *Geological Society of America Special Paper*, 267, 407pp.

PARRISH, J. T., DANIEL, I. L., KENNEDY, E. M. AND SPICER, R. A. (1998) 'Palaeoclimate significance of mid-Cretaceous floras from the middle Clarence Valley, New Zealand, **13**, 149–158.

PHILIP, J., MASSE, J.-P. AND CAMOIN, G. (1995) 'Tethyan carbonate platforms', in A. E. M. NAIRN *et al.* (eds) *The Ocean Basins and Margins, Volume 8: The Tethys Ocean*, Plenum Press, New York, pp. 239–265.

SKELTON, P. W., SPICER, R. A. AND REES, P. A. (1997) *S269 Earth and Life: Evolving Life and the Earth*, Open University, Milton Keynes.

STEUBER, T. (2002) 'Plate tectonic control on the evolution of Cretaceous platform-carbonate production', *Geology*, **30**, 259–262.

WEISS, R. F. (1974) 'Carbon dioxide in water and seawater: the solubility of a non-ideal gas', *Marine Chemistry*, **2**, 203–215.

WEISSERT, H., LINI, A., FÖLLMI, K. B. AND KUHN, O. (1998) 'Correlation of Early Cretaceous isotope stratigraphy and platform drowning events: a possible link?', *Palaeogeogr., Palaeoclim., Palaeoecol.*, **137**, 189–203.

9 The Lost World rediscovered

Bob Spicer

That the Cretaceous world was significantly different from our present one is undisputed, given the rich array of geological evidence discussed in previous Chapters. The most noticeable difference between then and now was the climate: the Cretaceous global mean surface air temperature (GMST) may have been at times up to 15 °C higher than now, according to some estimates. The Cretaceous therefore provides a useful case history showing how interactions within the Earth system function under extreme greenhouse conditions — conditions that we might experience again if some current predictions concerning the effects of burning fossil fuels prove correct. To try to fully understand how this world functioned, what the overriding controls were and which components of the Earth system were most sensitive to perturbation, is not an exercise that can be undertaken without the aid of complex numerical Earth system models that demand enormous computing power. In this Chapter, we review some of the components of the Cretaceous Earth system and see how these might have played a role in determining the overall pattern of the Cretaceous world environment. What we can cover here is, of course, not exhaustive and you might like to consider the interactions of other components of the Cretaceous world. What should be apparent is the methodology employed and the complexity of the problem. However, it is this complexity, and its relevance to managing our present world, that makes the topic so intellectually exciting.

The occurrence of fossil organisms, the distribution of different rock types, and palaeomagnetic evidence from both continental crustal blocks and the ocean floor tell us that the continents were configured differently during the Cretaceous (Chapter 2). However, while palaeogeography may have been a contributory factor to both global and regional climate, it cannot have been the primary cause for the differences between Cretaceous and contemporary climates. This is apparent from the simple observation that the configuration of the continents changed throughout the Cretaceous as Pangaea broke up and the Tethys Ocean developed without a simple corresponding climate pattern being obvious. Nevertheless, the possible role of continental configuration in respect of GMST was one of the first targets for the application of Atmospheric General Circulation Models in a palaeogeographical context. These models are far more complex than the geochemical box models we saw in Chapter 6 and so some background knowledge about how they work is necessary. Before dealing with this, it is worth considering how the simple box models and the more sophisticated General Circulation Models (GCMs) are complementary.

In Chapter 6, it was pointed out that advanced GCMs such as CCM-3 have been used to help inform the simpler geochemical box models such as GEOCARB III. To some this may seem like circular reasoning, particularly if GEOCARB III predictions are then used to constrain GCM parameterizations such as atmospheric CO_2 levels. Clearly, circularity is a danger but in dealing with complex systems such an approach has legitimacy if we are aware of what has been done and how the results help inform further reasoning and observation. Such an approach is taken in the hope that at the next turn of the cycle we get closer to a better understanding of the system. This iterative, or *spiral*, reasoning will — hopefully — converge eventually on a limited field of possibilities as the process progresses.

9.1   Climate modelling: a brief overview

The geological record of the Cretaceous is, as we have seen, rich and varied. However, because Cretaceous rocks are not preserved everywhere over the surface of the Earth, there are large gaps in our knowledge of the spatial and temporal patterns of the Cretaceous Earth system. In recent years, the development and application of climate models has helped us to think about how the Earth system works, integrate the geological data into more complete views of this fascinating period of Earth history, and provide a predictive tool for focusing new geological research activity. There are several types of climate models: qualitative and quantitative, simple and complex.

Qualitative models apply the generalized principles of atmospheric circulation to ancient palaeogeography and yield patterns of surface pressure that can be used to predict dominant wind directions on an annual and seasonal basis; and, from that, relative continental wetness and dryness as well as ocean circulation. However, predictions of past temperature fields are difficult. This is a severe limitation but the advantage of these models is that nothing more than a pencil and paper are required provided that the palaeogeography (patterns of land and sea and some indication of palaeotopography) is known.

A more quantitative approach to determining past air temperatures is by means of energy balance models which, as their name suggests, balance the incoming energy of the Sun with outgoing long-wave radiation. Surface characteristics like albedo are required, as is some estimate of the solar energy flux arriving per unit time over the area under consideration. In its simplest form, an energy balance model provides a single value for the mean global surface temperature. By dividing the surface of the Earth into latitude-parallel belts or even latitude-by-longitude grid boxes, the mean temperature within each belt or box can be calculated. The more complex energy balance models also divide a typical year up into smaller time slices so that temperature changes over the seasons or even typical days can be derived.

Clearly, the complexity of the real world is far greater than such energy balance models can reproduce. Consequently, GCMs have been developed to try to simulate the complex interactions of clouds, atmospheric composition, albedo and roughness of surfaces and even the feedbacks associated with different types of vegetation. Moreover, these models attempt to reproduce patterns of environmental change as model 'time' progresses hour by hour. These small time steps can be integrated over tens or even hundreds of model years.

GCMs work by dividing the surface of the Earth into a number of grid elements. Above each element, the atmosphere is divided into a number of layers. In this way, the entire atmosphere is divided into a set of contiguous boxes of air. For each box, such things as temperature, winds, humidity, air pressure, cloud types and heights etc., are calculated. Calculations are carried out at discrete model time steps (e.g. every 30 minutes) as the days and seasons progress. All the boxes also interact with their neighbours at each time step so the numbers and complexity of the equations involved are huge and usually require the most powerful computers available. Nevertheless, these equations include simplifications because to reproduce atmospheric processes with accuracy the models would need to simulate the movements of microscopic particles of air: this is obviously impossible. Unfortunately, these simplifications can lead to spurious results, but tuning of the models to reproduce the modern climate minimizes these problems. Such models are the basis for short- and medium-term weather forecasting as well as climate research.

When starting from cold, that is with nothing but the initial starting or boundary conditions as provided to the model, the model is often allowed to 'spin up' over several simulated years until a degree of stability or constancy is acquired. After this, averaging a number of simulated years provides an indication of average climate.

The preceding account only involved the atmosphere: such models are called Atmospheric General Circulation Models or AGCMs. These were the first form of GCMs because, until fairly recently, modelling the ocean system and coupling both atmosphere and ocean together involved computational demands that outstripped computing capacity. To understand why this is so, just consider the average size of an atmospheric low- or high-pressure cell. Typically, these are in the order of 1000 km across so the spatial resolution of the atmospheric grid need not be very high. By contrast, the higher density and viscosity of water results in equivalent oceanic eddy systems a few hundred kilometres across and so requires the ocean to be modelled at a much higher spatial resolution. Additional complications arise when coupling these two systems. This is partly because the atmosphere requires only several months to stabilize after a perturbation whereas the ocean system with greater mass and inertia may take many centuries to settle back to its stable state.

9.2 Modelling the Cretaceous

Because of the widespread evidence for global warmth and apparent equability in the Cretaceous, it was one of the first targets for computer-based palaeoclimate modelling. The early experiments were relatively crude and not only used a very coarse geographical grid, but also a simple 'wet carpet' ocean. This ocean had no depth or circulation and only the simplest of interactions with the air above (e.g. evaporation), but these simplifications were necessary due to the limitations imposed by the available computing power as well as the scientific understanding at that time. The experiments were still valuable in that they provided some insights into the various contributions to warming provided by, for example, geographical configuration, atmospheric composition, and, more latterly, vegetation/climate feedbacks.

Figure 9.1 shows the results of a series of early experiments using a model with a coarse spatial resolution (4.5° latitude by 7.5° longitude and nine levels in the atmosphere) and a simple wet carpet ocean. This ocean had no depth, no currents or heat capacity but was a moisture source. Moreover, there was no realistic vegetation and no annual cycle. Computational load in relation to the relatively limited computing resource then available demanded that many of these early experiments were run for 'perpetual spring', in which the model iterated about a repeated day at the spring equinox. Similarly, the models were often run for given days in July or December. Although not producing a realistic view of the Cretaceous world, these early experimental models emphasized the roles played by continental configuration and atmospheric CO_2 levels in determining average global temperature and precipitation. Figure 9.1 demonstrates clearly that mid-Cretaceous geographical configuration alone contributed to the greenhouse warmth of that time, even exceeding the effect of increasing by a factor of four CO_2 levels with present-day geography. When CO_2 is increased to four times present levels for mid-Cretaceous palaeogeography, average global precipitation and temperature are both raised even further.

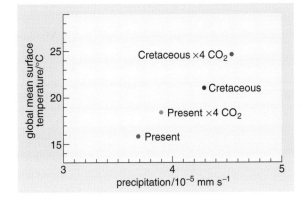

Figure 9.1 Results of an early climate modelling exercise using continental configurations of today ('Present') and for the Cretaceous using today's atmospheric CO_2 levels, and ×4 present CO_2 levels. *(Parrish and Barron, 1986.)* Further explanation in the text.

Figure 9.2a illustrates the effect on the zonal precipitation pattern of the gross distribution of land and sea. If all the continents were concentrated at the poles (Figure 9.2c), low- and mid-latitude precipitation would be high, even exceeding the levels that would result from a Cretaceous palaeogeography with four times present CO_2. A purely equatorial continent (Figure 9.2b) has the effect of drying the low latitudes. Basically, the message here is a simple one: the presence of land reduces precipitation. This result is entirely consistent with the Russian data compilation in Chapter 8, which shows that as the seaways penetrated the continental interiors, and in particular the Atlantic opened up, there was an increase in low-latitude humidity.

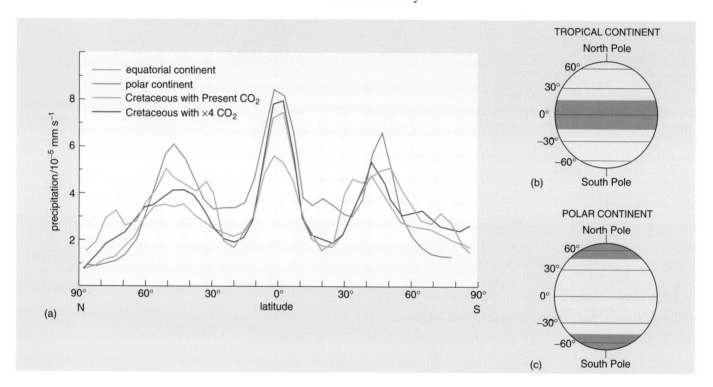

Figure 9.2 Modelled distributions of precipitation runs (the equatorial and polar continental configuration models used present-day CO_2 levels). *(Parrish and Barron, 1986.)*

These early experiments indicated that despite their limitations the models yielded results that were understandable in terms of the first-order observations. They also highlighted the effect of palaeogeography on climate and the model's sensitivity to such things as atmospheric composition. However, it proved extremely difficult to reproduce the observed pattern of Cretaceous climate, and data/model disparities were common. In particular, it proved difficult to reproduce the observed degree of polar warmth without overheating the tropics unless some very high, and possibly unrealistic, oceanic polar heat transport rates were imposed on the atmosphere models. The models did not automatically generate the necessary heat fluxes. On the contrary, the low equator-to-pole thermal gradients that arose as a consequence of having both low and high latitude warmth resulted in a very sluggish atmosphere. It was thought that perhaps part of the problem lay in the simplifications built into the models; more realistic models might be able to resolve these disparities.

Modern climate models are far more sophisticated and do approach a more realistic simulation of the Cretaceous world. The full seasonal cycle can be accommodated, the geographical resolution can be higher (typically 3.5° latitude by 3.5° longitude), and there are more layers in the atmosphere (19 is common). There is better cloud parameterization, vegetation can be specified either based on palaeobotanical data or other model results, and the oceans can be more realistic. Figures 9.3 and 9.4 illustrate the kinds of feedbacks in the real world and the interactions that are at the heart of a typical Earth system model in use in the 1990s.

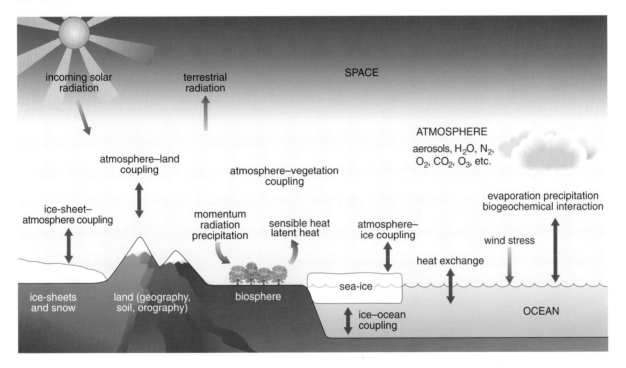

Figure 9.3   The primary components of the global climate system and the interactions between them.

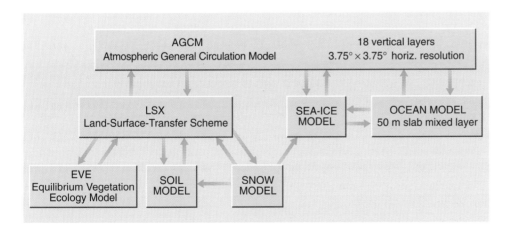

Figure 9.4   The components of the GENESIS Version 2.0 Global Climate Model. Bidirectional arrows indicate interactive communication between model components, allowing feedback mechanisms to operate in realistic ways. Further explanation in the text.

Until the year 2000, almost all Cretaceous model simulations used at best a mixed layer 'slab' ocean where the ocean was treated as being 50 m deep (the depth to which weather phenomena and the seasonal thermal cycles penetrate) and with limited predictions of latitudinal heat transport. This kind of ocean is a component of the Earth system model illustrated in Figure 9.4. At the time of writing (2002), Cretaceous models with fully coupled dynamic oceans are just beginning to be introduced. In these models, not only is the atmosphere divided into an array of contiguous 'boxes' of air, but so too is the ocean divided into a set of contiguous interacting parcels of water. Currents, temperature, and salinity variations due to ice formation and evaporation are all modelled. Unlike the atmosphere, which can return to a steady equilibrium state after perturbation within a few tens of years, the ocean system takes thousands of years to similarly settle down after any change. This means the models have to be run for long periods of simulated time, which makes the exercise computationally very expensive. Interestingly, the results of fully coupled ocean/atmosphere models are not significantly different to slab ocean results and both display similar disparities with observed geological data, particularly over land.

○ Why are fully dynamic ocean models for the Cretaceous difficult to circumscribe?

● The size and shape (bathymetry) of the ocean basins need to be specified at a reasonably high resolution because a small channel, such as a breach in the modern Panama isthmus, could have profound effects on overall circulation and heat transfer. Because much of the Cretaceous ocean basin system has been subducted, and therefore no longer exists, basin bathymetry has to be reconstructed using indirect means. This introduces large potential uncertainties into the model system. Another problem, although there are means of overcoming it, is that some key connections may be narrower than the grid size used in the model so it cannot properly 'see' the connections.

Interestingly, ocean circulation patterns and water body structure derived from geological data and the modelled behaviour of the Cretaceous oceans are in good agreement. This indicates that despite our uncertainties of Cretaceous bathymetry the first-order, large-scale, simulations are likely to be reliable.

Today, approximately half the heat transport is by the ocean and half by the atmosphere, but how this would have differed in a Cretaceous world is unclear. In the oceans, the weather may require much of the transport to have been accomplished by small-scale eddies; while in the atmosphere, higher temperatures may result in more water vapour and hence greater transport of energy via latent heat (Section 5.4).

○ What palaeogeographic factors may have limited oceanic heat transport to the Arctic during the Cretaceous?

● The Arctic Ocean was effectively an enclosed basin isolated from the deep-water global system by connections that were shallow (only a few hundred metres deep) (Section 2.2.1).

This may not be a significant problem as a lot of heat is transported in the uppermost few hundred metres of the water column. What remains necessary though is not only a poleward flow but also a return flow. The exact current pathways have yet to be determined and modelling this system is highly sensitive to precise widths, depths and positions of the connections.

The disparity between the data and model results even using coupled ocean/ atmosphere models can be quite marked. As we saw in Chapter 4, the differences between temperatures derived from geological proxies and those predicted by models for the interior of Asia exceed the combined uncertainties estimated for both. This indicates a genuine disparity that as yet cannot be explained. It may indicate problems in interpreting the proxies or in the way that the model attempts to reproduce the physics of the Cretaceous ocean/atmosphere system.

9.2.1 Volcanism in the Cretaceous

The Cretaceous also saw some of the most intense and prolonged periods of volcanism in the Phanerozoic. While this undoubtedly contributed CO_2 to the atmosphere, and therefore contributed to increasing the GMST, dust and aerosols may well have had the opposite effect, even if the period during which they remained aloft after each eruption was very short (a few tens of years at most). In Chapter 7, we discussed the possible effect of the Deccan eruptions and concluded that the environmental impact was difficult to predict with any certainty. This difficulty arose from our lack of knowledge of the height of the tropopause, the height of the eruptive columns, the volume and types of gases and aerosols emitted and the size and frequency of the eruptions. Whatever the global effect of such continental large igneous provinces (LIPs), one would perhaps expect that organisms living in close proximity to the eruptive centres might have been profoundly affected. In fact, the organisms whose demise effectively defined the end of the Cretaceous — the dinosaurs — are not only known to have lived in close proximity to the Deccan eruptive centres but actually chose that area as nesting sites for at least a million years after the onset of the volcanism! Clearly, the environmental impact of such large-scale volcanic activity, were it to happen today, would be devastating for a complex technology-dependent society of humans, but the overall long-term environmental impact of such an event has yet to be fully understood. Here again, modelling could offer a way forward.

One of the first imperatives that drove the development of global atmospheric climate modelling was not a wish to understand the natural variations in climate over time, but a need to understand the global impact of a possible full-scale nuclear war. The so-called 'nuclear winter' was a concept arising from such studies and similar analyses have subsequently been applied to estimating the environmental effects in the aftermath of an asteroid impact. To model such an event, not only do the complex physical interactions of the atmosphere and oceans need to be included, but also the chemical interactions within those environments. In current Earth system models, which are extensions of the GCMs, the interactions of as many as 92 chemical species are modelled in the atmosphere alone. To date, such computing power has not been applied in any comprehensive sense to the problems of understanding the Cretaceous world either as a steady-state system or one that was perturbed by volcanic eruptions for example, but within the next few years such exercises are inevitable. Such experiments would provide invaluable insights into how susceptible a possible future warm environmental system might be to potentially destabilizing influences.

The Deccan Traps erupted at low latitudes while India was still in the Southern Hemisphere. We might speculate that the direct immediate effects of such eruptions would largely be confined to the Southern Hemisphere and because the tropopause is high at low latitudes we might also suggest that stratospheric injection of aerosols may have been limited. The same is not true, however, for the volcanism that occurred in the Okhotsk–Chutotka Volcanogenic Belt. As you may recall from Chapter 4, north-eastern Russia hosts vast thicknesses of volcaniclastic material that was produced as a result of largely explosive volcanism throughout the Late Cretaceous. Ash plumes, as evidenced by bentonites, spread across northern Alaska and impacted the polar forests.

○ Is it likely that the eruptive plumes from these eruptions penetrated the stratosphere?

● Yes. You may recall from Chapter 7 that in the polar regions the tropopause is as low as 8 km so a vigorous explosive eruption is very likely to have injected large quantities of material into the stratosphere.

That the ash travelled long distances attests to the large size and energy of the eruptions. The effect of the aerosols might well have been to cool the polar regions by absorbing incoming solar radiation, but by how much and what the long-term effect at these critically sensitive latitudes might have been can only be properly assessed using Earth system models based on GCMs. Once again, these experiments have yet to be done.

9.2.2 Vegetation

One set of modelling experiments that have thrown light on the maintenance of warm polar temperatures involves 'planting' the polar regions with the kind of vegetation we know existed from the fossil record. When polar forests were imposed on an otherwise shrub-covered Cretaceous world, there was a marked warming, particularly in the spring and summer. In the control runs, the shrub-covered polar regions accumulated winter snow that lasted well into the summer; and over Antarctica, even with present-day orbital configurations, the snow only disappeared completely for a few weeks in December and January. When the shrubs were replaced with polar forests, the snow melted earlier.

○ Why do you think this was so?

● During the spring and summer months, low-angle sunlight would have been absorbed by the relatively dark, low albedo, trunks and branches of the trees, whereas the shrubs were buried under the snow and so little incoming solar radiation was absorbed. In the winter, with no sunlight, there is no albedo-effect warming.

Thus, the existence of the polar forests creates a positive feedback loop that helps maintain conditions conducive to their survival. The polar forests not only warmed the poles by having a low albedo, they also probably created clouds. We know that in the modern world deforestation leads to local increases in aridity because evapotranspirational feedback is curtailed. The opposite is also true: the development of forests increases atmospheric humidity. So the polar forests must have played a role in maintaining the wetness of the Cretaceous poles. Moreover, the high humidity would have also led to the creation of a blanket of polar cloud that would have helped limit the leakage of long-wave radiation to space and thereby warmed the atmosphere. This effect would have been particularly pronounced over, and adjacent to, an Arctic Ocean kept relatively warm by heat transported via ocean currents. In winter though, plant dormancy, especially when coupled with deciduousness, would have virtually eliminated the supply of moisture to the atmosphere through transpiration. Whether this would have made any significant difference to the cloud cover or whether the warm ocean offered enough evaporation to maintain the insulative blanket is another aspect of the Cretaceous polar environment best tackled through modelling.

9.2.3   Carbonate platforms

At low latitudes, there would also have been some interesting albedo issues. If you had been able to fly over the carbonate platforms of the Tethyan margins, you would probably have been dazzled by the high albedo of the carbonate deposits. The latter effect could well have mitigated the extent of heating at low latitudes to some extent, particularly because for most of the day the incoming solar radiation would have been at a high angle of incidence, so the shallow water above the platforms would have had little moderating effect. This albedo effect would have fluctuated with changing sea-level and would have been particularly strong as sea-level fell, exposing the carbonate surface directly to the atmosphere. At present, we can only speculate on the possible climatic consequences of such effects, which beg further, more sophisticated twists to the modelling.

9.2.4   What next?

The GCMs that we have been considering in this Chapter largely concern atmospheric circulation and its climatic consequences for given initial conditions. Later, more sophisticated versions integrate some dynamic oceanic influences (e.g. Figure 9.4), but still only in a relatively simplified fashion. Important forcing factors, such as the atmospheric content of CO_2, still have to be provided as 'given' inputs. Yet, as we have seen in Chapter 6 (Table 6.1), the atmosphere is a far smaller reservoir for carbon than is the network of oceans, so the composition of the former is very much at the mercy of the latter over the longer time-scales (100s to 1000s years) of oceanic circulation. On the other hand, biogeochemical box models such as GEOCARB (Section 6.3) set out precisely to address the longer-term influences on atmospheric composition. Although climatic feedbacks, mediated, of course, by the atmosphere, are already incorporated into the GEOCARB model (Section 6.3.3), in a rather broad way, the logical next step in modelling the system is to keep feeding the output of the latter directly back to the GCMs themselves. In other words, we can look forward to a new generation of models that will incorporate the interactions between the physical effects of both atmospheric and oceanic circulation, and the chemical consequences of biogeochemical cycles. Such virtual whole-Earth system simulations could provide us with a greatly increased potential both for modelling past conditions — for further testing against the geological record — and predicting those in the future.

9.3   Summary

- The Cretaceous world was sufficiently different from the present world that simple extrapolation of climate systems from present to past could be misleading.

- The Cretaceous rock record is globally incomplete, but models provide a way of integrating these data to build a more complete picture of what conditions may have been like at that time.

- There are various kinds of models of different degrees of complexity: coupling different models and using them to inform and evaluate each other provides an iterative approach for limiting uncertainties.

- GCMs, coupled to other models for components such as ice dynamics, vegetation and atmospheric and oceanic chemistry, allow the construction of Earth system models. These are essential for future research because of the different roles played by these components in the Cretaceous world.

- Earth system modelling of the Cretaceous is in its infancy but nonetheless has considerable benefits to offer in understanding how a possible future warm Earth might function under both stable and perturbed conditions.

9.4   Further reading

BARRON, E. J. AND WASHINGTON, W. M. (1982a) 'Cretaceous climate: a comparison of atmospheric simulations with the geologic record', *Palaeogeog., Palaeoclim., Palaeoecol.*, **40**, pp. 103–133.

BARRON, E. J. AND WASHINGTON, W. M. (1982b) 'Atmospheric circulation during warm geologic periods: is the equator-to-pole surface-temperature gradient the controlling factor?', *Geology*, **10**, pp. 633–636.

BARRON, E. J. AND WASHINGTON, W. M. (1984) 'The role of geographic variables in explaining palaeoclimates: results from Cretaceous climate model sensitivity studies', *Journal of Geophysical Research*, **89**, pp. 1267–1279.

BARRON, E. J., THOMPSON, S. L. AND SCHNEIDER, S. H. (1981) 'An ice-free Cretaceous? results from climate model simulations', *Science*, **212**, pp. 501–508.

BARRON, E. J., THOMPSON, S. L. AND HAY, W. W. (1984) 'Continental distribution as a forcing factor for global-scale temperature', *Nature*, **310**, pp. 574–575.

TRENBERTH, H. E. (1992) *Climate System Modelling*, Cambridge University Press, 783pp.

VALDES, R. J., SPICER, R. A., SELLWOOD, B. N. AND PALMER, D. C. (1999) *Understanding Past Climates: Modelling Ancient Weather*, CD-ROM, Routledge.

PART 3 THE END OF AN ERA

10 The end-Cretaceous mass extinction

Iain Gilmour

Extinction is the ultimate fate of virtually all species. As we saw in Chapter 1.2, the geological record of the Phanerozoic is punctuated by several abrupt changes when many of the species living at that time became extinct, and disappeared from the record over a relatively short time interval. In the 19th century, many early geologists believed that these periods represented gaps in the geological record. Today, many of these gaps are interpreted as short periods when the rate of extinction was much higher than normal: mass extinctions.

10.1 Mass extinctions in the geological record

Throughout your study of geology, you have probably become very familiar with the geological time-scale and its division into eons, eras and periods; indeed, few discussions in geology can occur without reference to it. We can recognize the subdivisions of the geological time-scale globally because most of them are defined by the fossils they contain. For example, if we made a detailed examination of the fossils from the bottom to the top of a particular succession of sedimentary rocks, the chances are that we would see a series of different fossils; some would disappear, to be replaced by new fossils as we moved forward in time. This is the principle of biotic succession established in the 19th century by William Smith (1769–1839). If we examined rocks from much of the last 500 Ma or so, the period for which we have a relatively detailed fossil record, this is exactly what we observe — a process of gradual change and evolution. However, the geological record is punctuated by several major episodes when between 25–90% of all the species living at the time became extinct, and disappeared from the record over a relatively short time interval. Such episodes of large-scale extinction, affecting many different species within a short time interval, are known as mass extinctions and, as you saw in Chapter 1.2, they have long been recognized as defining moments in the history of life on Earth.

This Chapter is concerned with the most recent of the five major mass extinction events, which occurred at the end of the Cretaceous and defines the end of the Mesozoic Era.

The boundary between the Cretaceous and Tertiary was first described by the 18th century French naturalist Georges Cuvier (1769–1832), who observed a sharp transition between a chalk unit and an overlying clay with distinctly different fossils (Figure 10.1). Cuvier's observations were made on the rocks of the Paris Basin and he believed they showed that the history of life was marked by episodes of creation, each followed by a period of equilibrium with little change, and then finally annihilation. Such a catastrophic view of the history of life was clearly at odds with the Principle of Uniformitarianism as developed by Charles Lyell (1797–1875) in his book *Principles of Geology*. Indeed, the recognition of this sharp boundary corresponding to the modern Cretaceous/Tertiary boundary preceded both the development of the modern concept of mass extinctions and Charles Darwin's theory of evolution by natural selection.

Figure 10.1 Cuvier's graphic representation of a section from the Paris Basin published in 1812. The Cretaceous beds end at the top of '1.Craie' printed at bottom left.

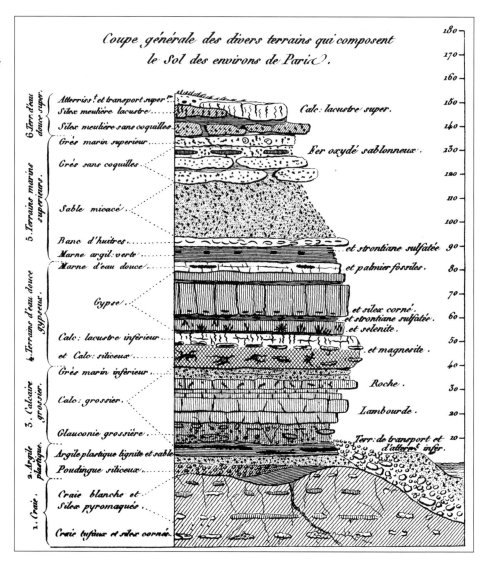

By the mid-19th century, geology was dominated by the recently established Principle of Uniformitarianism, which argued that very small changes, occurring over vast amounts of time, were responsible for the much greater changes observed in the geological record. Lyell had subdivided the Cainozoic using assemblages of fossil molluscs, but he also noted that there was an enormous change in fossil faunas between the uppermost Cretaceous and overlying Tertiary rocks that encompassed a wide range of both fauna and flora.

○ What inference can we draw about the nature of the Cretaceous/Tertiary boundary from the concept of gradual change envisaged in the Principle of Uniformitarianism and the sudden change in fossil assemblages?

● Since Uniformitarianism (at least as Lyell envisaged it) interprets the geological record in terms of a succession of minor changes, then there should be transition beds between the beds representing the fauna and flora of the Cretaceous and those of the Tertiary.

Lyell believed that the 'missing' sequences should occur somewhere. However, the expected transition beds for the Mesozoic/Cainozoic boundary were never found.

The success of Lyellian uniformitarianism led to a view that persisted for the next century, namely that interpretations of the geological record in terms of catastrophes were inaccurate, and there was little evidence that catastrophism was a valid approach. It was not until the 1960s–1970s, when palaeontologists began to compile large databases of fossil assemblages of different ages, that it became apparent there really were times of very high extinction rates. A more recent compilation of data for marine invertebrates is shown in Figure 10.2. This uses the same database that you first saw in Figure 1.24, and it indicates that there have been several times in the last 270 Ma when there were very high levels of extinction. The graph only shows the last 270 Ma because this is the better-preserved part of the fossil record, and uses abbreviations for geological periods. This is normally the first letter of the period such as 'J' for Jurassic, but in the case of the Cretaceous the letter 'K' is used, from the German '*kreide*' (chalk), to avoid confusion with the Carboniferous (C). 'T' is used for the Tertiary. The Cretaceous/Tertiary boundary is therefore commonly abbreviated to the K/T boundary. In this graph 'percent extinction' is defined as:

$$\text{Percent Extinction} = \frac{\text{number of extinctions of genera in an interval of time}}{\text{number of genera present in that time interval}}$$

There are three large peaks apparent: at the end of the Permian, at the end of the Triassic and at the end of the Cretaceous.

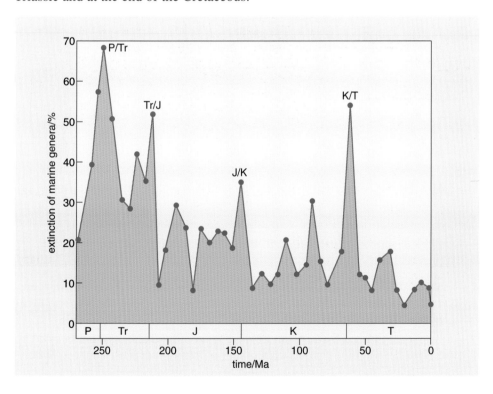

Figure 10.2 Percentage extinction of marine genera over the last 270 million years. The total number of genera represented is 10 383, of which 6350 are extinct. *(Sepkoski, 1990b.)*

○ What percentage of marine genera became extinct at the Cretaceous/Tertiary (labelled K/T) boundary?

● From Figure 10.2, it would appear that between 50% and 60% of marine genera became extinct at the K/T boundary.

In fact, as we shall see in Section 10.3, it is estimated that 80–90% of all marine species became extinct at the K/T boundary, with some groups of marine organisms, such as the ammonites, disappearing completely while others, such as corals, survived across the boundary and are still with us today.

With the recognition that there was undeniably a mass extinction marking the end of the Mesozoic Era, scientists began to turn their attention to its cause. While a much better understanding of the events surrounding the K/T boundary had emerged, there were still major uncertainties, which led to a wide spectrum of hypotheses on the mechanisms responsible for the mass extinction. Among the more common were sea-level changes, volcanism and global events in which deep-ocean waters became depleted in oxygen, all of which could have resulted in major environmental stresses and wiped out many species.

However, one idea was to lead to a major controversy about the K/T boundary that persists to this day. In 1980, two teams of scientists, one led by a Nobel Laureate, Luis Alvarez, together with his son Walter, and the other by a Dutch postgraduate student, Jan Smit, reached the same conclusion based on their studies of K/T boundary rocks in Italy and Spain. In papers published in the scientific journals *Science* and *Nature* within a few weeks of each other, both teams put forward the hypothesis that a large asteroid or comet impacted the Earth's surface at the time, triggering a catastrophic mass extinction. After nearly 200 years, catastrophism in the shape of what has become known as the impact–extinction hypothesis had returned to the forefront of geology.

10.2 An impartial record?

Our ability to interpret the fossil record and to examine the detail of the end-Cretaceous mass extinction depends upon the quality of the data we are able to assemble. Therefore, before examining the fossil record in detail across the K/T boundary, we need to assess how reliable these data are and what the potential for misinterpretation might be. We have to start by considering how we get the raw data for our investigation. The actual documentation of large changes in the fossil record, such as a mass extinction, relies on one of the oldest of all geological techniques: biostratigraphical sampling (Box 10.1).

○ What information do we need from the biostratigraphical record of a geological section to properly document the level of extinction that has occurred?

● To properly document a mass extinction event, we need to know the highest level — worldwide — at which particular taxonomic groups or taxa, such as genera or species, occur.

The difference between a biostratigraphical range chart constructed for a period of the geological record where background extinctions were the norm and one constructed across a mass extinction boundary could not be more marked. The example shown in Figure 10.4 is from a marine rock succession across the K/T boundary at Agost in southern Spain and, as with Figure 10.3, shows the appearance and disappearance of species of planktonic foraminifers. A large number of foraminifers that were abundant in the Cretaceous (shown in blue) disappear at the boundary; some 47 species that are common in the Cretaceous rocks of the

Box 10.1   Biostratigraphy

Biostratigraphy is the process of examining stratigraphy based on the information provided by fossils. Since the early 19th century, biostratigraphers have used the tried-and-tested technique of collecting fossil samples from measured sections, and then using checklists of species to document the ranges over which particular species are found.

One outcome of such a procedure might be the kind of chart shown in Figure 10.3, which indicates the presence or absence of a particular fossil (shown along the right by their Latin names), together with their first and last appearance in the geological period (shown by the solid orange horizontal lines). This type of chart is known as a biostratigraphical range chart. It was constructed by using data obtained from a marine rock succession, in this case using a particularly abundant group of marine microfossils, the calcareous single-celled planktonic foraminifers (see Figure 1.26c). In this particular example from the mid-Cretaceous, we can see that species appear and disappear consistent with William Smith's Principle of Faunal Succession. This continuous disappearance of some fossils from the record is the normal extinction of taxa that occurs as a result of local changes in environmental conditions or other factors and is referred to as background extinction.

Calcareous microfossils such as planktonic foraminifers are ideal biostratigraphical fossils because they occur in large quantities over wide areas, are well preserved in marine sediments and evolved fast enough over long periods of time to make them stratigraphically useful (Section 1.3.1).

In terrestrial sedimentary rocks, spores and pollen are the fossil group most commonly used as biostratigraphical fossils. Other fossil groups, such as vertebrates or other larger fossils, are generally not suitable because they occur in relatively low abundances and are not widespread. Spores and pollen can also be recovered as wind-blown fossils in marine sedimentary successions, making them important for correlating the marine and terrestrial records.

The region of the stratigraphical record defined by the first and last occurrences of given taxa is referred to as a biostratigraphical zone, or biozone. Biozones are usually based on assemblages of several taxa, but are named after a key constituent. They enable the stratigraphical record to be subdivided into relative time periods.

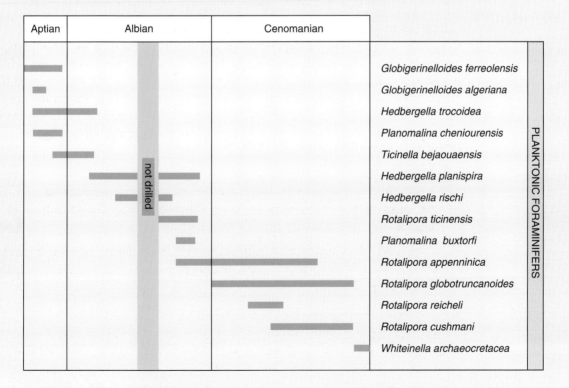

Figure 10.3   Biostratigraphical range chart for planktonic foraminifers from the mid-Cretaceous of the western North Atlantic. The appearance (origination) and disappearance (extinction) of individual species is part of a pattern of background extinction that occurs throughout the geological record. *(ODP.)*

area are not found in the Tertiary rocks. A few species of foraminifers (shown in red) appear to range a short way above the boundary into the Lower Tertiary before they too disappear, while other foraminifers (shown in brown) that were common in the Cretaceous seas appear to have survived intact well into the early Tertiary. However, life is resilient and, above the K/T boundary, we begin to see the appearance of new species of foraminifers (shown in green) that were not present in the Cretaceous; a process referred to as faunal turnover. An obvious conclusion to draw from Figure 10.4 is that large numbers of foraminifers became extinct at the K/T boundary.

○ Can we infer from the data in Figure 10.4 that there was a mass extinction at the K/T boundary?

● No, because while the data in Figure 10.4 are persuasive, they are data for a single location and for one group of organisms.

We will see in Section 10.3 that there is ample evidence for a widespread mass extinction of many groups of marine organisms across the K/T boundary. However, before we can reach such a conclusion, it is clearly important that we determine the degree to which the palaeontological record reflects actual extinction events, and to what extent it has been modified by gaps in the record, incomplete sampling, and diagenetic effects.

The major faunal turnover that occurred at the end of the Cretaceous has been known for a long time and, indeed, its very magnitude was used by 19th century geologists to separate the Mesozoic and Cainozoic Eras. However, what has been less clear is whether the end-Cretaceous mass extinction was an instantaneous and worldwide event, or whether it extended over a longer period of time. In any hypothesis proposed to explain the cause of the mass extinction, the *timing* of that extinction is the most critical issue.

We have already established that to properly document the magnitude of extinction, it is important to determine the highest stratigraphical level at which a particular taxon occurs. What are the chances that we will collect the *actual* last occurrence of a fossil, and on what factors does this depend?

○ Is the sedimentary rock record always complete?

● No, there are gaps or hiatuses in the record as a result of non-deposition of sediments or as a result of erosion. These gaps may be recorded as unconformities, but often there is no discordance in the orientation of strata above and below the hiatus.

A hiatus due to non-deposition or erosion can change the apparent pattern we observe in the fossil record for the disappearance of species (extinction) and the appearance of new species (origination). This is illustrated in the simplified biostratigraphical range charts shown in Figure 10.5. In Figure 10.5a, the ranges of particular species (shown as vertical lines) cross a boundary (shown as a horizontal line).

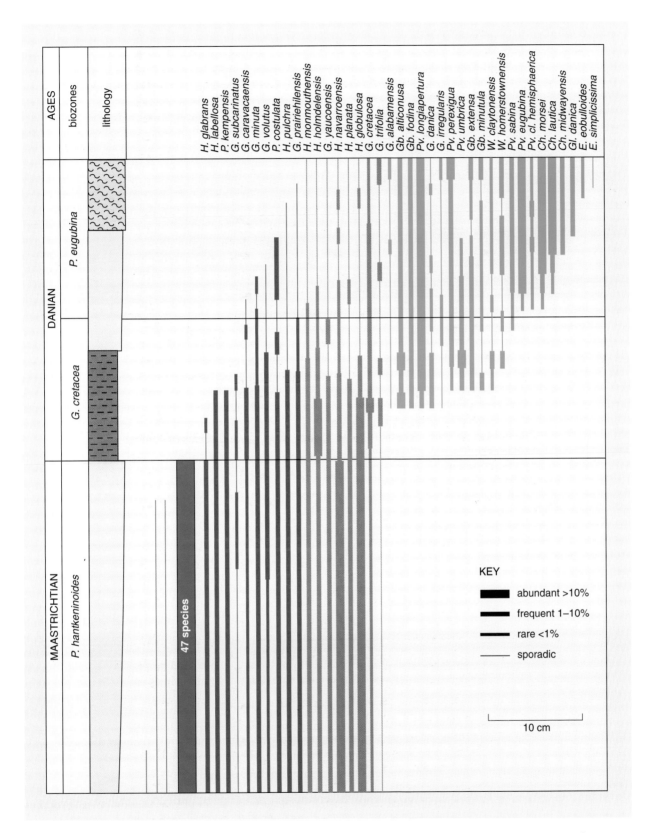

Figure 10.4 Biostratigraphical range chart for planktonic foraminifers across the K/T boundary at Agost in southern Spain. This locality has provided a particularly high-resolution record of the pattern of extinction of foraminifers across the K/T boundary, since there is little evidence of reworking or bioturbation in the section. (*Molina et al., 1996; data courtesy of Eustoquio Molina, University of Zaragoza.*)

Figure 10.5 Diagram showing how a hiatus in the stratigraphical record can affect the apparent pattern of extinction and origination. (a) The ranges of species cross a boundary without any extinctions, or originations, coinciding. (b) The introduction of a hiatus results in the apparent coincidence of extinctions and originations for several species. (c) The larger the hiatus, the greater the number of species with apparently coincident extinctions or originations. *(Ward, 1990.)*

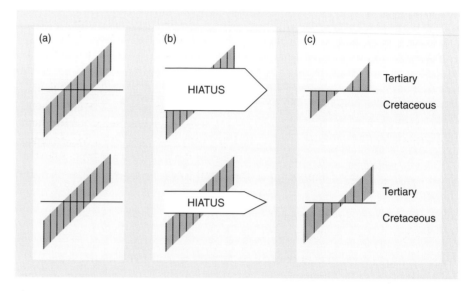

○ Do any of the extinctions or originations of species in Figure 10.5a coincide with each other?

● No, in this case none of the extinctions, or originations, coincide.

If we now introduce a hiatus, i.e. an interruption in chronological continuity such as a lack of deposition or a period of erosion, into our record (Figure 10.5b), then the pattern of disappearance or appearance can change and it can seem as if the extinction of several species was simultaneous. The larger the hiatus, the more pronounced is the apparent coincidence (Figure 10.5c). In this way, an unrecognized hiatus in the stratigraphical succession can create the impression of simultaneous extinction and origination where, in reality, none existed.

○ Are all species alive on Earth today present in equal abundances?

● No. For any particular community, both living and fossil, there are a few abundant species, a few common species, and a large number of rare species.

This uneven abundance of species throughout Earth's history can have a marked effect on the palaeontological sampling of the fossil record. It means that large numbers of species that are actually preserved in a fossil-rich succession of rocks will be missing from the sampling. Since we only ever collect, at most, a few kilograms of rock for each sample taken, it is therefore impossible to sample the entire record.

○ What effect will this uneven distribution of species have on the apparent abundance of rare species immediately below a mass-extinction boundary where these species disappear from the record?

● Samples taken immediately below the boundary will only contain a few of the rare species that we would find lower in the stratigraphical section, even if these particular species actually survived right up to the boundary. One consequence of this is that rarer species will require far more effort in collection than the more common species, if reliable biostratigraphical range charts are to be constructed.

The period of time over which a particular taxon was alive is given as the time from its first occurrence to its last occurrence (i.e. range), regardless of whether or not it is observed to occur at some intermediate position in the stratigraphical record. However, for most species, the last sampled occurrence is likely to occur *below* the stratigraphical position at which the species actually disappears. The effect of this sampling bias is that, since most species lived on beyond their last-known occurrences in the fossil record, the ranges we would record for them on a biostratigraphical range chart would be artificially truncated. As a consequence, rarer fossils can seem to disappear much earlier in the fossil record than more numerous ones, even if both groups survived until the same time.

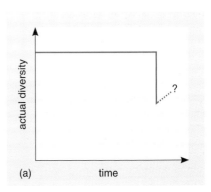

○   What effect might this sampling bias have on the record of a catastrophic extinction such as that shown in Figure 10.4?

●   It could make a sudden catastrophic extinction appear to be more gradual. Rarer species would apparently disappear before the main extinction horizon.

This effect of such sampling biases on the apparent pattern of an extinction event is known as the Signor–Lipps effect, after the palaeontologists Philip Signor and Jere Lipps (1982). They noted that catastrophic hypotheses for mass extinctions were commonly criticized because many species apparently gradually disappear from the fossil record prior to the extinction. The Signor–Lipps effect is illustrated in Figure 10.6 which shows how a sudden extinction involving the simultaneous disappearance of many taxa can appear to be a gradual extinction because of sampling problems. Figure 10.6 shows the effect on diversity (the variety of taxa present at a particular point in the geological record) of the presence of rare taxa or of incomplete sampling of the fossil record. In Figure 10.6a, the actual diversity of taxa in a stratigraphical record is constant through time and then undergoes a sudden and instantaneous reduction — a mass extinction. Figure 10.6b shows how the extinction may appear if sampling is incomplete or if the taxa are not commonly preserved.

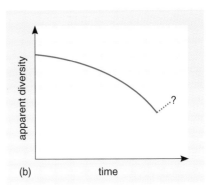

Figure 10.6   Diagram illustrating how a sudden extinction can appear to be gradual as a result of sampling biases. (a) The diversity remains constant through time and then shows a sharp reduction as a result of a mass extinction. However, (b) shows what might be observed if the sampling is incomplete or if many of the species being studied are rare. In this case, the decline in diversity may appear to be more gradual. *(Ward, 1990.)*

The recognition, during the late 1980s and 1990s, of the effects of hiatuses and of the Signor–Lipps effect on the pattern of change in the fossil record made palaeontologists more cautious when it came to interpreting their data. In particular, it became clear that extinction recorded by fossils in the rock record was not necessarily an accurate record of the actual event. It is worth noting that, in places where the fossil record is particularly sparse, the potential for misinterpretation due to the artificial truncation of a species' biostratigraphical range is greatly increased. A lack of data might, for example, result from a change in depositional environment. For some species, data may become sparse for other reasons: for instance, as we shall see in Section 10.3, there were considerably fewer genera of ammonites toward the end of the Cretaceous than there had been in the Early to mid-Cretaceous.

○   How might we attempt to overcome or reduce the effects of sampling bias?

●   If we were to consider sufficiently large numbers of species, then we would be less likely to misinterpret the early disappearance of a few rarer species.

○   What sort of properties would make a particular fossil a good one for studying levels of extinction?

- An ideal fossil would be geographically widespread, and would occur in large numbers and in a variety of environments. Size is also important, as smaller fossils are more likely to be sampled in relatively small samples of bulk sedimentary rock.

Perhaps the most useful organisms for studying extinction levels are microfossils. They can generally be found in large numbers and occur in sufficient quantities in small samples, so that they are easy to collect. Microfossils also yield reliable information about their biostratigraphical ranges, and their widespread geographical occurrence makes them useful tools for studying extinction levels on a global basis. They do have some drawbacks: many microfossils are most common in deep-sea sediments where other fossils are rare. This can make it difficult to correlate deep-sea sections with shallower marine facies where macrofossils may be more abundant. The small size of microfossils can also be a problem since they can be mobilized by bioturbation either upward or downward in the sediment soon after deposition, thus giving a false signal of their extinction levels.

In reading this Section, you may have started to question whether it is possible to obtain an accurate picture of an extinction event from the fossil record. This is the same question that palaeontologists found themselves asking not long after the impact–extinction hypothesis was first proposed for the K/T mass extinction. It quickly became apparent that a literal reading of the fossil record, the accepted palaeontological method for the previous 200 years, was both inaccurate and misleading. Overcoming the difficulties posed by the Signor–Lipps effect and hiatuses presented palaeontologists with quite a problem. Today, most palaeontologists studying extinction events use statistical methods to test for artificial extinction steps and to enable them to place a level of confidence on the stratigraphical range of a particular species. Nevertheless, different methodologies or discrepancies in the classification of some species can still lead to conflicting interpretations.

In Sections 10.3 and 10.4, you will examine the marine and terrestrial record of the mass extinction at the K/T boundary. From the preceding discussion, you should have a sense of the considerable care that has to be exercised in order to obtain good palaeontological data. However, as you will see, even good data can be open to more than one interpretation and there has been considerable scientific debate, and at times heated controversy, about the nature of the mass extinction at the K/T boundary.

10.3 The marine record of mass extinction at the K/T boundary

It may seem surprising that, given the magnitude of the faunal turnover at the K/T boundary, there should be so much controversy and debate surrounding the interpretation of the fossil record. However, the magnitude of the extinction itself has never been seriously questioned. As you have already seen, it is the timing of the extinction that is critical. For marine sedimentary facies, the best evidence of the changes that occurred in the diversity of life in the Late Cretaceous oceans comes from the fossil record of shelly marine invertebrates. Both microfossils and macrofossils register the effect of the extinction and, as they can be found in large numbers, microfossils are the most useful organisms for studying extinction levels worldwide.

10.3.1 The microfossil record

Six groups of microfossils have been the subject of extensive study in attempts to unravel the pattern of extinction at the K/T boundary. For three of these groups — the calcareous nanofossils known as coccoliths, the radiolaria which have silica skeletons, and the floating or planktonic foraminifers — more than 90% of their species, overall, became extinct at or near the K/T boundary (Figure 10.7). Other groups of microfossils such as the benthic foraminifers, dinoflagellates and the silica-shelled diatoms were less affected.

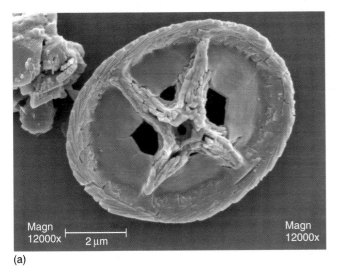

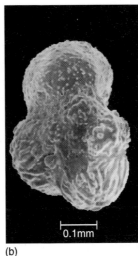

(a)   (b)

Figure 10.7 Two microfossil species from the chalk of southern England that become extinct at the K/T boundary. (a) A coccolith plate from *Eiffelithus turriseiffelii*. *(Courtesy of Jeremy Young, Natural History Museum, London.)* (b) The planktonic foraminifer *Rugoglobigerina macrocephala*. *(Courtesy of Kate Harcourt-Brown, University of Bristol.)*

The coccoliths are extremely small and widespread, making them an exceptionally useful fossil in biostratigraphy. Unfortunately, they are also easily moved around in sediments by bioturbation so that even a sudden extinction can look gradual. This makes them a problematic fossil for the study of the K/T boundary. The larger planktonic foraminifers, on the other hand, are much less affected by bioturbation. They apparently underwent one of the most catastrophic of extinctions of any group of organisms, with more than 97% of their species disappearing at or near the K/T boundary. Not surprisingly, therefore, the precise timing of the extinction of planktonic foraminifers has been the subject of intense scientific scrutiny. Two diametrically opposed interpretations have been proposed: a sudden and globally synchronous extinction event; and a gradual or stepwise extinction extending over a much longer period of time.

10.3.2 The macrofossil record

The idea that a gradual or a stepped pattern of extinction might exist at the end of the Cretaceous is worth further investigation since such a pattern would have profound implications for the possible causes of the mass extinction. In the previous Section, we noted that background extinctions are happening all the time, so that there would undoubtedly have been some extinctions occurring in the few million years prior to the K/T boundary. However, the concept of a stepped extinction was based on the hypothesis that different groups of Late Cretaceous organisms underwent extinctions in a stepwise fashion, with the first step starting well before the end of the Cretaceous. In this hypothesis, the K/T boundary corresponds to one of these extinction steps. If this hypothesis is correct, it would suggest that large-scale changes took place in the biosphere 2–3 Ma prior to the K/T event itself. The proponents of the stepwise extinction model cited as evidence the apparent simultaneous major decline of three major groups of marine invertebrates — the inoceramids, rudists (Figure 10.8), and ammonites — at around 67.5 Ma ago.

Figure 10.8 Rudist bivalves from the Upper Maastrichtian, Jebel Faiyah, United Arab Emirates. This group of large bivalves lived predominantly in shallow waters and apparently declined markedly in diversity some 2.5 Ma before the K/T boundary. *(Peter Skelton, Open University.)*

As we have seen, however, a literal interpretation of the fossil record can lead to misinterpretation of the pattern of an extinction event. The palaeontologist Peter Ward, from the University of Washington (2001), examined in some detail the extinction pattern of the rudists, inoceramids, and ammonites. It will not surprise you to learn that he came up with a different explanation from the proponents of the stepwise extinction model. Other researchers had previously noted that the disappearance of rudists coincided with the reduction in the area of shallow marine environments during the latest Cretaceous as a result of the regression of the sea. This means that it is difficult to distinguish between the effects of regression, which removed shallow marine facies from the geological record of the Late Cretaceous, and an extinction. Since rudists were organisms that occurred in large numbers in shallow waters, a regression in any given section could have resulted in only an apparent extinction, although the loss of their preferred habitat might indeed have caused their genuine extinction.

○ Since rudists were restricted to shallow waters, how easy would it be to correlate their extinction to the extinction of other organisms?

● Quite difficult. As we have seen, the best organisms for studying patterns of extinction are ones that not only occur in large numbers, but are also widely distributed across different environments and geographical areas.

Historically, dating rudist-bearing Campanian–Maastrichtian limestones has presented quite a problem, because of the paucity of other age-diagnostic fossils (e.g. ammonites or planktonic foraminifers) in them. Recently, however, new approaches, such as Sr-isotope correlation (Section 1.3.1), have led to a marked improvement. In particular, Thomas Steuber and colleagues demonstrated (2002) that diverse rudist associations in Jamaica persisted into the latest Maastrichtian (66–65 Ma ago), at least. So linkage of their extinction in the New World with the K/T event now appears highly probable. On the other hand, use of the same methods on Old World rudists (e.g. Figure 10.8) currently still suggests a more extended decline during Maastrichtian times.

The inoceramids, another group of large epifaunal marine bivalves, were widespread throughout the late Cretaceous. Unlike the rudists, the inoceramids were latitudinally widely distributed in a variety of environments and can be well correlated with the fossil record of other organisms, in particular, with planktonic foraminifers. There seems to be little doubt, therefore, about the interpretation of their fossil record and the evidence that they underwent worldwide decline well before the K/T boundary with only a few species surviving into the late Maastrichtian.

Ward made an important observation about the distribution and eventual extinction of ammonites in Late Cretaceous seas (Figure 10.9). The ammonites, abundant in the oceans throughout the Mesozoic, had been in decline for several million years prior to the K/T boundary. It was initially thought that they too had declined markedly along with the inoceramids 67.5 Ma ago, supporting the hypothesis of a gradual or stepwise extinction. This decline is reflected in Figure 10.9, which shows how the diversity of ammonites was decreasing throughout the Cretaceous.

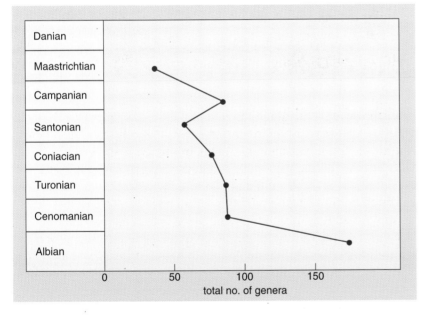

Figure 10.9 Ammonite diversity in the Cretaceous reached a peak in the Albian but then declined throughout the rest of the Cretaceous until they finally became extinct at the K/T boundary.

○ What effect might this decline in diversity have had on the apparent extinction pattern of ammonites?

● The decrease in diversity could have introduced a sampling bias, especially towards the end of the Cretaceous when ammonite diversity reached a minimum. An abrupt extinction might therefore appear as a gradual one.

Ward therefore wanted to know whether the remaining ammonites became extinct before the K/T boundary or abruptly at it. This entailed making an intensive collection of ammonite species in the uppermost Cretaceous in the Bay of Biscay in an effort to determine which ammonite species survived right until the K/T boundary (Figure 10.10). This was a much larger set of data than had been previously used. Following extensive statistical analysis, it was concluded that, of the few remaining ammonites, most had probably become extinct precisely at the K/T boundary and were therefore victims of a single K/T mass extinction.

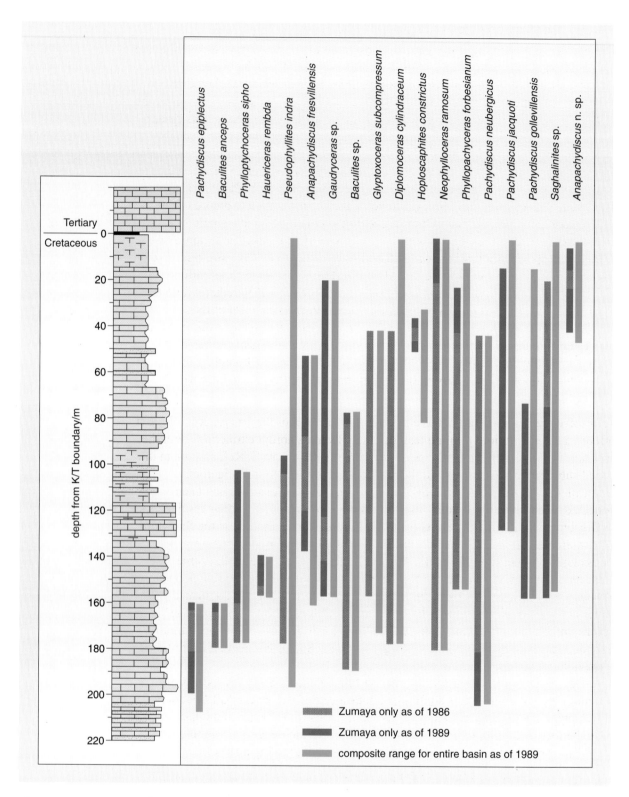

Figure 10.10   Ranges of ammonite species near the K/T boundary in Zumaya, Spain, and composite ranges for the entire sedimentary basin. Three datasets are shown. An initial collection made at Zumaya in 1986 (shown in grey), a subsequent collection at Zumaya in 1989 that extended the ranges of several ammonite species (blue) and a composite range for each species derived from data for the entire sedimentary basin, encompassing several sampling sites (green). (Ward, 1990.)

10.3.3   Testing the fossil record

The problems of sampling were further highlighted in a controversy surrounding the nature of the mass extinction of planktonic foraminifers. As we have already seen, these are ideal fossils with which to study patterns of extinction and should provide a reliable measure for the pace of extinction around the K/T boundary. However, the reality of the fossil record can result in intense scientific debate that, in this case, resulted in a somewhat unusual proposition to resolve a controversy between the two main protagonists: Jan Smit, from the Free University of Amsterdam and also one of the co-proponents of the impact–extinction hypothesis; and Gerta Keller, a palaeontologist from Princeton University. After analysing the El Kef section from Tunisia, an expanded section thought to contain the most complete record of the K/T planktonic foraminiferan extinction, Smit had concluded that a single catastrophic event at the K/T boundary had eliminated all planktonic foraminifers with the exception of a single species, *Guembelitria cretacea*. This single survivor species, he argued, then gave rise to all Early Tertiary planktonic foraminifers. Keller, on the other hand, looking at the same section at El Kef together with one from Brazos River in Texas, estimated that approximately one-third of the Cretaceous planktonic foraminiferal species in both sections had survived into the lower Tertiary. Keller concluded that there had been four successive extinction steps: two in the Cretaceous; an extensive extinction at the K/T boundary that eliminated 42–58% of planktonic foraminiferan species; and one in the earliest Tertiary.

To resolve the dispute, sedimentologist Robert Ginsburg of the University of Miami offered a novel solution: a 'blind test' of gradual versus abrupt extinction. With the assistance of Smit and Keller, he collected new samples from El Kef, split them into coded subsamples and distributed them to four foraminiferan investigators. Unaware of how far below or above the K/T boundary each sample had been collected, each investigator identified the species present and sent their results back to Ginsburg in Miami. The results of the blind tests were presented at the Snowbird III conference (see Box 10.2), in Houston in 1994. Perhaps not surprisingly, both sides claimed a measure of victory. The blind investigations found that some Cretaceous foraminiferan species apparently disappeared before the K/T boundary, although the estimates ranged from 2–21%. Keller argued that this confirmed a pattern of gradual extinctions with seven species of foraminifers disappearing before the K/T boundary. Smit viewed the blind investigations differently, arguing that the Signor–Lipps effect was biasing the results. Smit combined all four of the blind investigations, including only those species that two or more of the blind investigators spotted in their sample set. He found that, of the seven species of foraminifers that, by Keller's analysis, had disappeared before the K/T boundary, one or other of the blind investigators had found all seven species in the last sample taken before the boundary. Smit's argument was convincing, and combining the results was an effective means of reducing the risk of sampling biases.

Subsequent study of the blind test datasets have led other investigators to conclude that, while the gradual pattern of extinction advocated by Keller is indeed sample-biased, more Cretaceous planktonic foraminifers survived into the Tertiary than the single species observed by Smit (see also Figure 10.4). However, data from many more K/T sections have provided evidence that, with the exception of a few species, most species of planktonic foraminifers present in the latest Cretaceous had perished at the K/T boundary. Palaeontologists, for the moment, have reached a consensus that the combination of the extinction pattern and subsequent radiation of planktonic foraminifers indicates that an event of very large magnitude had occurred at the boundary.

Box 10.2 The Snowbird conferences: debating the impact–extinction hypothesis

Most of the scientific debate concerning the impact–extinction hypothesis has taken place in the published scientific literature over the last 20 years, the original 1980 papers having been cited by more than 1000 other scientific articles. It has been an extraordinarily wide debate, encompassing fields as diverse as palaeontology, geology, geochemistry, geophysics, planetary science and astronomy. Some of the landmark papers in the debate have been published in the broad international science journals such as *Nature* and *Science*. However, much of the essential detail exists in the more specialist publications of the individual fields. Every six years or so, therefore, a scientific conference has been held to bring together scientists from all over the world and from a wide range of disciplines. Dubbed the Snowbird conferences (because the first two were held in Snowbird, Utah, USA), they represent defining moments in the history of the debate (see Section 13.7 for references to the published proceedings of these conferences). Each has seen a sea change in how the wider geological community has viewed the impact–extinction hypothesis.

Snowbird I, entitled 'Conference on Large Body Impacts and Terrestrial Evolution: Geological, Climatological, and Biological Implications' was held at Snowbird, Utah in 1981. At the time, a sizeable number of scientists rejected the reality of an impact, or of a mass extinction, or both. Scientists were still coming to terms with the concept of a large impact and how it might have caused extinctions, as well as beginning to accumulate evidence for it. It was generally accepted that the chemical traces at the K/T boundary were probably those of an impact and that there was global evidence for it in both marine and terrestrial rock successions. However, there was considerable scepticism about its consequences, their severity, and whether they were enough to have caused a mass extinction.

Snowbird II, entitled 'Global Catastrophes in Earth History: An Interdisciplinary Conference on Impacts, Volcanism, and Mass Mortality', was held in 1988, again in Snowbird, Utah. By now there had been several significant developments in the debate. One was the recognition in boundary clays of shocked-quartz and the high-pressure mineral stishovite (a form of quartz only produced by very high pressures), both seen as diagnostic criteria of impact by many scientists. Many boundary deposits had also been found that contained silicate melt droplets or remnants of them. Two main themes dominated the conference. The first was whether the evidence favoured an impact at the K/T boundary or could be explained by terrestrial volcanism, a hypothesis that had several vocal proponents. By the end of the conference, a consensus had emerged that volcanism was inconsistent with the evidence. The second theme concerned the nature of the faunal transition itself — what had happened and when, and how sharp was the boundary? The palaeontological record around the K/T boundary was being subjected to ever-greater scrutiny but there were still widely differing opinions as to which species had become extinct abruptly at the K/T boundary.

Two dramatic developments occurred after the Snowbird II Conference. The first was the recognition of some very unusual deposits at the K/T boundary in the regions around the Gulf of Mexico. These deposits that were postulated by some to be the result of tremendous tsunamis caused by an impact. The second was the recognition and identification of an extremely large crater buried in the Yucatan Peninsula of Mexico. The stratigraphical age of this crater, Chicxulub, was consistent with the age of the Cretaceous/ Tertiary boundary. Snowbird III, entitled 'New Developments Regarding the K/T Event and Other Catastrophes in Earth History', held at the University of Houston, Texas, in 1994, was preceded by a field trip to north-eastern Mexico by some 50 scientists and postgraduate students to examine some of the unusual K/T boundary deposits. By this time, there was an overwhelming consensus that an impact had occurred at the K/T boundary with very few arguments in favour of a volcanic influence on boundary events. The debate focused on whether the extinctions were caused by the impact. Some scientists argued that the extinctions were 'stepped' and of a pattern inconsistent with impact. Similarly, not everyone agreed that the boundary deposits in the Gulf of Mexico region were related to an impact. At issue was a question central to all of the Snowbird meetings: the precise timing and pattern of the extinction. The results of a 'blind test' of the abruptness of the planktonic foraminiferan extinction presented at the meeting fuelled the controversy over the interpretation of the fossil record.

Snowbird IV, the most recent conference, entitled 'Catastrophic Events and Mass Extinctions: Impacts and Beyond', was held at the University of Vienna, Austria, in 2000. The title reflected the maturity of the debate, with few scientists now disputing that the Chicxulub impact and its global ejecta layer mark the end of the Mesozoic Era. This conference focused on the effects of the impact and on other periods of rapid environmental change in the history of the Earth. The palaeontological record is considerably more refined than it was 20 years ago, although there are still areas of uncertainty. Some scientists, while accepting that there was an impact, argued that it had little effect, or that the mass extinction was not that significant, or even both. A more widespread view was that the Chicxulub impact was the final straw that tipped the balance for many species already stressed by climatic and environmental changes during the Late Cretaceous. A new tool available to geologists, powerful computer programs that can model both the impact process and its effects on climate, provided evidence in support of the latter view.

10.3.4   Geochemical evidence for a collapse of primary productivity in the oceans

Evidence in support of an abrupt extinction of planktonic foraminifers and in favour of a major disruption of the carbon cycle in the oceans also comes from geochemical evidence. When scientists measured the carbon-isotope compositions of planktonic foraminifers across the K/T boundary, they noticed something very unusual: there was a marked change in the $^{13}C/^{12}C$ ratio in the carbonate skeletons of the foraminifers at the boundary. If we look at the carbon-isotope record across the K/T boundary (Figure 10.11), this change is very apparent. The Figure shows the carbon-isotopic composition across the K/T boundary for seven different K/T boundary sections: El Kef in Tunisia; Caravaca in south-eastern Spain; Biarritz in southern France; Lattengebirge in Bavaria; Stevns Klint in Denmark; and deep sea sites in the South Atlantic and Pacific Oceans.

At these sites, and many others around the world, we can see that as we move through the Late Cretaceous the background carbon-isotope values (shown as $\delta^{13}C$ values on the Figure; see Box 1.1) remain fairly constant. However, at the K/T boundary, the $\delta^{13}C$ suddenly becomes more negative. Eventually, a few centimetres beyond the boundary the isotope values return to values similar to those observed in the Cretaceous.

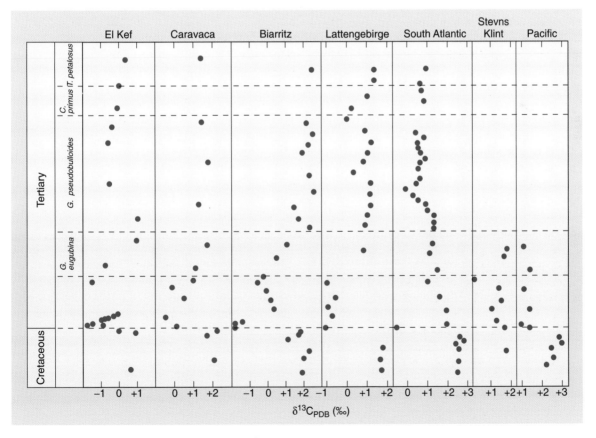

Figure 10.11   Carbon-isotope compositions ($\delta^{13}C$) of carbonate from marine planktonic foraminifers across the K/T boundary at seven locations worldwide. *(Perch-Nielsen et al., 1982.)*

Under normal conditions of photosynthesis in ocean surface waters, the organic or living part of phytoplankton preferentially use ^{12}C over ^{13}C when they fix CO_2 or bicarbonate during the first stages of photosynthesis that result in production of simple organic compounds (see Box 1.1). As a result, relatively more ^{13}C finds its way into the inorganic calcium carbonate that makes up the foraminifer skeletons. However, at the K/T boundary we have already seen there is evidence that more than 95% of surface water planktonic foraminifers became extinct.

○　If there were fewer surface water plankton, what would have been the effect on the level of photosynthetic activity in ocean surface waters?

●　The level of photosynthesis would have been much less. In effect, there would have been much less primary production in ocean surface waters.

With considerably fewer organisms fixing carbon during photosynthesis, therefore, much less ^{12}C would have been removed resulting in a relative excess of this isotope in the surface waters of the ocean. At this point, any carbonate formed in the skeletons of surviving plankton, such as those in the *eugubina* zone of the Lower Tertiary, would then contain more ^{12}C than normal. Geochemist Ken Hsü has called this a Strangelove ocean (after the mad scientist in the film *Dr Strangelove*), to emphasize the fact that the biological pump of surface water primary productivity was, in effect, shut down at the K/T boundary.

The abrupt change in the carbon-isotope composition of planktonic foraminifers is observed globally in K/T successions and it provides evidence for a major disruption of the global carbon cycle at the time of the K/T boundary. It also provides a stringent test for any hypothesis put forward to explain the K/T mass extinctions. In addition to explaining the decrease in diversity at the K/T boundary, i.e. the mass extinction itself, any plausible explanation must also be able to account for the apparent sudden and globally synchronous loss of primary productivity in the oceans that the abrupt change in carbon-isotope composition appears to represent.

10.3.5 Pattern of extinction of marine life at the K/T boundary

The pattern of extinction of components of marine ecosystems at the K/T boundary is summarized in Figure 10.16. The consensus of scientific opinion that this Figure represents is not unchallenged. Some scientists still argue that the end-Cretaceous mass extinction was gradual, and that more species survived the K/T boundary. However, the general consensus is that the marine palaeontological record contains evidence for a sudden and abrupt extinction of much of marine life at the K/T boundary 65 Ma ago. The more abrupt the mass extinction, the more likely a catastrophic event was responsible for it. However, such a conclusion cannot be drawn from the study of the marine fossil record alone.

10.4 The fossil record of terrestrial fauna extinction at the K/T boundary

10.4.1 The record from terrestrial vertebrates

The idea that a large asteroid impact 65 Ma ago wiped out the dinosaurs is certainly one that has all the attributes of a good story, and has led to some creative reconstructions of the event (e.g. Figure 10.12). But what is the scientific evidence that the dinosaurs and other vertebrates became extinct at the K/T boundary, and was their extinction abrupt?

Figure 10.12 Artistic reconstruction of the demise of the dinosaurs. *(© NASA Ames Space Science Division.)*

Our evidence will have to come from the fossil record of terrestrial vertebrates across the K/T boundary. However, at the outset, we should bear in mind that this record is far less complete than that of marine microfossils. In part, this is due to the nature of the vertebrate fossil record.

○ How abundant and widespread are vertebrate species in the world today?

● Many vertebrates today, e.g. the mammals, have evolved to occupy particular habitats and are not geographically widespread. Compared with shallow marine invertebrates, they are also much fewer in number.

We are therefore dealing with large fossils of species that were not abundant and were often geographically restricted. We also have to contend with a lack of information concerning vertebrates across the K/T boundary in many parts of the world, because preserved terrestrial sedimentary facies are less common than marine ones. While the K/T boundary can be readily identified worldwide in marine sections, in non-marine sections the K/T transition has only been located with any accuracy in extremely well-studied sections in Mexico, the Western Interior of the United States and Canada.

The dinosaurs are undoubtedly the most spectacular of the various creatures that became extinct at the end of the Cretaceous. However, they represent only a small percentage of vertebrate species in existence at the time. Determining the precise stratigraphical level at which the dinosaurs and other vertebrates actually disappeared from the geological record is not without difficulties. To do so requires a sufficiently continuous vertebrate faunal succession across the K/T boundary, and there are very few locations where such a record exists. Some of the best information comes from a database of terrestrial vertebrates compiled from the Hell Creek Formation which lies to the east of the Rocky Mountains in Montana, USA, and is amongst the most-studied of terrestrial K/T sections in the world. During the Late Cretaceous, the area was a broad, low-lying plain, crossed by several narrow rivers and bordered by an inland sea to the east. Hell Creek's most famous dinosaur is probably *Tyrannosaurus*, which was described for the first time in 1906 by Henry Fairfield Osborn, who also coined the name.

Initial investigations of the Montana database indicated that 53–64% of terrestrial vertebrates survived the K/T boundary.

○ How does this pattern of extinction compare with that for marine invertebrates?

● It suggests that the K/T mass extinction was not as catastrophic amongst terrestrial fauna as it was in the marine realm.

The initial investigation had examined the database as a whole but when in further studies the differences in extinction patterns between land-dwelling vertebrates and freshwater aquatic species were scrutinized, a dramatically selective pattern of extinction emerged. It revealed that around 90% of freshwater species, such as crocodiles, survived into the Tertiary, compared with only 12% of land-dwelling vertebrates, the most notable survivors in the latter group being the mammals. While scientists have debated the precise methods used to arrive at this conclusion, the dramatic selectivity in the extinction patterns of terrestrial vertebrates remains one of the more startling features of the end-Cretaceous mass extinction.

○ What does the selectivity of the extinction patterns imply for any explanation of why they occurred?

● The selective character of the terrestrial vertebrate extinctions requires that any plausible explanation of the extinction should be able to justify why more freshwater fauna were able to survive compared to land-dwelling fauna, and why mammals survived.

One suggestion is that the selectivity was related to a primary productivity drop on land. As we shall see in Section 10.5, there is evidence of a major perturbation in the plant fossil record coincident with the K/T boundary. A variety of short-term effects might have disrupted food chains that were dependent on living plants, while food chains dependent on detritus, both on land and in the water, were little affected. However, given the very small number of detailed field observations, it is difficult to draw any general conclusions about the pattern of terrestrial vertebrate extinctions at the K/T boundary.

10.4.2 The problems of the dinosaur extinction

The extinction of the dinosaurs is particularly problematic, and we still do not know the pattern of dinosaur extinction on a global scale. The main reason for this is the sparse and intermittent nature of the dinosaur fossil record. We saw in Section 10.2 how the Signor–Lipps effect and hiatuses can affect the pattern of extinction and the very detailed sampling that was required to resolve questions about the extinction of the ammonites and other marine invertebrates. In contrast to this level of detail, the total number of remains of individual dinosaurs identifiable to the level of genus that has been globally collected from deposits representing a time interval of 160 Ma is only around 2100. While the Late Cretaceous is perhaps the best-studied period of the dinosaur fossil record, it is likely that only a few of the total number of dinosaur genera in existence at the time have been identified. This makes it difficult to assess any short-term (i.e. a few million years) changes in their diversity.

As with the marine invertebrate record, there has been debate as to whether the dinosaur extinction was a gradual or an abrupt one. The difficulty of analysing the dinosaur fossil record is illustrated in Figure 10.13, which shows the minimum number of individual dinosaurs recorded in the upper 60 m of the Upper Cretaceous Hell Creek Formation in Montana. The data were recorded in 1.5 m intervals through the section, which extended laterally some 35 km. This record was interpreted as evidence for a gradual decline in both the species of dinosaurs and the number of individuals lasting perhaps 7 Ma over the end of the Cretaceous. However, other studies of the same section have found gaps in the record at around 73–70 Ma ago during which no dinosaur remains were found at all, although we know that the dinosaurs were still around during that time period. Such gaps illustrate the paucity of the dinosaur fossil record, the problems of sampling that record, and the difficulties in its interpretation.

In 1991, David Fastovsky and co-workers presented the results of a three-year field study that looked at the family level diversity of dinosaurs in the Hell Creek Formation. Their fieldwork had focused on assessing the ecological diversity rather than taxonomic diversity of dinosaurs. By examining the ecological diversity, in effect by seeing how dinosaurs that lived in a range of habitats coped with the end of the Cretaceous, Fastovsky argued that his approach was more useful in examining patterns of extinction. An important aspect of his study was that it considered the various factors that might have affected the preservation of fossils in the Hell Creek Formation. For example, dinosaur bones that were found in stream deposits resembled a death assemblage, one that had probably been derived from different habitats upstream of their final resting place. Similarly the preservation of large dinosaur bones was favoured over small mammalian bones in fluvial systems, as the smaller bones were more mobile. Taking these and other factors into account, Fastovsky concluded that the dinosaur extinction was sudden and that there was no evidence of a gradual decline in dinosaur diversity during the Late Cretaceous. In fact, they found that all eight of the dinosaur families studied ranged into the upper units of the Hell Creek Formation, immediately below the K/T boundary.

As with the marine invertebrate record, intense scrutiny of the dinosaur fossil record appears to indicate that they became extinct abruptly at the K/T boundary. However, the advocates of a gradual extinction scenario have proposed a final provocative argument: that dinosaurs survived into the Tertiary. This would be an important finding because it would reduce the biological magnitude of the K/T boundary extinction event. This idea is based mainly on stratigraphical arguments: dinosaur fossils have been recovered from Lower Tertiary river channel deposits cut into the Hell Creek Formation. However, whether these fossils are *in situ*, or reworked Cretaceous fossils, remains unclear. Until this question is settled, the idea that dinosaurs may have survived the K/T boundary has not gained wide acceptance among palaeontologists. A catastrophic scenario to explain the end-Cretaceous extinction of terrestrial vertebrates is therefore a distinct possibility.

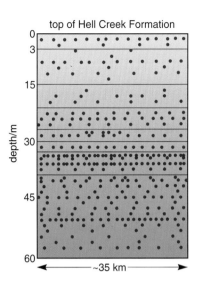

Figure 10.13 The minimum number of individual dinosaurs recorded in the upper 60 m of the Hell Creek Formation, Montana. Each line of dots represents the minimum number of individuals recorded at 1.5 m intervals. A total of 650 fossil specimens were recorded. *(Archibald et al., 2000.)*

10.5 The fossil record of terrestrial flora at the K/T boundary

10.5.1 The evidence from fossil spores and pollen

Plants in the Late Cretaceous were dominated by flowering plants, the angiosperms, and the gymnosperms, especially conifers, which did not have true flowers (Box 4.1). As with the terrestrial fauna record, the best evidence of changes in flora associated with the K/T boundary comes from the North American continent. There have been relatively few investigations of other locations, but those that have been undertaken can give information on the extent and magnitude of the changes associated with the K/T boundary. North American land plants appear to have been overwhelmed by the events surrounding the K/T boundary. Below the boundary, Cretaceous sedimentary rocks contain pollen of gymnosperms and angiosperms, together with the spores of ferns and related plants. In most of these rocks, fern spores make up between 10–30% of the pollen and spore total. However, in rocks immediately above the K/T boundary, angiosperm pollen disappears and fern spores suddenly peak at 70–100% of the total pollen and the spore count, then the spore count gradually drops to the levels seen in the Upper Cretaceous (Figure 10.14).

Figure 10.14 The ratio of angiosperm pollen to fern spores across the K/T boundary in the Raton Basin, New Mexico. *(Orth et al., 1981.)*

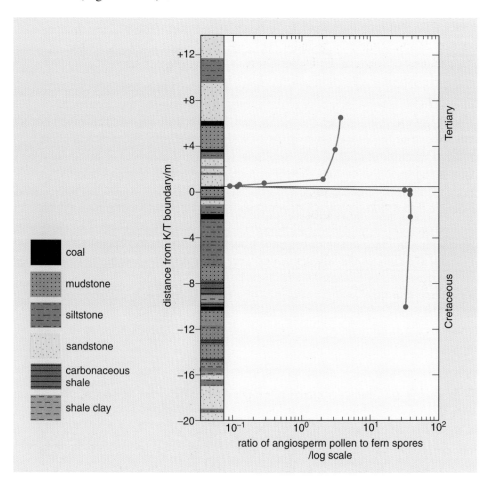

This sudden increase in the frequency of fern spores immediately above the K/T boundary is often referred to as the 'fern spore spike' or the 'fern spore abundance anomaly'. It was first recognized in 1984 by Robert Tschudy, who suggested that it represented a floral assemblage associated with an ecological disaster. He proposed that the dominance of fern spores indicated that recolonization of the land following the K/T boundary event began with regrowth dominated by pioneer species, which were able to rapidly exploit a major change in environmental conditions. In other areas, e.g. in parts of Western Canada, the post-extinction flora is dominated by particular species of angiosperms. In the Netherlands, an enrichment in bryophyte spores (from plants such as mosses, liverworts and hornworts) immediately above the K/T boundary has been interpreted as local domination of the terrestrial ecosystems by this group of plants.

As we have seen, much of the information about the plant fossil record comes from the study of plant microfossils — pollen and spores — referred to as palynology. For North America, the region for which we have most data, several biotic changes have been identified in this record, which provide an important insight into the pattern of extinction for plants at the K/T boundary. Some plant taxa do disappear abruptly at the K/T boundary, while others apparently disappear only to re-appear further up in the Lower Tertiary — a sort of pseudoextinction. However, other groups of pollen types show little change through the K/T boundary suggesting that, on the whole, plants were much less affected by the perturbation around the K/T boundary than were animals. However, plant microfossils are not our only source of data. A large database comprising some 25 000 specimens of plant megafossils (mainly leaves) has been compiled from the Upper Cretaceous and Lower Tertiary in the northern Rocky Mountains and Great Plains of North America. Analyses of these data indicate that there were major floral changes both before and at the K/T boundary (Figure 10.15), characterized by the disappearance of most of the Upper Cretaceous angiosperm taxa.

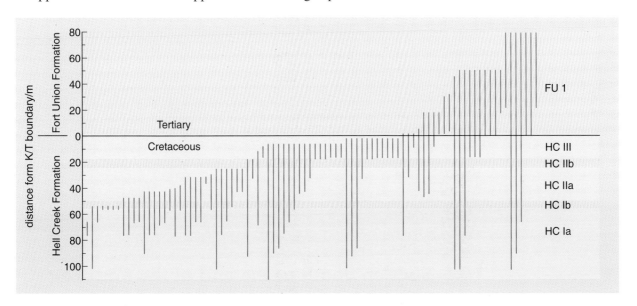

10.5.2 Plants as indicators of climate change at the K/T boundary

Studies on fossil leaf assemblages have also provided some interesting, if at times controversial, information on the K/T transition and on possible climate changes associated with it. Palaeobotanist Jack Wolfe (1986, 1987) examined the morphological and anatomical adaptations of land plants in various North American K/T sections employing the CLAMP technique you first met in Section 4.5.2. Recall that plants are highly sensitive indicators of past environmental changes for a relatively simple reason: they are spatially fixed

Figure 10.15 Biostratigraphical range chart constructed from plant megafossils (mainly leaves) showing the ranges of plant taxa in a composite section across the K/T boundary at Marmath, North Dakota. A high level of floral turnover is evident during the latest Cretaceous, culminating in the K/T boundary with only 21% of the latest Cretaceous flora persisting across it. (Johnson and Hickey, 1990.)

while animals can move around. Consequently, plants need a well-defined regime of light, humidity and temperature to survive. Significant climate changes affect these regimes enabling the characteristics of plants and their distribution to be used to infer ancient land climates. Wolfe concluded that an extinction event at the K/T boundary had indeed produced a major disruption of terrestrial ecosystems. He noted that there appeared to be higher levels of plant extinctions in the southern regions of the North American continent than in the northern regions, and that there was a higher survival rate among deciduous plants compared to evergreen plants. Deciduous plants are able to survive lower temperatures better because of their ability to become dormant. Perhaps the most controversial results of Wolfe's study came from changes he observed in the size and shape of leaves above the K/T boundary. These were generally larger and had shapes more typical of species that were acclimatized to wetter conditions (see Section 4.5.2). This led Wolfe to suggest that there had been an increase in both temperature and precipitation in the Early Tertiary before both returned to levels just slightly higher than those in the Late Cretaceous.

○ Are Wolfe's observations of the selective survival of deciduous plants over evergreen plants consistent with a warmer and wetter climate?

● No, because deciduous plants are better adapted to cooler conditions (recall Section 4.1.1).

Hence, Wolfe suggested there had been a brief low-temperature excursion or even a period of freezing before the warming, based on the selective survival of plants that could enter dormancy over evergreen plants. By itself, warming accompanied by a major precipitation increase would not favour dormancy.

Are Wolfe's interpretations based on the North American record confirmed elsewhere in the world? Unfortunately, outside of North America, there is sparse and, at times, conflicting evidence of floral changes across the K/T boundary. Minor changes in the distribution of plants between the Late Cretaceous and Early Tertiary have been found in New Zealand, Australia, and Antarctica, indicating that, if a climate change did occur at the end of the Cretaceous, it may have been much less severe at high latitudes of the Southern Hemisphere. In other parts of the world, for example Japan and China, a Late Cretaceous flora typical of a relatively warm climate appears to have been replaced by evergreen conifers in the Early Tertiary, suggesting a cooling of the climate.

10.6 Sudden death at the end of the Mesozoic

We have seen that the end-Cretaceous mass extinction was a relatively complex one. However, putting together the evidence from the faunal and floral fossil record, a less equivocal picture begins to emerge. In the marine realm, it is apparent that a major event occurred; the advocates of both catastrophic and gradual extinction hypotheses generally agree that something significant happened at the K/T boundary. Whether this boundary event is the only thing responsible for the mass extinction, or one of several episodes occurring over an extended period of time, is still a source of debate. The K/T boundary event itself would have to have been severe enough (and to have lasted long enough) to cause the mass extinction. We will examine the geological and geochemical evidence that this event was a large asteroid or comet impact in the next Section.

However, any plausible explanation for the K/T boundary event must be able to explain both the timing and pattern of the end-Cretaceous mass extinction.

In the marine environment, the main animals that became extinct were those that lived in the water column and would have relied on phytoplankton as their primary food source. In contrast, animals that lived on the sea-floor, and were predominantly deposit feeders, seem to have preferentially survived the K/T boundary event. The marine food chain in the Late Cretaceous is illustrated in Figure 10.16, which summarizes the interdependence of different groups of organisms — referred to as a food web. One thing apparent from Figure 10.16 is that not all animals living in the water column and in food chains dependent on phytoplankton became extinct — there were also survivors. For example, the nautiloids survived with only a minor extinction in contrast to the ammonites, which became extinct. Yet, both groups of organisms lived in the water column, though with some likely differences in living and feeding habits. Some of this selectivity may be explained by biological differences; the larval stages of ammonites were dependent on phytoplankton for nutrition, while nautiloid eggs have large yolks which could provide nutrition. Similarly, other molluscs that had larval stages which depended on phytoplankton also became extinct while other invertebrates, such as brachiopods, which are known to be relatively resistant to starvation, survived. Many organisms have the ability to switch food sources, an ability which may have allowed some groups to feed on detritus during the K/T event. Some animals may simply have been unfortunate enough to have had adaptations that hindered them. While most brachiopods survived the K/T boundary event, those that were specialized to live in the chalk seas of north-western Europe were totally devastated. The Chalk was composed primarily of the skeletons of coccolithophores and foraminifers so that when these became extinct, production of the Chalk temporarily ceased. This had disastrous consequences for the Chalk brachiopods that were specialized for life on the soft chalk sea-floor.

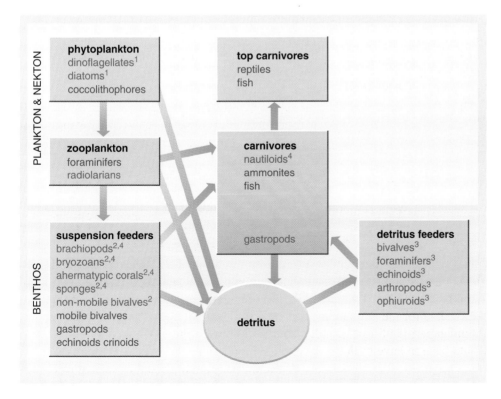

Figure 10.16 Late Cretaceous marine food chains. 1 = cyst or resting spore stages; 2 = starvation resistant; 3 = detritus feeding; 4 = development independent of planktonic food. Green labelling indicates preferential survivors; red, those suffering significant extinction. (Sheehan et al., 1996.)

On land, extinction appears to have been concentrated in food chains that relied on primary production, including both herbivorous and carnivorous dinosaurs (Figure 10.17). Other land animals may have preferentially survived because they lived in detritus-based food chains, such as forest-soil food chains based on leaf litter, decaying wood, roots, branches and fungi. Detritus-based food chains would have included protozoans, molluscs, many insects, amphibians, lizards and small mammals. Animals in this food chain were not tied directly to primary productivity and appear to have survived a period of reduced primary productivity before radiating in the Early Tertiary. Within the resolution of the Signor–Lipps effect, the extinction event on land was abrupt. The best resolution comes from fossil pollen and spores, which show an abrupt change at the K/T boundary. Other groups provide poorer resolution, but even the fossil record of dinosaurs appears to be consistent with an abrupt, and not a gradual, extinction.

The apparent collapse of terrestrial food chains and the subsequent extinction of land animals is reinforced by inferences that can be drawn from the North American terrestrial flora record. This shows an episode of climate change with a short-term cooling episode followed by a slow warming trend that re-established pre-K/T boundary conditions. An increase in rainfall also seemed to characterize the climate change in the North American plant fossil record. A short cooling event by itself was probably not enough to kill terrestrial fauna directly, but it may have initiated a collapse in the food chain in terrestrial ecosystems (Figure 10.17). Any drop in primary production would have affected the survival of herbivores and, in turn, lowered the ratio of prey to predators resulting in an increase in mortality among carnivores.

Figure 10.17 Comparison of Late Cretaceous and Early Tertiary terrestrial food chains. *(Sheehan et al., 1996.)*

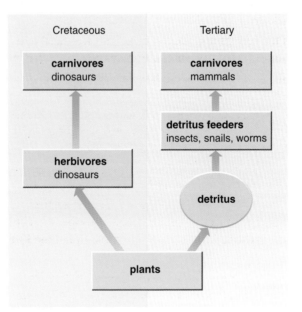

Following the K/T boundary event, primary productivity gradually resumed. However, many animals previously dependent on the primary production of plants had become extinct, and there were now no large-bodied herbivores or carnivores on land. It would take several million years before large-bodied mammals and birds evolved to replace the dinosaurs.

10.7 Summary

- Extinction is a normal part of evolution, and background extinctions have occurred throughout the geological record. However, the Cretaceous/Tertiary (K/T) boundary has long been recognized as representing a major faunal turnover in the geological record, an episode when the rate of extinction was much higher than normal. It is one of five major mass extinctions that have occurred during the Phanerozoic.

- Examining the detail of the K/T mass extinction depends on our ability to accurately interpret the fossil record. To properly document the magnitude of the K/T extinction, it is important to determine the highest stratigraphical level at which a particular taxon occurs. However, the apparent pattern of extinction obtained from studying the fossil record can be affected by sampling biases, known as the Signor–Lipps effect. These biases result because we can only ever collect, at most, a few kilograms of rock for each sample taken, so that large numbers of species which are actually preserved in a fossil-rich succession of rocks may be missing from the sample. Similarly, a hiatus due to non-deposition or erosion can change the apparent pattern we observe in the fossil record.

- Marine microorganisms experienced the most catastrophic of extinctions of any groups of organisms at the K/T boundary. For some, such as the coccolithophores, radiolaria and planktonic foraminifers, more than 90% of species became extinct at or near the K/T boundary. The planktonic foraminifers are ideal fossils to study as they are abundant and widespread, and are large enough so that they are less susceptible to bioturbation. Consequently, the precise timing of the extinction of planktonic foraminifers has been the subject of intense scientific scrutiny.

- Some marine animals such as the rudists and inoceramids declined markedly in diversity before the K/T boundary. Others, such as the ammonites, were in decline during much of the Late Cretaceous but apparently finally became extinct abruptly at the K/T boundary.

- An abrupt change in the carbon-isotope composition of planktonic foraminifers is observed globally in K/T successions, and it provides evidence for a major disruption of the global carbon cycle at the time of the K/T boundary.

- The pattern of extinction for the marine environment at the K/T boundary remains a controversial subject for scientific debate. Some scientists argue that the end-Cretaceous mass extinction was a gradual one. However, the general consensus that has emerged is that the marine palaeontological record contains evidence for a sudden and abrupt extinction of marine life at the K/T boundary 65 Ma ago.

- The disappearance of the dinosaurs is the most famous of the K/T extinctions. However, we do not know the pattern of dinosaur extinction on a global scale due to the sparse and intermittent nature of the dinosaur fossil record. As with the marine fossil record, there has been debate as to whether the dinosaur extinction was a gradual or an abrupt one. However, intense scrutiny of the dinosaur fossil record appears to indicate that they became extinct abruptly at the K/T boundary. A catastrophic scenario to explain the end-Cretaceous extinction of terrestrial vertebrates is, therefore, a distinct possibility.

- The best evidence of changes in flora associated with the K/T boundary comes from the North American continent where land plants appear to have been overwhelmed by the events surrounding the K/T boundary. Below the boundary, Cretaceous sedimentary rocks contain fern spores that make up 10–30% of the pollen and spore total. However, in rocks immediately above the K/T boundary, fern spores suddenly peak at 70–100% of the total, and the spore frequency then gradually drops to the levels seen in the Upper Cretaceous. This suggests a floral assemblage associated with an ecological disaster. The predominance of fern spores has been interpreted as evidence that recolonization of the land following the K/T boundary event began with regrowth dominated by pioneer species that were able to rapidly exploit a major change in environmental conditions.

10.8 References

ALVAREZ, L., ALVAREZ, W., ASARO, F. AND MICHEL, H. V. (1980) 'Extraterrestrial cause for the Cretaceous–Tertiary extinction', *Science*, **208**, 1095–1108.

SIGNOR, P. J. AND LIPPS, J. H. (1982) 'Sampling bias, gradual extinction patterns, and catastrophes in the fossil record', *Geological Society of America Special Paper*, **190**, 291–296.

SMIT, J. AND HERTOGEN, J. (1980) 'An extraterrestrial event at the Cretaceous–Tertiary boundary', *Nature*, **285**, 198–200.

STEUBER, T., MITCHELL, S. F., BUHL, D., GUNTER, G. AND KASPER, H. U. (2002) 'Catastrophic extinction of Caribbean rudist bivalves at the Cretaceous–Tertiary boundary', *Geology*, **30**, 999–1002.

WARD, P. D. (2001) *Rivers in Time: The Search for Clues to Earth's Mass Extinctions*, Columbia University Press, 320pp.

WOLFE, J. A. AND UPCHURCH, G. R. (1986) 'Vegetation, climatic and floral changes at the Cretaceous–Tertiary boundary', *Nature*, **324**, 148–152.

WOLFE, J. A. AND UPCHURCH, G. R. (1987) 'Leaf Assemblages Across The Cretaceous Tertiary Boundary In The Raton Basin, New Mexico And Colorado', *Proc. Nat. Acad. Sci. USA*, **84**, 5096–5100.

10.9 Further reading

GLEN, W. (1994) *The Mass-Extinction Debates: How Science Works in a Crisis*, Stanford University Press, 370pp.

SMIT, J. (1999) 'The global stratigraphy of the Cretaceous–Tertiary boundary impact ejecta', *Annual Review of Earth and Planetary Sciences*, **27**, 75–113.

11   Seeking an explanation

Iain Gilmour

Hypotheses that attempt to explain mass extinctions, including some involving meteorite impacts, reach back to the French astronomer and Fellow of the Royal Society of London, Pierre Louis de Maupertuis (1698–1759) (Figure 11.1). In his *Essai de cosmologie* published in 1750, Maupertuis suggested that comets had repeatedly struck the Earth and snuffed out life *en masse*. This hypothesis was explored by other scientists in the 18th and 19th centuries, including the French mathematician Pierre-Simon Laplace (1749–1827) in his *Exposition du systeme du monde* published in 1796:

> '… the small probability of collision of the Earth and a comet can become very great in adding over a long sequence of centuries. It is easy to picture the effects of this impact on the Earth. The axis and the motion of rotation have changed, the seas abandoning their old position …, a large part of men and animals drowned in this universal deluge, or destroyed by the violent tremor imparted to the terrestrial globe.'

Laplace's view may have been modern in its concept, if somewhat severe in its portrayal of an impact's effects, but these early hypotheses were essentially untestable. In the 20th century, the German palaeontologist M. W. de Laubenfels suggested that a meteorite impact was responsible for the K/T mass extinction in a paper published in 1956. Other scientists also explored the idea, including the Nobel Laureate Harold C. Urey who, in a paper published in 1973, discussed the wide range of environmental effects that would follow from an asteroid or comet impact, suggesting that it was:

> '… even probable that [such a collision] destroyed the dinosaurs and initiated the Tertiary division of geologic time.'

Figure 11.1   The French astronomer Pierre Louis de Maupertuis was one of the first to propose that the impact of comets or asteroids could have catastrophic consequences for life on Earth.

11.1   The state of play prior to the impact–extinction hypothesis

Throughout the 1960s and 1970s, hypotheses invoking exogenous causes of mass extinctions, such as solar flares, supernovae and meteorite impacts, were generally disregarded by the wider geological community — as, for that matter, were hypotheses invoking volcanism as a cause of mass extinctions. In contrast, hypotheses of gradual, endogenous causes, such as climate and sea-level change, or changing atmospheric carbon dioxide and oxygen levels, were methodically pursued. In the 1970s, plate tectonic theory was ascendant: as this new paradigm was being fitted to Earth history, hypotheses began to emerge that invoked plate tectonics as the causal mechanism behind major faunal turnovers in the fossil record. The proponents of these hypotheses argued that the splitting and reassembly of the continents explained changing biological diversity, including mass extinctions. All invoked a plate-tectonic-driven mechanism to change climates, alter ocean currents, modify continental shelf areas, or increase competition, which accorded perfectly with the then widely accepted notion that the major extinction boundaries had been brought into being within a time-frame of several million years. The widespread acceptance in the 1970s of the plate-driven mass extinction hypothesis was, in large part, due to its accommodation of uniformitarianism: the resultant environmental changes were readily accepted within the context of well-documented sea-level changes.

Figure 11.2 Luis Alvarez (left) and his son Walter (right) at a limestone exposure near Gubbio, Italy. Walter Alvarez's right hand is resting on a whitish band of limestone, overlain by a thin clay layer that marks the K/T boundary. Together with colleagues Frank Asaro and Helen Michel, they proposed that the end-Cretaceous mass extinction was caused by an asteroid or comet impact. *(Copyright © Lawrence Berkeley National Laboratory, USA.)*

Given the importance of extinction to studies of evolution and the history of life on Earth, you may have been surprised at how little attention was paid to the question of the nature of extinction or mass extinction. Most of the material you studied in Chapter 10 is based on research conducted in the last 20 years. While the question received greater prominence during the 1970s, it was the publication of the impact–extinction hypothesis in 1980 that fuelled research and scientific debate on the subject.

11.2 The first clues: an extraterrestrial cause for the Cretaceous–Tertiary extinction?

We saw in the previous Section that the idea that an asteroid or comet impact could cause a mass extinction was not new. So why was the impact–extinction hypothesis, published in 1980 to explain the K/T mass extinction, not rejected out of hand, as those before it had been? The answer should become increasingly apparent as we review the evidence for an impact at the K/T boundary but, in essence, the difference this time was that the hypothesis was testable. In fact, the initial reception given to the paper published by Luis and Walter Alvarez and their colleagues (Figure 11.2) from the University of Berkeley was highly sceptical. The geological community charged with evaluating the merits of their ideas was generally very unfamiliar with the phenomenon of impact cratering and its effects. Indeed, to many geologists, the instantaneous and worldwide effects of the proposed impact seemed to contravene the Principle of Uniformitarianism.

Since 1980, a long series of 'clues' have been followed, culminating with the discovery of the 65-million-year-old Chicxulub impact structure in the Yucatan peninsula of Mexico precisely at the K/T boundary. The evidence is substantial and the occurrence of an extraterrestrial impact event at the K/T boundary is now widely accepted as a fact. In this Section, we will explore this evidence and how it was obtained. You should bear in mind, however, that we are looking primarily at the evidence for an impact and *not* for evidence that the impact caused a mass extinction. We will examine that debate instead in Chapter 12.

11.3 An element out of place: the geochemical evidence for an impact

We saw in Chapter 10 how the original distinction of the Mesozoic from the Cainozoic was founded on fossil sequences that included the presence of ammonites, rudists and other macrofossils in the Cretaceous that were quite distinct from the marine fossils in the overlying Tertiary. It was recognized that mammals were fairly common in Tertiary rocks but rare in Mesozoic ones while the dinosaurs that had dominated the large Mesozoic faunas were completely absent from the Cainozoic. Lyell had postulated that uniformitarianism required that there should be passage beds between the two, but these were never found.

Today, the K/T boundary in marine successions can be readily identified from its microfossils, largely based on detailed biostratigraphical studies of the large successions of pelagic carbonates that are common in the Upper Cretaceous and Lower Tertiary of southern Europe. One region in particular, the Umbria–Marche Basin of north-eastern Italy, has been the subject of numerous interdisciplinary studies for several decades. Here, there is a nearly continuous and complete

Upper Cretaceous–Lower Tertiary succession of pelagic limestones, one of the best exposures being a section in the Bottaccione Gorge near the town of Gubbio. This section had been used in the 1960s to establish a simple and precise criterion for recognizing the K/T boundary in any pelagic carbonate sequence in the world: the appearance of a specific planktonic foraminifer species *Globigerina eugubina*, the first and last occurrences of which define the *Globigerina eugubina* biozone (Box 10.1), often abbreviated simply to the *eugubina* zone. Similarly, the last biozone of the Cretaceous is defined by the planktonic foraminifer *Abathomphalus mayaroensis*: the *mayaroensis* zone. The biostratigraphy of the section across the K/T boundary in the Bottaccione Gorge is summarized in Figure 11.3. This Figure contains a lot of information in addition to the biostratigraphy. On the left is the lithostratigraphy denoting the different lithologies of the section. For example, below the K/T boundary we have a thinly bedded pink-coloured limestone. This can be seen in Figure 11.4, which shows the exposure of the K/T boundary in the Bottaccione Gorge near Gubbio. The top of the limestone is shown by the whitish band. The change in colour from pink to white is due to a reduction in the iron content of the limestone. This is a measured section, so also on the left of Figure 11.3 is a metre scale. The scale highlights the completeness of the succession in this part of Italy since it is some 400 m thick and covers a time-span of almost 50 Ma from the beginning of the Late Cretaceous until the end of the Palaeocene. The lithostratigraphical log also contains information on the extent of bioturbation of the sections and marks the occurrence of a thin volcaniclastic layer. The latter is an important source of time-scale information as the biotite crystals it contains can be dated using radiometric dating techniques providing absolute ages.

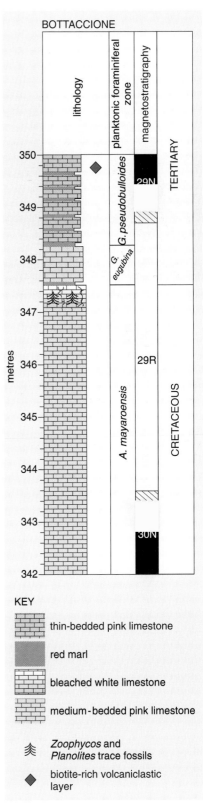

Figure 11.3 A stratigraphical synthesis of the K/T boundary interval in the Bottaccione Gorge, Gubbio, Italy. (Montanari and Koeberl, 2000.)

Figure 11.4 The K/T boundary in the Bottaccione Gorge section near Gubbio, Italy. The pale pink rock below the K/T boundary is the Scaglia Rossa Formation which was deposited during the Maastrichtian. At this locality, the top of this limestone is white in colour, immediately below the K/T boundary. The boundary itself is marked by a thin clay layer some 1–2 cm thick. (Iain Gilmour, Open University.)

The final piece of information on the Figure also concerns time: it is a record of the magnetostratigraphy of the section and provides an important mechanism by which the absolute ages obtained from dating the few volcanic layers that occur are correlated with the relative ages derived from biostratigraphy (Section 1.3.1). As with most sedimentary rocks, the limestones contain small amounts of magnetic iron-rich minerals that record the orientation of the Earth's magnetic field at the time the sediments were deposited. It is therefore possible to reconstruct a stratigraphical record of the periodic reversals of the Earth's magnetic field. These are shown in Figure 11.3 as the black and white bars on the right of the Figure; periods when the Earth's magnetic field has the same orientation as today are shown in black and the reversals are shown in white. Each magnetic period, referred to as a 'magnetochron', is given a number followed by the letter N or R denoting whether the Earth's magnetic field was normal or reversed.

○ During which magnetic period or magnetochron does the K/T boundary occur?

● The K/T boundary occurs within magnetochron 29R.

The difference in the microfossil fauna between the limestone layers immediately above and below the K/T boundary in Italian K/T sections is striking. Figure 11.5 shows two thin section photomicrographs of the planktonic foraminiferal assemblages above and below the boundary, taken at the same magnification. In the lower photograph, showing the last Cretaceous foraminifers, we can see large numbers of individuals up to about 0.6 mm across. In contrast, in the lowest Tertiary, the foraminifers are more sparse and distinctly smaller in size. At Gubbio, and in most other marine K/T boundary sections in the world, the boundary itself is marked by a thin (0.5–2.5 cm) claystone layer (Figure 11.6) that is sandwiched between the Cretaceous limestone, with its large and abundant foraminifers which are usually visible with a hand lens, and the first Tertiary limestone containing the tiny examples of *Globigerina eugubina* shown in Figure 11.5. Did this claystone layer perhaps represent the missing passage bed looked for by Lyell? The claystone layer was more than a puzzling curiosity to the scientists working on these sections. Palaeontologists argued that it represented a hiatus with little or no sedimentation occurring. Remember from Figure 10.5 how such a hiatus might create the impression of a sudden and abrupt extinction. What was needed was some indication of how much time was represented by the claystone. However, determining absolute time in the sedimentary rock record is not easy. Sedimentary rocks are, in general, not dateable using radiometric techniques and the magnetostratigraphical time-scale tends to give more accurate dates only at the point of a reversal, i.e. the change from normal to reverse or *vice versa*, and not for the period in between.

A possible solution to the problem came from a quite different field of study: meteoritics, the study of meteorites. In the early 1960s, cosmochemist Edward Anders and his colleagues from the University of Chicago, had suggested that the continuous accretion of micrometeorites to the surface of the Earth could be used to calculate the rates at which sediments were deposited. They had found that, by measuring the abundance of an element common in meteorites but rare on the Earth's surface, they could estimate sedimentation rates in deep-sea sediments. The elements in question were those in the Periodic Table referred to as the platinum group elements: platinum (Pt), palladium (Pd), iridium (Ir),

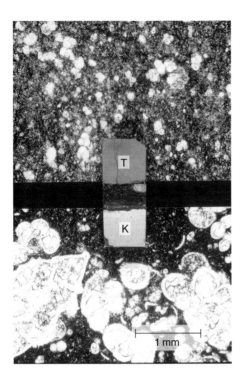

(a)

(c)

Figure 11.5 Photomicrographs of planktonic foraminiferal assemblages in the last Cretaceous limestone (lower photograph) and the first Tertiary limestone (upper photograph) in the Petriccio section, Italy. (Repeat of Figure 1.20.)

Figure 11.6 The clay layer that marks the K/T boundary in most pelagic marine successions in the world. (a) The boundary claystone at Stevns Klint, Denmark. (b) The boundary claystone at Agost, S.E. Spain. (c) The boundary claystone at Sumbar, Turkmenistan. (d) The boundary claystone at Petriccio, Italy. (a, Peter Skelton; b, d Iain Gilmour, Open University; c, Michael Nazarov.)

(b)

(d)

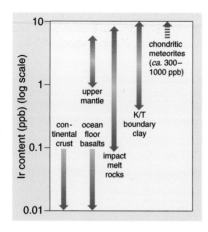

Figure 11.7 Ranges in Ir contents of the Earth's crust, mantle, K/T boundary claystones and meteorites. (Montanari and Koeberl, 2000.)

osmium (Os), ruthenium (Ru) and rhodium (Rh). Their method was based on the assumption that elements such as Ir accumulated in sediments at a more or less steady rate. This idea was picked up by Luis Alvarez, father of Walter Alvarez, one of the geologists working on the Gubbio section. His basic idea was to measure the concentration of the iridium in the boundary claystone and in the limestones above and below it through a known time-span. It followed that the concentration of Ir observed in the claystone would be proportional to the length of time of the alleged hiatus. The Alvarez team found levels of Ir in the Cretaceous and Tertiary limestones of around 0.3×10^{-9} g of Ir for each gram of rock (0.3 parts per billion, ppb) very similar to the levels measured by Anders for deep-sea sediments. This was as they expected, if that Ir represented the steady accretion of material from micrometeorites.

However, when they measured the boundary claystone they found that it contained much more Ir, reaching a maximum concentration of 8 ppb, 30 times the level in the Cretaceous and Tertiary limestones. This presented the team with a paradox: when they looked at the amounts of Ir in the boundary claystone and the first 20 cm of the Tertiary, the apparent amount of time that it would have represented due to the steady accretion of micrometeoritic material was much longer than the amount of time that they had estimated this 20 cm interval represented on the basis of magnetostratigraphy: there was an anomalously high amount of Ir in the boundary claystone. What had caused this Ir anomaly? Figure 11.7 shows the ranges in Ir concentrations (in ppb) for the Earth's crust and mantle, the K/T boundary and meteorites. Some meteorites contain particularly high levels of Ir so the Alvarez team explained the Ir anomaly with the hypothesis that there had been a massive, catastrophic accretion of an Ir-rich extraterrestrial object such as a comet or asteroid. From the amount of Ir present at Gubbio and at a second K/T section in Denmark, they estimated that the object must have had a diameter of around 10 km.

The Alvarez discovery was soon supported by reports of similar enrichments in Ir at other K/T sections in Spain and Denmark. Within a matter of weeks, two papers had been published citing anomalous Ir contents in K/T boundary claystones as evidence for a large asteroid or comet impact at the end of the Cretaceous. Both also suggested that this impact was responsible for the end-Cretaceous mass extinction, a hypothesis that became known as the impact–extinction hypothesis.

However, the impact–extinction hypothesis was not generally well received by the geological community. Many geologists had difficulty in accepting the presence of Ir as evidence for an asteroid or comet impact and various competing hypotheses were proposed to explain the Ir anomaly. Some scientists suggested that marine processes such as sudden changes in ocean chemistry precipitating Ir from seawater were responsible. Others suggested that the boundary claystone represented a substantial sequence of limestone from which carbonate had been removed, thereby enriching the amount of Ir. Both were shown to be wrong when, a year later, an Ir anomaly 300 times higher than background levels was measured in K/T boundary samples from a drill core, this time in continental sedimentary rocks of the Raton Basin, New Mexico, USA.

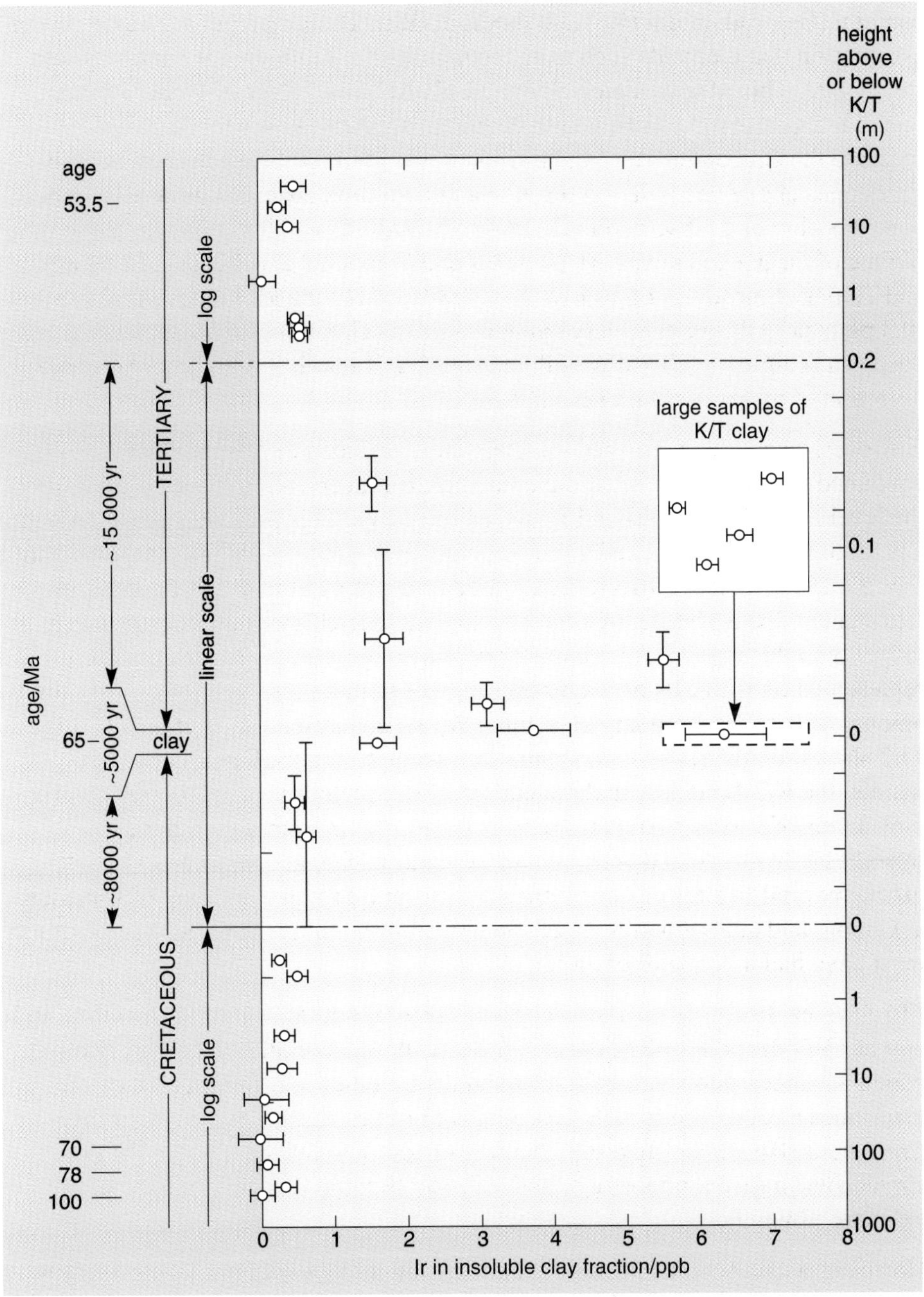

Figure 11.8   The abundance of Ir across the K/T boundary at Bottaccione Gorge, Gubbio, Italy. *(Alvarez et al., 1980.)*

The profile of Ir measured by the Alvarez team in 1980 for the Bottaccione Gorge and other K/T sections in the same region is shown in Figure 11.8. One problem with the Ir anomaly at the K/T boundary is the presence of Ir tails above the main peak. Occasionally, small secondary peaks have also been reported for some K/T sections. This led some scientists to suggest that prolonged volcanism was responsible for the Ir enrichments.

○   Look again at Figure 11.7. What terrestrial sources might contain enough Ir to explain the Ir anomaly at the K/T boundary?

●   The only possible terrestrial source is the upper mantle, which contains amounts of Ir similar to those found in K/T boundary sections across the world.

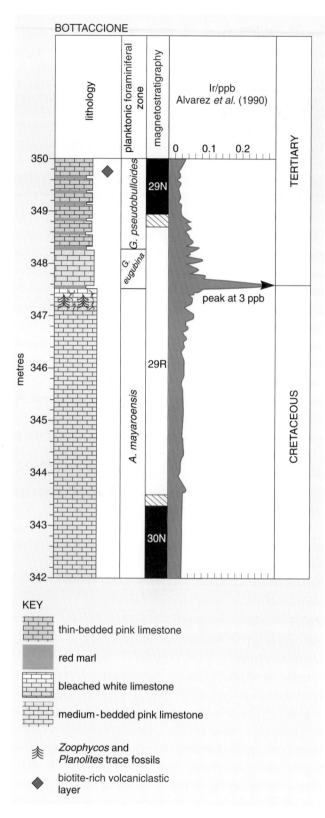

KEY

thin-bedded pink limestone

red marl

bleached white limestone

medium-bedded pink limestone

Zoophycos and
Planolites trace fossils

biotite-rich volcaniclastic
layer

Figure 11.9 The results of a study of the Ir abundance
through a 10-million-year-long section across the K/T boundary
at Bottaccione Gorge, Gubbio, Italy. *(Montanari and Koeberl,
2000.)*

We have already seen in Section 7.3 that there was substantial volcanic activity occurring around the K/T boundary: the Deccan flood basalts in India had begun erupting some 1 million years previously. Could these have been the source of the Ir anomaly? The debate between scientists who wanted an extraterrestrial source for the Ir in the form of an asteroid or comet impact versus those who wanted a terrestrial source in the form of volcanism went on for several years. More detailed analyses of the Bottaccione K/T section were done in 1990, this time analysing the Ir content of a 57-m-thick, 10-million-year-duration section; the results are shown in Figure 11.9. They showed that there was only one Ir anomaly at the K/T boundary, dominating over an intricate fine structure. Such a sharp spike in the Ir content argued against a volcanic origin for the Ir from Deccan volcanism, which lasted for several million years. There were other problems with a volcanic source for the Ir. Detailed analyses of the Ir contents from many volcanic ash beds, lavas and large basaltic flows, including the Deccan flood basalt, were done to assess the plausibility of a volcanic origin. The results indicated that the Ir abundances in ash from silicic eruptions, and from basalt samples from the Deccan basalt, were too low to account for the K/T boundary anomalies. In fact, the apparent spread observed in the Ir anomaly at some K/T sections is the result of more mundane geological processes such as bioturbation, local reworking of the sedimentary rocks or possibly the fluvial transport of Ir from the continents. Some scientists, however, continued to argue that volcanism could explain the Ir enrichments. What was needed was unambiguous proof of an impact. Critics of the impact hypothesis also cited the lack of an impact crater; a 10-km-diameter asteroid would have produced a crater almost 200 km in diameter.

Before leaving our discussion of the Ir anomaly at the K/T boundary, it is worth looking at the current state of the *geochemical* evidence. To date, an Ir anomaly has been measured in over 80 K/T sections worldwide ranging in concentration from 0.5 to more than 50 ppb. These data are summarized in Figure 11.10 which shows the maximum Ir concentration measured in K/T boundary sites worldwide. The distribution of K/T sections with relatively high concentrations of Ir appears to be random with no one geographical area showing a strong grouping of high Ir concentrations as might possibly be expected, for example, in the vicinity of an impact crater.

The other platinum group elements also show enrichments similar to Ir, and some of these, in particular ruthenium and osmium, have been used to try to identify the type of object that impacted. The results seem to indicate that it was an asteroid, rather than a comet, with a composition similar to two known groups of meteorites: the iron meteorites and the chondritic meteorites.

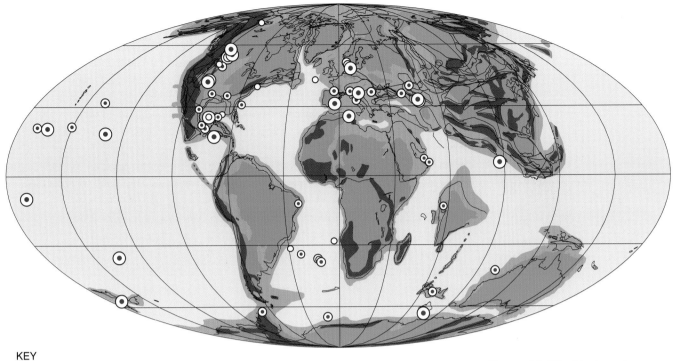

KEY

■ 1   ■ 2   ■ 3   □ 4   ○ 5   ⊙ 6   ⊙ 7   ⊙ 8

Figure 11.10 Maximum Ir concentration measured in K/T boundary sites. The data are shown on a palaeogeographical reconstruction of the Earth at the time of the K/T boundary. Key: 1 mountains; 2 lowland; 3 shelf sea; 4 ocean; 5 Ir anomaly indicated but no precise data available; 6 <1.5 ppb; 7 1.5–10 ppb; 8 >10 ppb. *(Buffetaut and Koeberl, 2002.)*

11.4 Traces of an impact: the mineralogical evidence

When an asteroid or comet impacts the Earth, typically at velocities between 14 km s^{-1} and 25 km s^{-1}, the rocks that the object hits are subjected to tremendous pressures and temperatures, considerably beyond those that normally occur during metamorphic processes. Extremely high amounts of energy are liberated, almost instantaneously, when the object hits the ground. This energy is equal to the kinetic energy of the impactor:

kinetic energy $= \frac{1}{2}mv^2$, where m is the mass of the object and v its velocity.

During the early stages of an impact, this kinetic energy ($\frac{1}{2}mv^2$) is transformed into heat and to supersonic shock waves that penetrate into the projectile and the target. The shockwaves momentarily compress the target rocks to extremely high pressures. If the pressures are high enough, then the rocks and the minerals they contain undergo change. These can include changes in the crystal structure of minerals or the formation of a new phase of a mineral, for example at high enough pressures graphite can be transformed into diamond. The pressures generated by impact shock waves are enormous and far exceed the pressures normally achieved in geological processes. The maximum pressures experienced in metamorphic events are typically around 4–5 × 10^9 pascals or 4–5 gigapascals, abbreviated to GPa (Box 11.1). In contrast, the pressures experienced by target rocks in an impact can exceed 100 GPa and temperatures can reach several thousand degrees Celsius. This gives rise to a whole new field of metamorphism known as shock metamorphism. The study of how different materials respond to shock is not new but was the subject of thorough research for much of the 20th century, in part stimulated by nuclear weapons research. Shortly after the first atomic bombs were detonated in the late 1940s, scientists investigating the explosion craters noticed that quartz grains in the rocks that had been subjected to the shock waves from the nuclear explosion looked different under the microscope from 'normal' quartz grains. They contained small intersecting sets of parallel lines on their surfaces that gave the impression of fractures within the quartz. In fact, the features were planar, ran through the quartz crystals and were the result of dislocations within the crystal lattice of quartz on an atomic scale; they are known as planar deformation features, abbreviated to PDFs. Today, these shocked quartz grains, displaying multiple intersecting sets of PDFs, are considered to indicate that the rock containing them experienced an extremely high pressure for a short time. They have only ever been observed in laboratory shock experiments, at the site of nuclear explosions, and at the sites of known impact craters. It is worth noting that no quartz or other minerals showing genuine shock metamorphic characteristics have ever been found in a volcanic environment.

Box 11.1 Units of pressure

The SI unit of pressure is the pascal, abbreviated to Pa:

$$1\,Pa = 1\,N\,m^{-2} = 1\,kg\,m^{-1}\,s^{-2}.$$

The maximum pressures experienced in metamorphic events are typically around $4-5 \times 10^9$ pascals (Pa), 4000–5000 megapascals (MPa) or 4–5 gigapascals (GPa). You may be more familiar with pressure measured in kilobars (kbar) — 1 kbar is around 1000 times atmospheric pressure at sea-level and the conversion to gigapascals is given by:

$$10\ kilobar\ (kbar) = 1\ gigapascal\ (GPa).$$

In 1984, Bruce Bohor, a geologist working for the US Geological Survey, discovered shocked quartz grains in the K/T boundary claystone at Brownie Butte in Montana, USA. This was quickly followed by their discovery in other K/T boundary sections around the world. An example of one of these grains is shown in Figure 11.11, which shows a scanning electron microscope image of a quartz grain from the K/T boundary claystone in a drill core from the south-west Pacific Ocean. The diagnostic indicator of an impact, the intersecting sets of PDFs are visible on the surfaces and penetrating through the grain.

Figure 11.11 Scanning electron microscope image of a quartz grain from the K/T boundary. The intersecting sets of planar deformation features are visible on the surface of the grain and penetrate through it. *(Courtesy of Bruce Bohor.)*

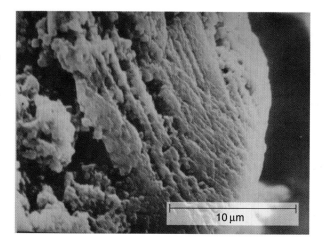

10 μm

Bohor also noticed that the size of the quartz grains was not the same at different localities around the world. He suggested that this was related to how far the grains had travelled in the material ejected from the impact site, with smaller grains indicating a greater distance from the crater. A modern compilation of the grain-size distribution is shown on the map in Figure 11.12. The largest shocked quartz grains, with diameters ranging from 500 to 1250 µm, are found in the vicinity of North America, the Gulf of Mexico, and the Caribbean, while less abundant and much smaller grains with sizes typically less than 190 µm were found at European sites, in the Pacific, and in New Zealand. A few years later, scientists examining the K/T boundary sediments of the Beloc Formation on the island of Haiti, found that between 54% and 70% of the quartz grains present at the K/T boundary had been shocked. Haiti, it seemed, must have been even closer to the impact structure.

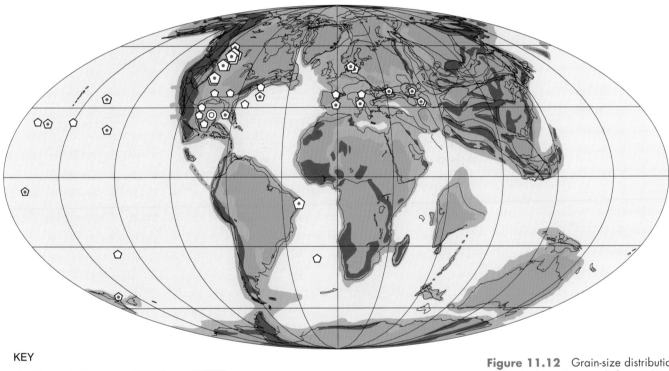

KEY

1 ☐ 2 ☐ 3 ☐ 4 ⬠ 5 ⬠ 6 ⬠ 7 ⬠ 8 ◎ 9

Figure 11.12 Grain-size distribution of shocked quartz crystals from K/T boundary sites around the world. The data are superimposed on a palaeogeographical reconstruction of the Earth at the time of the K/T boundary. Key: 1 mountains; 2 lowland; 3 shelf sea; 4 ocean; 5 shocked quartz present, but no maximum size indicated; 6 maximum diameter <150 µm; 7 maximum diameter 150–300 µm; 8 maximum diameter >300 µm; 9 Chicxulub impact site. (Buffetaut and Koeberl, 2002.)

Bohor's discovery was the key piece of evidence that many geologists needed in order to persuade themselves that there had indeed been a massive impact at the K/T boundary. It also established that the boundary claystone, the thin layer typically 1–2 cm thick found around the world, contained material ejected from the impact crater and, for the first time, pointed to where that crater might be: somewhere near the North American continent.

Since the original discovery of shocked quartz, several other mineralogical indicators of an impact have also been found, considerably strengthening the case for an impact. Figure 11.13 shows the pressures and temperatures at which various structural changes can occur due to shock metamorphism. One such change is the transformation of quartz from one crystal structure to another, shown on Figure 11.13 by dotted lines separating the regions where these transformations occur. Two high-pressure forms of silica are known: coesite, which forms at pressures above about 3 GPa and occurs in some metamorphic rocks; and stishovite, which requires pressures in excess of 10 GPa and is only known from impact craters. Stishovite was found in K/T boundary claystones from the Raton Basin in New Mexico in 1989. Also shown in Figure 11.13 is the transformation from graphite to diamond that occurs during shock metamorphism at a pressure of around 5 GPa. Diamonds produced by shock are also well-known diagnostic features of an impact, having first been observed in fragments of the Canyon Diablo meteorite that formed Meteor Crater in Arizona, USA, and in rocks from the 37-Ma-old Popigai impact crater in Siberia. Shock-produced diamond crystals up to 30 μm in size were found in K/T boundary claystones from Montana, Colorado and north-eastern Mexico, and an example is shown in Figure 11.14.

Figure 11.13 A comparison of the pressure and temperature ranges for normal metamorphic processes with those of shock metamorphism. Also shown are the pressures at which various structural changes occur in rocks and minerals due to shock metamorphism.

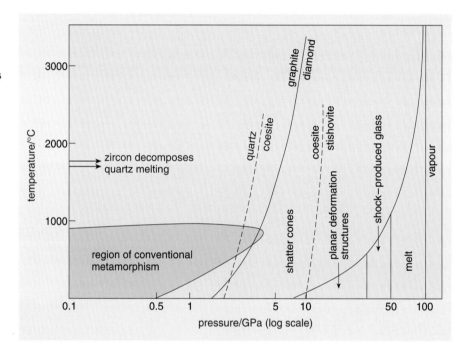

Other shocked minerals have also been found, including the mineral zircon, which provided an important clue to the impact site as zircons are common constituents of continental felsic rocks but not oceanic mafic rocks. A particularly important group of minerals common in K/T boundary claystones are crystals of magnetite (Fe_3O_4) and magnesioferrite ($MgFe_2O_4$), collectively known as spinels. However, the spinels found in K/T boundary claystones are nickel-rich, similar to those found in the fusion crust formed around meteorites when they enter the Earth's atmosphere. Ni-rich spinels do not occur in terrestrial igneous or metamorphic rocks. Ni-rich spinels have now been well documented in K/T sections around the world where they occur and coincide with the maximum Ir concentration.

There are some important differences between the boundary claystones found at continental sites in the Western Interior of North America and those of marine successions. At undisturbed K/T sections in North America, the layer is up to 3 cm thick and has a dual-layered nature (Figure 11.15), whereas in most marine successions there is just a single layer. Bohor noted that at the Western Interior locations there was a lower layer, up to 2.5 cm thick, composed mainly of clay

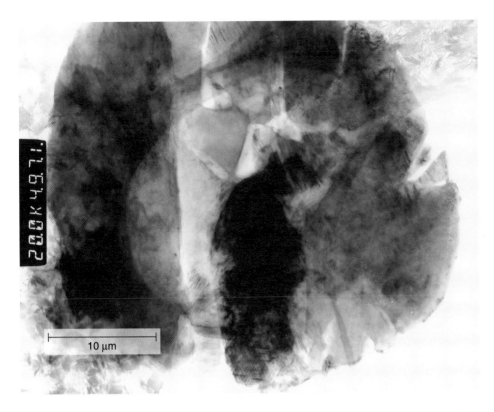

10 μm

Figure 11.14 An electron microscope image of a 30 μm diameter shock-produced diamond crystal from the K/T boundary at the Arroyo El Mimbral section in north-eastern Mexico. *(Iain Gilmour, Open University.)*

minerals that contained hollow spheres up to 1 mm in diameter (Figure 11.16). Some of these spheres were composed of a clay mineral called goyazite that can be formed by the alteration of silicate glass. The upper layer is generally thinner, around 5 mm, is also composed of clay minerals, and contains the shocked quartz and Ir anomaly. The hollow spheres were interpreted by Bohor as representing altered glass spheres produced by the impact and they have been found in a large number of K/T sites in North America, Mexico and on the island of Haiti. Glasses, produced by the rapid quenching of molten rock, are a well-known feature of impact craters. They are produced by the melting of target rocks which are then ejected large distances from the crater.

T
K

Figure 11.15 The K/T boundary claystone at the Clear Creek North K/T site in the Raton Pass, Colorado. At K/T sites in North America, the boundary clay consists of two layers. In this image the lower layer, some 2.5 cm thick, is the pale-coloured layer running though the middle of the photograph. This is overlain by a thinner darker-coloured layer about 5 mm thick, which in turn is overlain by a thin coal around 3 cm thick. The rest of the exposure is composed of Late Cretaceous and Early Tertiary mudstones and shales. *(Iain Gilmour, Open University.)*

Figure 11.16 Spherules of the mineral goyazite in the lower layer of the K/T boundary from Teapot Dome, Wyoming, USA. Goyazite is a common alteration product and the spherules, which are up to 1 mm in size, have probably formed by the alteration of spheres of impact glass similar to microtektites. *(Courtesy of Bruce Bohor.)*

1mm

The spheres found in the K/T claystone are similar in many respects to objects known as microtektites. These are small (up to 1 mm in size) impact-produced silicate glass particles formed by melting rocks in the uppermost layer of the target surface of an impact, and are found in a geographically extended distribution around an impact crater. The interpretation that the K/T spheres were the altered remains of impact glasses similar to microtektites was reinforced when fragments of preserved glass were found inside the cores of clay spherules from the K/T boundary at Beloc in Haiti and at two K/T sites in north-eastern Mexico (Figure 11.17). Geochemical analyses of the preserved glasses indicated that they contained very little water, again very similar to microtektites but quite different from glasses formed in terrestrial processes such as volcanism. Other characteristics of the glasses also ruled out a volcanic origin: they were generally very homogeneous and entirely crystal-free, unlike volcanic glasses, which invariably contain small phenocrysts of high-temperature minerals due to fractional crystallization of magma. The preserved glasses were to prove critical in one other respect: the fact that they were produced by melting meant that their age could be determined using radiometric dating (using the $^{40}Ar/^{39}Ar$ method; Section 1.3.2). High-precision age determinations made on glasses from the Haitian K/T boundary site gave an age of 65.07 ± 0.11 Ma, indistinguishable from the K/T boundary and, as we shall see in Section 12.3, indistinguishable from the age of the Chicxulub impact structure itself.

Figure 11.17 A fragment of an unaltered glassy spherule recovered from the K/T boundary at Beloc in Haiti. When dated, the glass was found to have an age of 65.07 ± 0.11 Ma, indistinguishable from the age of the K/T boundary. *(Iain Gilmour, Open University.)*

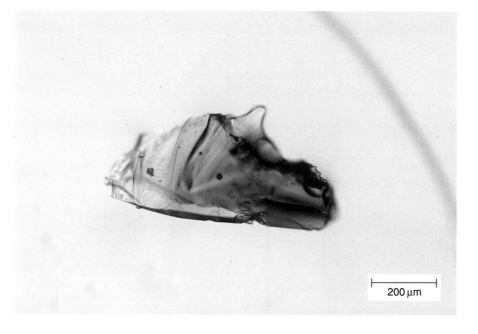

200 μm

The Ir anomaly and enrichments in other platinum group elements, the occurrence of shocked minerals such as quartz and diamond, the discovery of impact-derived glasses and the presence of spinel crystals of possible extraterrestrial origin constituted a pretty substantial body of evidence that the K/T boundary clay layer found worldwide represented the ejecta deposit from a large asteroid or comet impact at the K/T boundary. Nevertheless, in the late 1980s and early 1990s, a few geologists held out against the idea of an impact at the K/T boundary, citing a variety of relatively rare terrestrial geological processes that could explain some of the observations. These included mantle-derived volcanic processes to explain the Ir enrichment together with massive continental explosive volcanism to explain shocked minerals, although the levels of shock required have never been observed in the latter. However, while some of the alternative explanations offered were able to explain one or two of the features associated with the K/T boundary, they singularly failed to provide a complete explanation for all of the evidence we have looked at in this Section.

11.5 Summary

- The first evidence for a large impact event at the K/T boundary came from studies of the geochemistry of Cretaceous–Tertiary boundary successions in Northern Italy. Detailed studies of the biostratigraphy and magnetostratigraphy of the section at Gubbio had been undertaken prior to 1980. In this section, the K/T boundary is marked by a thin clay layer some 0.5–2.5 cm thick. It was in this layer that the Alvarez group from the University of Berkeley found anomalous levels of the element iridium.

- Iridium is a rare element in the Earth's crust but does occur in meteorites and in rocks derived from the Earth's mantle. High resolution studies, spanning some 10 Ma, of the section at Gubbio indicated that there was just a single sharp peak in the Ir abundance in these rocks, precisely at the K/T boundary.

- The amounts of iridium observed were much higher than would have been expected for the steady accretion of this element in sediments due to the continuous infall of micrometeorites. This led the Alvarez team to suggest that there had been a large impact of an asteroid or comet at the K/T boundary.

- While Ir was not incontrovertible evidence of an impact event because it could also have come from large-scale mantle-derived volcanism, the subsequent discovery of shocked minerals led most scientists to accept that there had indeed been a large impact at the K/T boundary. In particular, the occurrence of planar deformation features in quartz crystals was consistent with those crystals having experienced pressures in excess of 7–10 GPa. These are pressures well in excess of those common in terrestrial metamorphic processes.

- Subsequent to the discovery of Ir and shocked quartz, other diagnostic features of an impact event have also been observed at the K/T boundary. These include other shocked minerals such as diamonds and zircons, spherules of preserved glass with chemistries very similar to impact-derived microtektites and with ages indistinguishable from the K/T boundary.

11.6 References

ALVAREZ, L. W., ALVAREZ, W., ASARO, F. AND MICHEL, H. V. (1980) 'Extraterrestrial cause for the Cretaceous–Tertiary extinction', *Science*, **208**, 1095–1108.

BOHOR, B. F., FOORD, E. E., MODRESKI, P. J. AND TRIPLEHORN, D. M. (1984) 'Mineralogic evidence for an impact event at the Cretaceous–Tertiary boundary', *Science*, **224**, 867–869.

UREY, H. C. (1973) 'Cometary collisions and geological periods', *Nature*, **242**, 32–33.

12   The 'smoking gun'

Iain Gilmour

The proponents of an impact did have one apparent flaw in their argument: they had no crater. If the estimates of the impactor's size, 10 km, calculated from the amount of Ir present at the K/T boundary were right, then it would have produced a crater of between 100 and 200 km diameter. Such craters are rare: the largest-known impact structure on any planet in the inner Solar System formed during the past 500 million years is the 270-km-diameter Mead crater on Venus (Figure 12.1a). Large craters are known on Earth: for example, the Vredefort impact structure in South Africa is nearly 250 km in diameter, but it was formed almost two billion years ago at a time when large impacts may have been more frequent.

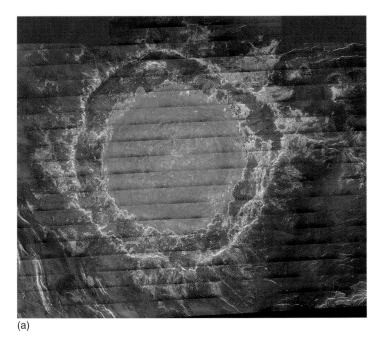

(a)

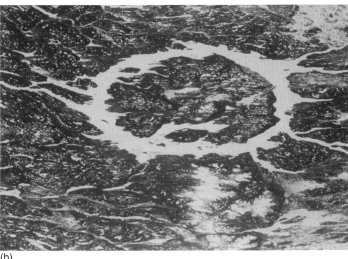

(b)

Figure 12.1   (a) The 270-km-diameter Mead crater on Venus is the largest-known impact crater on any planet in the inner Solar System formed during the past 500 Ma *(NSSDC)*. (b) The 100-km-diameter, 214-Ma-old Manicouagan impact structure in Canada. *(© NASA/JPL.)*

○ What evidence do we have as to where the K/T crater might be located?

● The size-distribution of shocked-quartz grains (Figure 11.12) points to an impact in the region of the North American continent. Larger grains are found there than in K/T sites in Europe or elsewhere. Since finer-grained material would be easier to transport over large distances, it suggests the impact was on or near North America.

12.1 Crater? What crater?

The Earth is not short of impact craters (Figure 12.2); to date, some 160 have been identified despite the problems of erosion and the geologically active surface of the Earth that have reworked the planet's surface for the last 4.5 billion years. There are some large impact structures on the North American continent, for example the 100-km-diameter Manicouagan crater in Canada (Figure 12.1b), but this too is the wrong age since it is 214 million years old. In the late 1980s, there were only two known impact structures that appeared to be the correct age and therefore contenders for the title 'K/T boundary crater'. The first of these was the 65–80-km-diameter Kara crater in Russia, which was thought to have an age of 65 Ma. However, when rocks from the crater were dated using radiometric techniques, it was found to have an age of around 70.3 Ma. The second contender, the Manson impact structure in Iowa, was on the North American continent itself, although at the time little was known about it. The area was covered in Pleistocene glacial drift so there was no surface expression of the crater, but it had been known as an area of anomalous geology since the early 20th century when a well was drilled as a water supply for the small town of Manson. Early attempts to date the rocks of the structure suggested that it might be of K/T boundary age. The crater was therefore the subject of an extensive research program in 1991–1992 during which over 1200 m of drill cores were recovered. When the rocks from the crater were dated using the ^{40}Ar/^{39}Ar technique, an age of 74.1 ± 0.1 Ma was obtained, nine million years prior to the age of the K/T boundary. The drill cores also revealed that the crater was only 35 to 37.8 km in diameter; it was also too small to be the K/T crater.

Figure 12.2 Distribution of known impact craters on the Earth, as of the year 2000. Crater diameters are given in the key. *(National Geophysical Data Center, USA.)*

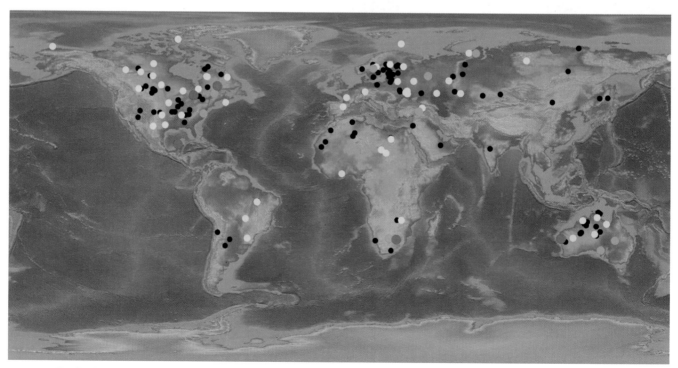

KEY   ● >100 km   ● 50–100 km   ○ 10–50 km   • 0–10 km

12.2 Chicxulub rediscovered

At the same time as the Manson crater was being explored, an unusual structure in Mexico was also being looked at in more detail. Some ten years previously, two geologists working for the Mexican petroleum company, Petróleos Mexicanos (Glen Penfield and Antonio Camargo), had identified a circular gravity anomaly of about 180 to 200 km diameter on the Yucatan Peninsula of north-eastern Mexico, centred on the small town of Chicxulub. In 1981, they presented a paper at a meeting of economic geologists suggesting that the anomaly was a large impact crater, but it seems that their audience was not interested in impact craters or extinctions! It was not until 1990, however, that the gravity data were re-examined and new gravity anomaly maps compiled (Hildebrand *et al.*, 1990). The Chicxulub structure, as it became known, was marked by an extensive circular negative gravity anomaly of about 180 to 200 km in diameter with a small central positive gravity anomaly (Figure 12.3). Circular negative gravity anomalies are common features of impact craters and correspond to the relatively lower densities of the brecciated impact rocks and sedimentary fill within the crater. The gravity high in the centre is also a feature of large complex impact craters and reflects the fact that large craters have an uplifted area at their centres (Box 12.1).

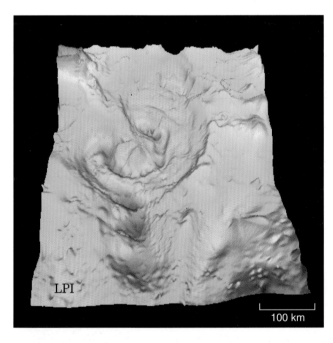

Figure 12.3 Computer-generated 3-D Bouguer gravity anomaly map over the Chicxulub crater, represented by the near-circular gravity low, shown in blue, at the centre. This image does not show the crater's shape. The negative gravity anomaly of the crater corresponds to the relatively low densities of the rocks within it, breccias and the melt sheet, and the Tertiary sedimentary rocks that fill it. The double-humped central gravity high, shown in green, corresponds to the crater's central uplift now buried deep inside. The blue area to the bottom is part of the regional gravity anomalies of the area. Shortly after the impact, the Chicxulub crater probably looked similar to the larger 270-km-diameter Mead crater on Venus shown in Figure 12.1a. *(Courtesy of Buck Sharpton.)*

Box 12.1 Geological characteristics of impact craters

Impact craters are divided into two groups based on morphology: simple craters and complex craters. Simple craters are relatively small with depth-to-diameter ratios of about 1 : 5 to 1 : 7 and a smooth bowl shape (Figures 12.4a, 12.5a). In larger craters, however, gravity causes the initially steep crater walls to collapse downward and inward, forming a complex structure with a central peak or peak ring (Figures 12.4b, 12.5b) and a shallower depth compared to diameter (1 : 10 to 1 : 20). The diameter at which craters become complex depends on the surface gravity of the planet. The greater the gravity, the smaller the diameter that will produce a complex structure. On Earth, this transition diameter is 2 to 4 km depending on the properties of the target rocks; on the Moon, at one-sixth Earth's gravity, the transition diameter is 15 to 20 km.

The central peak or peak ring of the complex crater is formed as the initial (transient) deep crater floor rebounds from the compressional shock of impact. Slumping of the rim further modifies and enlarges the final crater. Complex structures in crystalline rock targets will also contain coherent sheets of melted rock (impact melt) on top of the shocked and fragmented rocks of the crater floor. On the geologically inactive lunar surface, this complex crater form will be preserved until subsequent impact events alter it. On Earth, weathering and erosion of the target rocks quickly alter the surface expression of the structure; despite the crater's initial morphology, crater rims and ejecta blankets are quickly eroded and concentric ring structures can be produced or enhanced as weaker rocks of the crater floor are removed.

On other planets and satellites (e.g. the Moon, Mercury, Mars, Ganymede and Callisto), larger structures than the simple or complex craters have been identified. These are the so-called multi-ring impact basins, which have diameters of a few hundred to at least 2000 km (Figure 12.5c).

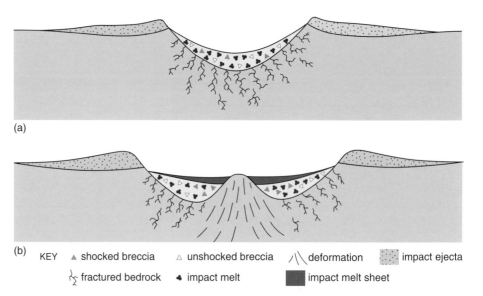

(a)

(b) KEY ▲ shocked breccia   △ unshocked breccia   /\ deformation   ▓ impact ejecta
       fractured bedrock   ◄ impact melt   ▓ impact melt sheet

Figure 12.4 Schematic cross-section of (a) simple and (b) complex impact craters. *(Montanari and Koeberl, 2000.)*

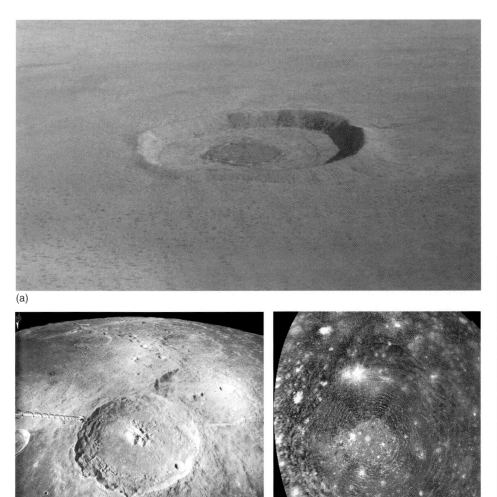

(a)

(b)   (c)

Figure 12.5   Examples of a simple crater, a complex crater and a multi-ring impact basin. (a) The simple terrestrial crater, Wolfe Creek, in Australia *(courtesy Buck Sharpton)*; (b) the complex lunar crater Theophilus seen from the *Apollo 16* spacecraft; (c) the multi-ring impact basin Valhalla on Jupiter's moon, Callisto. *(b, c NSSDC.)*

A combination of computer-based modelling and new geophysical data led to the suggestion that the Chicxulub structure was a multi-ring impact basin, similar to large impact structures on the Moon and other bodies in the Solar System (Box 12.1). A three-dimensional computer-generated image of the circular gravity anomalies at Chicxulub is shown in Figure 12.3, which shows a well-defined inner ring and several outer rings. If Chicxulub is a large multi-ring impact basin, it would be the only well-preserved one on Earth. However, there was a problem: the structure was buried under almost 1 km of Tertiary sediments. The only surface expression of the structure appears to be a circular distribution of sinkholes known as '*cenotes*' in Yucatan (Figure 12.6) some 120 km from the structure's centre. Cenotes form due to the collapse of underground caves created by the dissolution of limestone, and it is thought that the cenote ring at Chicxulub reflects part of the buried structure of the crater.

Figure 12.6   Radar image of a portion of the buried Chicxulub impact structure in the Yucatan peninsula, Mexico taken by the Space Shuttle *Endeavor* on April 14, 1994. The 10-km-wide band of yellow and pink with blue patches along the right is a mangrove swamp. Blue patches are islands of tropical forests created by freshwater springs (*cenotes*) that emerge through fractures in the limestone bedrock and are most abundant in the vicinity of the buried crater rim. *(© NASA/JPL.)*

Fortunately for us, Petróleos Mexicanos drilled the structure in the 1950s and 1960s and some of the core samples were still in existence almost 40 years later. When samples from two of these drill cores near the centre of the structure were examined, two rock types characteristic of an impact structure were found. The first was a dark crystalline rock containing fragments of glass indicating that parts of the rock had been melted and cooled very quickly; the second was a breccia that contained clasts of other rocks, including some that were hundreds of millions of years older than the Cretaceous and Tertiary sediments of the area. The most important findings for both rocks, however, were that they contained clear evidence for the presence of shock metamorphism including abundant multiple sets of planar deformation features in quartz crystals (Figure 12.7). Elevated levels of iridium were also found in the rocks. Chicxulub was indeed a very large buried impact crater, but was it the K/T boundary crater?

Figure 12.7 Photomicrograph (between crossed polars) of a fine-grained impact breccia recovered from a drill core near the centre of the Chicxulub impact crater in the Yucatan peninsula of Mexico. There are clear indications of shock metamorphism in the form of multiple sets of planar deformation features in the labelled quartz crystal. (*Courtesy of Buck Sharpton.*)

0.1 mm

foraminifer   quartz

To try to correlate the Chicxulub structure with the K/T boundary, we could look at the stratigraphy of sedimentary rocks inside the crater; if we found Cretaceous sedimentary rocks, then this might indicate that the crater was too old. A schematic model of the crater constructed from data from a variety of boreholes across the structure is shown in Figure 12.8. Tertiary sedimentary rocks, labelled Cainozoic basin fill, overlie the impact-produced rocks, labelled in the Figure as melt + breccia and Chicxulub breccia. The Chicxulub breccia contains fragments of Cretaceous limestone; in fact, a Cretaceous foraminiferan can be seen in Figure 12.7 just to the left of the shocked quartz crystal, confirming that Cretaceous rocks were part of the target for the impactor. The oldest rocks on top of the Chicxulub breccia were identified as Early Tertiary in age, based on the microfossils they contained.

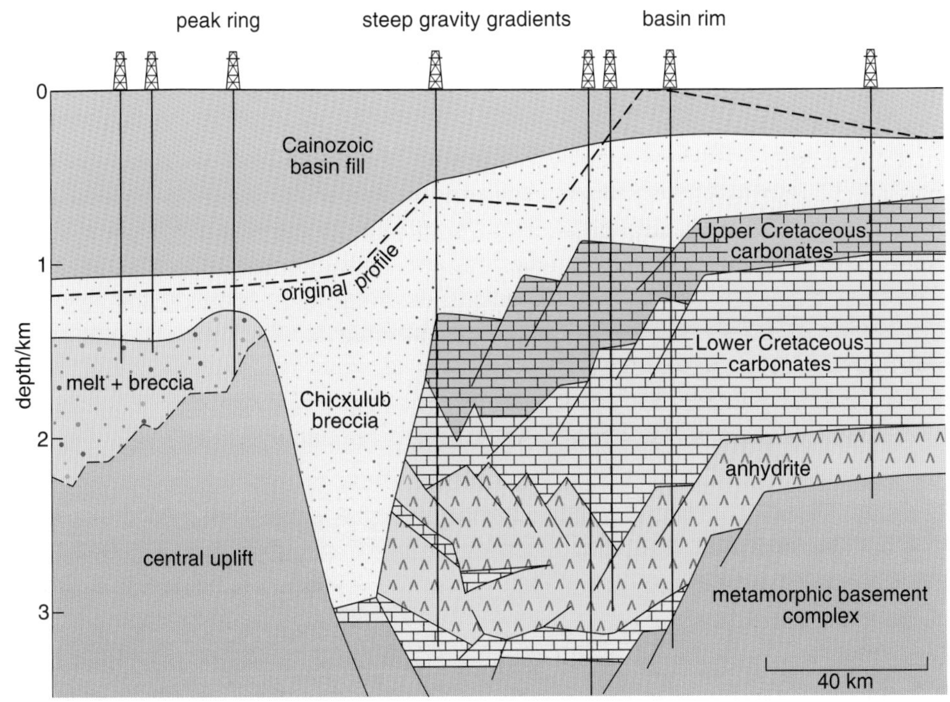

Figure 12.8 Schematic cross-section compiled from drill core and seismic data across the Chicxulub impact crater. (Sharpton et al., 1996.)

○ Does this tell us that the Chicxulub crater was formed at the K/T boundary?

● Unfortunately, it does not. While the microfossils found in the rocks above the impact breccias were Tertiary, they were not specific enough to identify the rock as being from the very base of the Tertiary. All we can deduce is that the crater was formed before these Lower Tertiary rocks were deposited.

Similarly, the underlying Cretaceous limestones were identified as Late Cretaceous in age. The stratigraphical evidence allows us to constrain the age of the Chicxulub impact to some time between the Late Cretaceous and Early Tertiary, but we cannot be more specific than that. However, there is another important rock type shown in Figure 12.8, the melt + breccia. Large impact structures often contain coherent sheets of impact melt on top of the shocked and fragmented rocks of the crater floor (Box 12.1). Since the melting of these rocks took place at the time of the impact, their radiometric clocks would have been reset. Bear in mind that impact glasses from the Beloc K/T boundary site on Haiti had given an age of 65.07 ± 0.11 Ma; subsequent measurements of impact glasses from the Arroyo El Mimbral site in Mexico gave the same age. When samples of impact melt glass from one of the Chicxulub boreholes were dated using the $^{40}Ar/^{39}Ar$ technique (Swisher *et al.*, 1992), they gave ages of 65.00 ± 0.08 Ma and 64.98 ± 0.11 Ma, indistinguishable from the ages for the Beloc and Arroyo El Mimbral K/T boundary layers. Geochemical data also linked the K/T boundary layer glasses to the Chicxulub melt rocks. Studies using the uranium–lead-isotope system not only gave identical ages for crystals of the mineral zircon from Chicxulub melt rock and K/T boundary layers in Colorado but also indicated that these zircons had come from the 544-Ma-old metamorphic basement rocks underlying the crater. Other isotopic geochemical studies using rubidium–strontium, samarium–neodymium and oxygen-isotopic data also tied the Chicxulub melt rock to the K/T boundary layer. The geochemical and age data confirmed that Chicxulub was the long-sought K/T impact structure. It was of the right size, location, and age, and rocks from the structure had similar geochemistry to the ejecta found in the K/T boundary layer.

12.3 Summary

- Prior to 1991, several impact craters had been suggested as potential structures for the K/T boundary impact event. However, none was of both the right size and right age.

- The Chicxulub structure in the Yucatan Peninsula, Mexico, was identified as a possible Late Cretaceous impact structure on the basis of geophysical measurements. A circular negative gravity anomaly some 180 to 200 km in diameter, together with a small central positive gravity anomaly, matched the known features of large complex impact structures, which have an uplifted area at their centres.

- Extensive geophysical investigations of the Chicxulub structure in the early 1990s led to it being confirmed as an impact structure similar to the large multi-ring impact basins that occur on the Moon. Drill core samples indicated that the structure was now buried under almost 1 km of Tertiary sediments but also provided positive verification of the structure's impact origin in the form of planar deformation features in quartz crystals, evidence of shock metamorphism.

- The Chicxulub impact structure was dated at 65 Ma using ^{40}Ar/^{39}Ar dating of impact melt glasses from drill cores. This age is indistinguishable from the age of the K/T boundary and the age of impact-produced spherules found in K/T boundary deposits in Haiti.

12.4 References

HILDEBRAND, A. R., PENFIELD, G. T., KRING, D. A., PILKINGTON, D. A., CAMARGO, A., JACOBSEN, S. B. AND BOYNTON, W. V. (1990) 'Chicxulub crater — a possible Cretaceous–Tertiary boundary impact crater on the Yucatan peninsula, Mexico', *Geology*, **19**, 867–871.

SWISHER, C. C., GRAJALES-NISHIMURA, J. M., MONTANARI, A., MARGOLIS, S. V., CLAEYS, P., ALVAREZ, W., RENNE, P., CEDILLO-PARDO, E., MAURRASSE, F. J. M. R., CURTIS, G. H., SMIT, J. AND McWILLIAMS, M. O. (1992) 'Coeval Ar^{39}/Ar^{40} ages of 65.0 million years ago from Chicxulub crater melt rock and Cretaceous–Tertiary boundary tektites', *Science*, **257**, 954–958.

13 The effects of the Chicxulub impact

Iain Gilmour

To most scientists, the discovery, or more accurately rediscovery, of the Chicxulub impact crater was the 'smoking gun' needed to settle the debate about whether there was an impact at the K/T boundary. For many, it was time to go back to the beginning of the whole story of the impact ejecta layer and the search for the crater: the mass extinction that marked the end of the Mesozoic Era. Was the Chicxulub impact responsible for the mass extinction? Or was it simply one of several contributory factors acting on already stressed ecosystems? Alternatively, did the impact of a 10-km-diameter asteroid have no appreciable consequences for life on Earth? Establishing the effects that the Chicxulub impact had on the Earth's climate and environment is perhaps the only means by which we can try to assess what role the impact played in the mass extinction.

13.1 Local devastation

The immediate devastation in the area surrounding the Chicxulub crater would undoubtedly have been catastrophic and widespread; there is evidence from exposures in the Gulf of Mexico region that attests to some of this devastation. The palaeogeography of the area at the time of the Chicxulub impact is shown in Figure 13.1. The crater itself is situated on a relatively shallow water (100 to 200 m) carbonate platform with deeper water off to the north, and beyond that, land, which had gradually emerged as the sea that had occupied the Western Interior of North America receded during the Late Cretaceous. Examination of field exposures has built up a picture of the geological upheaval that resulted from the impact.

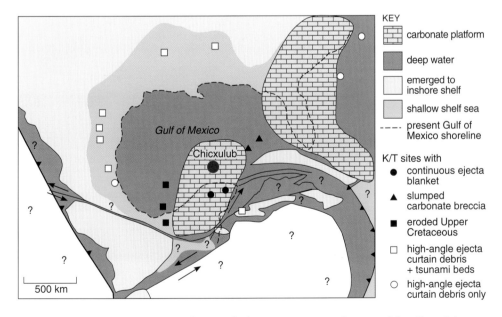

Figure 13.1 Palaeogeography of the Gulf of Mexico region at the time of the Chicxulub impact. (See Figures 2.13 and 2.14 for geodynamic context.)

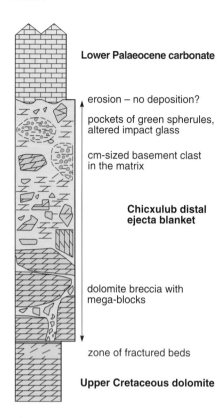

Figure 13.2 Graphic log of the impact breccia sequence from Albion Island, Belize.

To the south of the crater, an unusual deposit of breccia-like material with a thickness of several tens of metres was found in the region of the Mexico–Belize–Guatemala borders, some 250 to 350 km from the crater centre. These deposits (shown in Figure 13.1 as ●) have been interpreted as part of the continuous ejecta blanket from the crater and have since been found all over the Yucatan peninsula. The ejecta blanket is represented by more than 20 m of a breccia that contains large blocks of dolomite up to 10 m in size as well as small pockets of mm- to cm-size rounded to elongated green clay spheres. These closely resemble the clay spheres observed at other K/T boundary sites and are thought to represent altered impact glass. The breccia sits on top of a Late Cretaceous dolomite — similar to the blocks forming most of the breccia components — and is overlain by Tertiary rocks; a graphic log of the sequence is shown in Figure 13.2.

We have already seen that around the world the K/T boundary is marked by a millimetre- to centimetre-thick claystone layer (e.g. Figure 11.6) enriched in Ir and other impact debris. However, around the Gulf of Mexico, from Alabama to Guatemala, the K/T boundary is marked by a 2- to 4-m-thick sequence of clastic sandstones characteristic of high-energy depositional environments (shown in Figure 13.1 as □). One such site, a deep-water deposit at Arroyo El Mimbral in north-eastern Mexico (Figure 13.3), was deposited in several hundred metres of water. A graphic log for the section is shown in Figure 13.4. The base of this section is composed mainly of Chicxulub ejecta and contains shocked minerals, spherules and impact glass together with large rip-up clasts from the underlying Late Cretaceous marls. A high-energy massive and laminated coarse sandstone, which contains terrigenous material such as abundant plant fragments (Figure 13.5), overlies the ejecta layer. The sandstone grades into a succession of fine sandstone and siltstone layers (Unit IV in Figure 13.4), the latter being enriched in Ir, impact-produced diamonds and magnesioferrite spinels. Three interpretations have been suggested for these K/T successions:

Figure 13.3 Exposure of a K/T boundary sequence at Arroyo El Mimbral in north-eastern Mexico, marked by a coarse clastic sequence. *(Iain Gilmour, Open University.)*

1 Their high-energy nature led to the suggestion that these layers were the
   result of tsunamis generated by the impact.

2 They represent a Bouma sequence as a result of turbidite flows initiated by
   the impact.

3 As with most controversies surrounding the K/T boundary, there was also a
   non-impact scenario that proposed that these horizons were deposited over a
   longer period of time, i.e. thousands of years, as a result of sea-level changes.

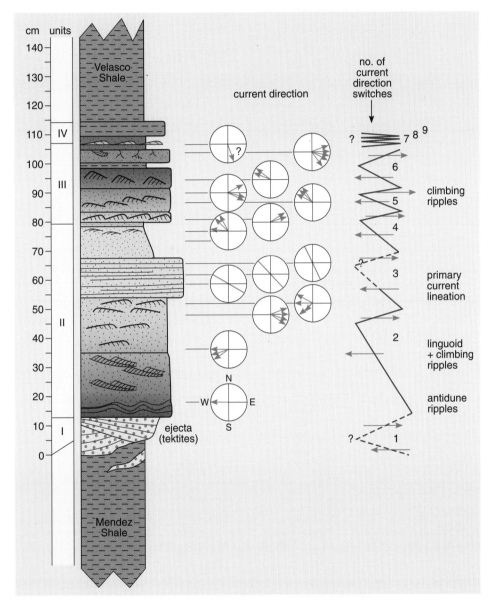

Figure 13.4 Graphic log of the postulated tsunami deposit at Arroyo El Mimbral, north-eastern Mexico. *(Smit, 1999.)*

However, there are several pieces of evidence that make the latter two suggestions
unlikely. The juxtaposition of the massive sandstone unit between two horizons
containing abundant evidence of shock metamorphism (Units I and IV in Figure 13.4)
would require two massive impacts in the region within several thousand years of
each other, which, while not entirely impossible, would seem to stretch the bounds
of probability. There is also evidence within the sandstone itself that indicates rapid
deposition: coarse sandstones deposited in deep water require strong currents and
rapid flows to move them. Thus, for the non-impact scenario, this would require a
series of rapid flows over a long period of time. Between these flows there would
have been ample time for life to colonize the sea-floor and therefore there should be
evidence of bioturbation, for example the trace fossils of burrowing animals, within

Figure 13.5 Plant fragments in the base of the massive sandstone unit at Arroyo El Mimbral, north-eastern Mexico. (Iain Gilmour, Open University.)

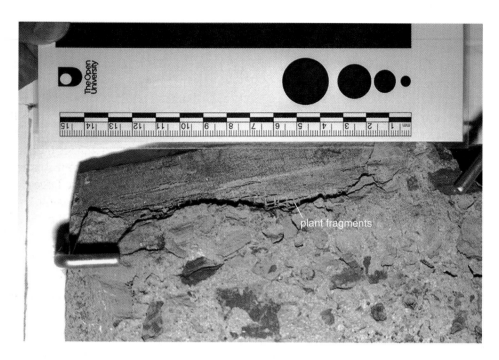

the massive sandstone. However, there is no evidence within the sandstone of trace fossils of any burrowing animals suggesting that the whole deposit was formed in days rather than millennia. If you look at the graphic log in Figure 13.4, you can see another crucial piece of evidence: the ripple marks at the top of the sequence.

○ What do the palaeocurrent directions determined from these ripple marks and shown in Figure 13.4 suggest about the nature of the current that deposited these sandstones?

● They indicate a series of different current direction flows, at times in opposite directions.

In fact, when the palaeocurrent directions indicated by these ripples were determined, some 200 different directions were found. This was evidence consistent with the effects of large waves interacting with each other as they hit land but not dense bottom currents, which generally only flow in one direction.

Other sedimentary sequences in the Gulf of Mexico region also seem to indicate some of the localized effects of the Chicxulub impact. At Blake Nose, off the coast of Florida in the Atlantic Ocean, both geophysical studies and examination of drill cores indicate extensive folding and slumping of the sedimentary rocks directly underlying the K/T boundary ejecta. This has been interpreted as resulting from a massive failure of the North Atlantic continental margin that was triggered by the impact. It has been speculated that such extensive disruption of the sea-floor could have resulted in the release of large quantities of methane gas, often trapped in some oceanic sediments, which could have had a significant effect on the Earth's climate.

13.2 Global effects

There is little direct evidence of the effects of a large asteroid or comet impact. The 1994 impact of a series of 2-km-sized fragments of comet Shoemaker-Levy 9 into the atmosphere of Jupiter provided some information on the kind of processes that can happen. For the Chicxulub impact, however, much of what is postulated to have happened is based on our understanding of the impact process from two sources. The first are laboratory experiments involving the high-velocity impact of small projectiles of known composition into targets of known composition. Typically, the projectiles are a few millimetres in diameter and the velocities achieved in the experiment range up to around $9 \, \text{km s}^{-1}$. From these experiments, it is possible to scale the process up to larger impacts, a process that provides useful information on various aspects of the impact event including the effects of high-velocity impact on various types of rock and the trajectories of ejecta from a crater. A second source of information comes from hydrocodes. Hydrocodes are large computer programs that can be used to simulate numerically a highly dynamic event, especially those that include shocks. While they cannot supersede the information provided by real impact experiments, hydrocodes are an invaluable tool for extending to planetary scales the limited results from laboratory experiments.

○   Look again at Figure 12.8. What are the three main rock types that would have comprised the target rocks for the Chicxulub impact?

●   The underlying metamorphic basement rocks, which were probably highly crystalline, an evaporitic sequence with abundant anyhdrite ($CaSO_4$) and gypsum ($CaSO_4.2H_2O$) and a sequence of limestones that would predominantly have been composed of calcium carbonate ($CaCO_3$).

Hydrocode modelling of the Chicxulub impact event suggests the impact vaporized a large quantity of sedimentary rock, producing gases whose physical and chemical properties had an effect on the Earth's climate due to adsorption of radiation or chemical reactions. During the impact event, thousands of cubic kilometres of rock and dust were ejected from the crater, and modelling suggests large amounts of water, CO_2 and SO_2 would have been released into the atmosphere. Estimates of around $300–2000 \, \text{km}^3$ of vaporized sediments yield sulfur masses of around $100–200 \, \text{Gt}$ (1 gigatonne (Gt) $= 1 \times 10^9$ tonnes), and about $1000–1500 \, \text{Gt}$ of CO_2 that were released almost instantaneously into the atmosphere. The quantity of gases released varies between different model calculations. In addition, the marine sediments that comprised the Chicxulub target rocks also contained halogens (chlorine, bromine), which, when released into the upper atmosphere, destroy the ozone layer. This would have led to a substantial increase in the flux of ultraviolet radiation at the Earth's surface.

13.3 Climatic implications

Before the discovery of the Chicxulub structure, studies of the impact-related climatic effects focused on the short-term climate change produced by dust in the atmosphere. Indeed, this was the 'killing mechanism' that the Alvarez group proposed in their 1980 paper. Calculations on the effect of this dust in the atmosphere suggested that there would have been a drastic cooling event due to interception of sunlight by high-altitude dust, lasting for a few months to about a year. However, once it was realized that the target contained carbonate and sulfate sedimentary rocks, scientists began exploring the possible climatic effects of a massive release of CO_2 and sulfur-bearing gases to the atmosphere.

As noted above, the hydrocode simulations of the Chicxulub impact event indicate a production of $1000–1500 \, \text{Gt}$ of CO_2.

○ For comparison, calculate the total mass of CO_2 in the present-day atmosphere from the mass of C given in Table 6.1. Note that $1 \, Gt = 10^{12} \, kg$, and assume (for ease of calculation) that the atomic masses of C and O are 12 and 16, respectively.

● According to Table 6.1, the present-day atmosphere contains $0.76 \times 10^{15} \, kg$ of C. The mass of CO_2 is therefore $[(12 + 32)/12] \times 0.76 \times 10^{15} \, kg = 2.79 \times 10^{15} \, kg$, or 2790 Gt. Hence the estimated yield of CO_2 from the Chicxulub impact amounts to rather more than a third to a half of that present in the atmosphere today — a rather alarming thought, should such an event recur.

These estimates of impact-released CO_2 should, however, be compared to the inferred global atmospheric inventory of CO_2 at the end of the Cretaceous, which is estimated at around 9000 Gt. Impact degassing of the target rocks at Chicxulub would then produce an increase of 20% in the CO_2 atmospheric inventory. It has been estimated that this amount may increase to as much as 40% if the asteroid impacted at a lower angle since laboratory experiments indicate that there is an increase in the vaporization of the surface layers of a target as the impact angle, measured from the surface, decreases. Was this enough of an increase in CO_2 to cause a catastrophic climate change? Probably not: climate models indicate that this level of increase in the atmospheric inventory of CO_2 might lead to a global temperature increase of around 1.2 °C.

We can examine the climatic effect of the injection of sulfur into the stratosphere by looking at the climatic effects associated with volcanic eruptions. Long-lived stratospheric aerosols, such as SO_2, tend to cool the Earth's atmosphere system by reflecting an additional amount of energy back into space (Section 7.1.3). In terms of the amount of gases injected into the stratosphere, the largest-known explosive volcanic event in the late Quaternary was the eruption of Toba, in Sumatra, about 73 500 years ago. Estimates indicate that it lofted about 1 Gt of SO_2 into the stratosphere for a predicted cooling of about 3.5 °C in the year following the eruption. The amount of sulfur added to the stratosphere by the Chicxulub impact, estimated by the hydrocode simulations, would have been between 100 and 200 Gt; this would correspond to around 200–400 Gt of SO_2, more than two orders of magnitude greater than any volcanic eruption in the geological record.

○ Why is it hard to infer the effect that such a large load would have had on the Earth's climate?

● The magnitude of any climate forcing is not a linear function of the amount of sulfur-bearing gases injected into the upper atmosphere and, as we saw in Section 7.1, is limited by the presence of H_2O in the stratosphere.

However, larger amounts of sulfur-bearing gases will take longer to decay, therefore prolonging the climate forcing in time. It is evident, though, that a Chicxulub-type injection of sulfur-bearing gases (and water vapour) in the stratosphere must have produced devastating changes to the end-Cretaceous global climate lasting for perhaps several years or even decades.

13.4 A question of cause and effect

The present consensus of scientific opinion is that the impact–extinction hypothesis provides the only coherent explanation for the origin of the K/T boundary, relating features as diverse as the dinosaur extinction in the Hell Creek Formation in Montana, the extinction of planktonic foraminifers at Gubbio in Italy, a negative shift in carbon-isotope values at the K/T boundary in New Zealand, the disappearance of ammonites at Zumaya in Spain, an iridium anomaly at Poty in Brazil, impact diamonds at the K/T boundary in Colorado, shocked zircon crystals from Saskatchewan, Canada, and shocked quartz crystals in the South Pacific. It is the global nature of the evidence that is impressive: the same evidence being found time and again around the world.

The consensus is by no means unchallenged: hypotheses requiring more gradual processes such as climate change are vociferously argued. Those more sceptical of the impact–extinction hypothesis have suggested that the selectivity of the pattern of extinctions, for example the survival of particular faunal groups, is more compatible with a gradual extinction scenario than an impact-related one. Similarly, the survival of some species or even genera only to become extinct shortly after the boundary, and the disappearance of some groups of organisms beforehand such as the inoceramids, has been cited as evidence for gradual or stepwise extinction.

The impact–extinction hypothesis does demand that the extinctions that define the end of the Mesozoic Era took place fairly rapidly as a result of, and thus not before, the impact. However, the fact that the inoceramids or some species of foraminifers went extinct before the K/T boundary does not contradict a later catastrophic extinction. Similarly, the persistence of some groups of animals through the boundary, even if they went extinct soon after, does not contradict the possibility of impact-caused extinctions either. While many effects of the Chicxulub impact were immediate, longer-term effects seem unavoidable as well.

We do not have a sufficient understanding of the complexity of Late Cretaceous ecosystems to reach firm conclusions about extinction mechanisms for both gradual and abrupt scenarios. Even present-day ecosystems are poorly understood and those of the Late Cretaceous are far removed from observation and analysis. Whatever the extinction scenario, it seems therefore inappropriate to claim that a selective pattern of extinction is compatible with one mechanism but incompatible with another.

13.5 The end of an era

The impact–extinction hypothesis also brought about the end of another era. Not, as many thought, the central role played by uniformitarianism in the Earth sciences. There never was any real conflict between the impact–extinction hypothesis and uniformitarianism. Many scientists had simply failed to make the connection between small meteorites that are observed to fall from the sky all the time and large impact events. Rather, it was recognized that strict adherence to a principle that invoked slow, gradual uniform rates of change to explain the Earth's history was inappropriate in light of the increasing evidence for periods of rapid environmental change throughout the Earth's history.

The controversy surrounding the K/T boundary continues. Some scientists remain firmly opposed to the idea that an impact played any role in the faunal turnover at the end of the Mesozoic, de-emphasizing the abruptness of the boundary and its importance. The irony of this position was not lost on Graham Ryder, a lunar geologist, who surmised it thus:

> '… [opponents] give the impression that it is those who invoke an impact who have required a particular paleontological significance, for instance that impact proponents *claim* abrupt extinction. Yet it has never been the case that an impact was inferred and that then there was a search for associated extinctions. It is an ironic reversal that some paleontologists chose to reduce the significance of the boundary after the impact was inferred. Rather than evaluate the record in the light of an impact, they chose to construct inappropriate straw men.'

13.6 Summary

- The immediate, local effects of the Chicxulub impact would have been severe. The impact occurred in relatively shallow water (100 to 200 m deep) on a carbonate platform, and an ejecta blanket tens of metres thick has been found 250 to 350 km from the crater centre in the region of the Mexico–Belize–Guatemala borders.

- In north-eastern Mexico, high-energy, massive coarse-grained sandstone deposits that grade upwards into finer sandstone and siltstone layers, the latter enriched in Ir, impact diamonds and spinels, have been interpreted as tsunami deposits produced by tidal waves generated by the impact.

- Hydrocode modelling, computer-based numerical simulations of the impact, are one means by which we can assess the possible global effects of such a large-scale impact event.

- The impact occurred in a thick sedimentary sequence of carbonates and evaporites and hydrocode modelling suggests that large quantities of the target rocks were vaporized by the impact, releasing large amounts of water, CO_2 and SO_2 into the atmosphere.

- Initial studies on the effects of the Chicxulub impact focused on the role of dust ejected from the crater and calculations suggested that there would have been a marked lowering of temperatures following the impact due to the interception of sunlight by high-altitude dust. Once it was realized, however, that the target contained carbonate and sulfate sedimentary rocks, the possible climatic effects of a massive release of CO_2 and sulfur-bearing gases to the atmosphere also had to be considered. Long-lived stratospheric aerosols, such as SO_2, tend to cool the Earth–atmosphere system by reflecting an additional amount of energy back to space.

13.7 Further reading

ALVAREZ, W. (1997) 'T. rex and the crater of doom', Princeton University Press, Princeton, 185pp.

GLEN, W. (1998) The Impact–Extinction Debates, Stanford University Press, Stanford, California, 371pp.

SMIT, J. (1999) 'The global stratigraphy of the Cretaceous–Tertiary boundary impact ejecta', Annual Review of Earth and Planetary Sciences, 27, 75–113.

The research papers presented at the 'Snowbird' series of conferences on the K/T boundary mass extinction are published in the following Geological Society of America Special Papers, listed chronologically:

SILVER, L. T. AND SCHULTZ, P. H. (eds) (1982) 'Geological implications of impacts of large asteroids and comets on the Earth', Geological Society of America Special Paper 190, 528pp.

SHARPTON, V. L. AND WARD, P. D. (eds) (1990) 'Global catastrophes in Earth History', Geological Society of America Special Paper 247, 631pp.

RYDER, G., FASTOVSKY, D. AND GARTNER, S. (eds) (1996) 'The Cretaceous–Tertiary boundary and other catastrophes in Earth History', Geological Society of America Special Paper 307, 576pp.

KOEBERL, C. AND MACLEOD, K. G. (eds) (2002) 'Catastrophic events and mass extinctions: impacts and beyond', Geological Society of America Special Paper 356, 579pp.

Epilogue

Peter W. Skelton

In this final Section, we stand back from considering the Cretaceous world itself to reflect on three more general issues that emerge from the case study. These concern, respectively, the limitations of uniformitarianism, how historical science is done, and lessons for the future.

Uniformitarianism

The Scottish 'father of geology', James Hutton (1726–1797), argued that geological phenomena should be interpreted only in terms of natural processes that can be seen to operate today — an approach later encapsulated in the aphorism 'the present is the key to the past'. Hutton's ideas were developed further by his compatriot Charles Lyell (1797–1875), in his highly influential *Principles of Geology* (1830–3). However, Lyell's version of uniformitarianism, as the approach came to be called, incorporated two further assertions. Besides explanation in terms only of known causes, he also insisted on their gradual (or incremental) operation, as well as a long-term overall steady state in global conditions (though allowing for some regional fluctuation in climate and sea-level). These embellishments were intended to counter speculative catastrophic explanations, such as deluges of continental scale, or even more menacing-sounding 'convulsions', which some geologists had advanced for such dramatic geological phenomena as mass extinctions, and which Lyell regarded as unscientific.

The main challenge posed by Lyell's assertions was the task of documenting credible evidence for conditions wholly unfamiliar to (and thus verifiable from) human experience. Yet, even when his *Principles of Geology* was published, Lyell's steady-state hypothesis sat uncomfortably with the existing fossil evidence for faunal and floral progression, while critics pointed out that the short span of human history was unlikely to have sampled all kinds of environmental catastrophe. Within a few decades, evidence for biotic turnover through evolution and extinction was well established, but Lyellian gradualism and a reluctance to invoke markedly exotic conditions prevailed largely through the want of contrary evidence, bolstered by the understandable conservatism of geologists who were wary of being associated with mere 'speculation'.

Over recent decades, however, the technological and methodological advances in all areas of the Earth sciences have at last begun to yield the credible evidence against Lyell's additional strictures. As you have seen in this book, for example, there is now compelling evidence, ranging from oxygen-isotope ratios in skeletal materials to multivariate analyses of leaf assemblages, that the Earth's mean surface temperature rose to significantly higher levels in the Cretaceous than at any time during the past few million years. Evaluation of the major geological sources and sinks of carbon, together with computer modelling of the feedbacks involved, implicate levels of atmospheric carbon dioxide significantly above those of today in the establishment of that greenhouse climatic regime. Palaeontological and sedimentological evidence reveals the existence of flourishing Cretaceous ecosystems, such as polar forests and tropical carbonate platform communities, radically different in character from their counterparts today. Moreover, catastrophism is firmly back on the scientific agenda, thanks to specific geochemical clues, linked with refined stratigraphical evidence for mass extinction. Hence the Earth's past was evidently a lot stranger than geologists could have believed just half a century ago, and only the original core of uniformitarianism — in effect, conformity of explanation with the currently known laws of physics and chemistry — is generally adopted by them today.

Historical science

Our exploration of the Earth's exotic and eventful Cretaceous history has also illustrated how it is possible to test hypotheses concerning the past.

It is often asserted that science progresses by successively falsifying competing hypotheses. Falsification can be achieved through experiment, designed to see whether the outcome predicted by a given hypothesis occurs or not: if not, the hypothesis can be rejected. Lacking the ability to travel back in time, however, we obviously cannot experimentally replicate past circumstances in all their complexity in order to test predicted outcomes. So a sceptic who is used to the practice of experimental physics or chemistry might well ask how historical hypotheses can be tested. As you have seen in this book, it is possible to run computer simulations of past climates, for example, in which years can be compressed to mere moments of computer time. Yet, despite their current (and growing) sophistication, these are still highly simplified models of reality, which beg the question as to whether the assumptions concerning the initial conditions for the runs, as well as the networks of feedbacks involved, are correct. In Chapter 9, it was stressed that the results of computer modelling have themselves to be repeatedly tested against geological data. The computer models are, in effect, elaborate hypotheses in their own right, and testing is still based on observation of the geological record. While such iterative verification and refinement of computer simulations should improve their reliability, the observable residue of the past will always remain the ultimate arbiter for rejecting or accepting hypotheses that have been proposed to explain it — but how?

As any devotee of crime literature knows, the key to getting the story right is corroboration. The English philosopher of science, William Whewell (1794–1866), who, incidentally, also invented the term 'uniformitarianism', referred to this method of scientific verification as *consilience*, meaning (from the Latin) a 'jumping together' of inferences. Consilience occurs when the same causal explanation is independently inferred from two or more unrelated observations. The greater the number of instances of such agreement, the more likely the explanatory hypothesis is to be correct, or, to put it another way, the more improbable it is that the concordance of the evidence could be due to the chance coincidence of unrelated effects. With a sufficient weight of corroborative evidence, a hypothesis may eventually be regarded as a well-founded theory, in the same way that a defendant would be found guilty or not guilty in a criminal trial. Perhaps the most celebrated exponent of this method was the fictional detective, Sherlock Holmes, so much so that in October 2002 he was granted an Extraordinary Honorary Fellowship of the Royal Society of Chemistry, in the UK, in recognition of his exposition of forensic science. One feels he might have appreciated the consilience of the combined geochemical, mineralogical, sedimentological and geophysical evidence for a devastatingly huge impact at Chicxulub at the end of the Cretaceous. His creator, Sir Arthur Conan Doyle, whose enthusiasm for the 'Lost World' of the dinosaurs is well corroborated (Figure 1.1), would surely have done so.

The need for as many independent lines of evidence as possible for testing hypotheses about past conditions on the Earth, especially where these resulted from the complex interactions and associated feedbacks of the Earth system, demands an interdisciplinary approach. We hope this message has also been abundantly illustrated in this book, as we have woven together such diverse topics as the physics and chemistry of the atmosphere and oceans, the rock cycle and stratigraphy, the palaeobiology of terrestrial and marine ecosystems, mantle dynamics and volcanism, and the biogeochemical cycles that are driven by these and other factors. Earth system science is indeed a demanding mental exercise, but that is amply rewarded by the insights gained into the workings of our rich and varied world.

Lessons for the future

Given the inherently chaotic nature of the Earth system, specific and reliable predictions about future global change, in terms of what will happen where and when, rapidly become more prone to error the further into the future that we try to look, largely because of the non-linear effects of feedbacks. Every instance of the weather not quite matching even the previous day's forecast reminds us of that limitation. Nevertheless, existing models can already predict some generalized trends of change, within limits of error governed by input assumptions. This approach is illustrated, for example, in the reports of the Intergovernmental Panel on Climate Change (IPCC), which was established 'to assess scientific, technical and socio-economic information relevant for the understanding of climate change, its potential impacts and options for adaptation and mitigation' (http://www.ipcc.ch/).

In this context, one might hope that past conditions on the Earth, when already in a greenhouse state, as in the Cretaceous, for example, might give us some clues as to the likely consequences of global warming today. To some extent this may be true, though the same reservations that apply to the maxim 'the present is the key to the past', which we noted above, have also to be borne in mind when applying the lessons of the past to the future. Neither the sustained rate at which climate warming is being forced today, nor the initial conditions for our current 'natural experiment', correspond well to the circumstances of the Cretaceous world, so we cannot expect too close a parallel in the outcomes, especially as far as life is concerned. One obvious point in this respect is that our possible journey towards a greenhouse world is starting out from what, for the past couple of million years, has been a predominantly glacial age, the like of which was never experienced (as far as we can tell) during the entire Mesozoic Era. On the other hand, iterative testing of the ability of complex models to replicate the conditions both of today and of contrasting periods such as the Cretaceous will help us to refine them further and so improve our ability to narrow down the range of possible futures. Already, Earth system science is of much more than just academic interest.

Acknowledgements

The production of this book involved a number of Open University staff, to whom we owe a considerable debt of thanks for their commitment and the high professional standards of their contributions, despite challenging deadlines. Glynda Easterbrook commented on draft versions, compiled the index and, together with Jessica Bartlett, tracked down figure copyrights, as well as helping to manage the project in many other ways. Jo Morris styled the text for handover to Gerry Bearman, who copy-edited it and kept the project running through the production process with tact and resourcefulness. Pam Owen prepared virtually all the graphic artwork with exemplary skill and efficiency, though we are also grateful to John Watson, of the Department of Earth Sciences, for his colourful dioramas of Cretaceous scenes. Ruth Drage was responsible for the excellent design and layout, with support from Liz Yeomans for revision of this co-published version. We are also very grateful to Giles Clark (Open University) and Susan Francis (Cambridge University Press) for their support and help with co-publication.

In addition, we wish to thank the following people who commented on earlier versions of our text: Stephen Hesselbo (University of Oxford), Angela Coe, Chris Wilson, Mark Sephton, Steve Killops (Associate Lecturer), and four anonymous referees appointed by Cambridge University Press. Many other individuals and organizations furnished and/or granted permission for us to use their diagrams or photographs, as acknowledged in the Figure References, and to them we also express our gratitude.

Figure references for this book

All of the figures in this book that are based on published work have been redrawn and the majority have been modified, some significantly. Every effort has been made to trace all copyright owners of these figures, but if any has been inadvertently overlooked, the publishers will be pleased to rectify any omissions in these acknowledgements when the book is reprinted.

ALVAREZ, L., ALVAREZ, W., ASARO, F. AND MICHEL, H. V. (1980) 'Extraterrestrial cause for the Cretaceous–Tertiary extinction', *Science*, **208**, 1095–1108. [**Figure 11.8** copyright © 1980 American Association for the Advancement of Science.]

ALVIN, K. L. (1971) 'The spore-bearing organs of the Cretaceous fern *Weichselia* Stiehler', *Botanical Journal of the Linnean Society*, **61**, 87–92. [**Figure 8.9**.]

ANON. (1978) *CIA Handbook, Polar Regions Atlas*, National Foreign Assessment Center, CIA, 66pp. [**Figure 4.6**.]

ARCHIBALD, J. D., SHEEHAN, P. M., FASTOVSKY, D. E., BARRETO, C. AND HOFFMANN, R. G. (2000) 'Dinosaur abundance was not declining in a "3 m gap" at the top of the Hell Creek Formation, Montana and North Dakota: Comment and Reply', *Geology*, **28**, 1150–1151, Geological Society of America. [**Figure 10.13**.]

ARTHUR, M. A. (2000) 'Volcanic contributions to the carbon and sulphur geochemical cycles and global change', pp. 1045–1056 in SIGURDSSON, H. (ed.), *Encyclopedia of Volcanoes*. [**Figures 6.4, 6.12** copyright © 2000 Elsevier Science (USA), reproduced by permission of the publisher.]

BERNER, R. A. (1999) 'A new look at the long-term carbon cycle', *GSA Today*, **9**(11), Nov., 1–6, Geological Society of America. [**Figures 6.3, 6.5**]

BERNER, R. A. AND CALDEIRA, K. (1997) 'The need for mass balance and feedback in the geochemical carbon cycle', *Geology*, **25**, 955–956, Geological Society of America. [**Figure 7.24**.]

BERNER, R. AND KOTHAVALA, Z. (2001) 'GEOCARB III, a revised model of atmospheric CO_2 over Phanerozoic time', *American Journal of Science*, **301**(2), 182–204, Yale University. [**Figures 6.6–6.10**.]

BIRD, K. J. (1987) 'The framework geology of the North Slope of Alaska as related to oil–source rock correlations', *Alaskan North Slope Geology*, **1** (1 Oct.), 138–139, Pacific Section, Society for Economic Paleontologists and Mineralogists. [**Figure 4.9**.]

BUFFETAUT, E. AND KOEBERL, C. (eds) (2002) *Geological and Biological Effects of Impact Events*, Springer-Verlag. [**Figures 11.10, 11.12** copyright © 2002 Springer-Verlag Berlin Heidelberg.]

CHUMAKOV, N. M. *et al.* (1995) 'Climatic zones in the middle of the Cretaceous Period', *Stratigraphy and Geological Correlation*, **3** , 3–14, International Academic Publishing Co. [**Figures 8.1–8**.]

CONAN DOYLE, A. (1960) *The Lost World*, paperback edition, John Murray Publishers Ltd., London. [**Figure 1.1**.]

COURTILLOT, V. (1999) *Evolutionary Catastrophes — The Science of Mass Extinctions*, trans. Joe McClinton, Cambridge University Press. [**Figure 7.15**.]

CROWLEY, T. J. AND BERNER, R. A. (2001) 'CO_2 and climate change', *Science*, **292**(4), 870–872. [**Figure 6.11** copyright © 2001 American Association for the Advancement of Science.]

DRAPER, G., JACKSON, T. A. AND DONOVAN, S. K. (1994) 'Geologic provinces of the Caribbean region', pp. 3–12 in DONOVAN, S. K. AND JACKSON, T. A. (eds), *Caribbean Geology — An Introduction*, University of the West Indies Publishers' Association, Kingston, Jamaica. [**Figure 2.12** © the authors.]

EAMES, A. J. AND MacDANIELS, L. H. (1947) *An introduction to plant anatomy*, 427pp., McGraw-Hill Book Co. [**Figure 4.41**.]

FRAKES, L. A. (1979) *Climates throughout geologic time*, 322pp., Elsevier. [**Figure 1.23**.]

FRAKES, L. A. (1999) 'Estimating the global thermal state from Cretaceous sea surface and continental temperature data', pp. 49–57 in BARRERA, E. AND JOHNSON, C. C. (eds) *Evolution of the Cretaceous Ocean–Climate System*, Geological Society of America Special Paper, **332**. [**Figure 5.11** © 1999 Geological Society of America.]

FRIIS, E. M. (1990) '*Silvianthemum suecicum* gen. et sp. nov…', *Biologiske Skrifter*, **36**, 5–21. [**Figure 4.28**.]

GALE, A. S., SMITH, A. B., MONKS, N. E. A., YOUNG, J. A., HOWARD, A., WRAY, D. S. AND HUGGETT, J. M. (2000) 'Marine biodiversity through the Late Cenomanian–Early Turonian: palaeoceanographic controls and sequence stratigraphic biases', *Journal of the Geological Society*, **157**, 745–757, Geological Society Publishing House. [**Figure 5.10**.]

GRADSTEIN, F. M. AND OGG, J. G. (1996) 'A Phanerozoic timescale', *Episodes*, **19**, 1–2, International Union of Geological Sciences. [**Figure 1.25**.]

GURNIS, M. (2001) 'Sculpting the Earth from inside out', *Scientific American*, **284**(3), March 2001, pp. 34–41. [**Figure 3.6** © David Fierstein.]

HANCOCK, J. M. AND KAUFFMAN, E. G. (1979) 'The great transgressions of the late Cretaceous', *Journal of the Geological Society of London*, **136**, 175–186, Geological Society Publishing House. [**Figure 3.4**.]

HAQ, B. U., HARDENBOL, J. AND VAIL, P. R. (1988) 'Mesozoic and Cenozoic chronostratigraphy and cycles of sea-level change', pp. 71–108 in WILGUS, C. K., HASTINGS, B. S., KENDALL, C. G. ST. C., POSAMENTIER, H. W., ROSS, C. A. AND VAN WAGONER, J. C. (eds), *Sea-level changes: an integrated approach, Society of Economic Paleontologists and Mineralogists Special Publ.*, **42**. [**Figures 1.22a, 3.10.**]

HARDARSON, B. S., FITTON, J. G., ELLAM, R. M. AND PRINGLE, M. S. (1997) 'Rift relocation — a geochemical and geochronological investigation of a palaeo-rift in northwest Iceland', *Earth and Planetary Science Letters*, **153**, 181–196. [**Figure 7.22** copyright © 1997 by permission of Elsevier Science.]

HARRIS, T. M. (1944) 'A revision of *Williamsoniella*', *Phil. Trans. Roy. Soc., Lond.*, **231B**, 313–328. [**Figure 4.29.**]

HAUGEN, R. K. (1982) 'Climate of Remote Areas in North-Central Alaska: 1975–1979 Summary', in *U.S. Army Cold Regions Research and Engineering Laboratory Report*, **82-35**, 1–114. [**Figure 4.3.**]

HAY, W. H. AND DECONTO, R. M. (1999) 'Comparison of modern and Late Cretaceous meridional energy transport and oceanology', pp. 283–300 in BARRERA, E. AND JOHNSON, C. C. (eds) *Evolution of the Cretaceous Ocean-Climate System, Geological Society of America Special Paper*, **332**. [**Figure 5.13** © 1999 Geological Society of America.]

HAY, W. W., DECONTO, R., WOLD, C. N., WILSON, K. M., VOIGT, S., SCHULZ, M., WOLD-ROSSBY, A., DULLO, W.-C., RONOV, A. B., BALUKHOVSKY, A. N. AND SOEDING, E. (1999) 'Alternative global Cretaceous Paleogeography', BARRERA, E. AND JOHNSON, C. C. (eds), *The Evolution of Cretaceous Ocean/Climate Systems, Geological Society of America Special Paper*, **332**, 1–47. [**Figures 2.15, 2.18** © 1999 Geological Society of America.]

HAYS, J. D. AND PITMAN, W. C. III. (1973) 'Lithospheric plate motion, sea level changes and climatic and ecological consequences', *Nature*, **246**, 16–22, Macmillan. [**Figure 3.3.**]

HOUGHTON, B. F., WILSON, C. J. N. AND PYLE, P. M. (2000) 'Explosive volcanism in pyroclastic fall deposits', in SIGURDSSON, H. (ed.), *Encyclopedia of Volcanoes*. [**Figure 7.7** copyright © 2000 Elsevier Science (USA), reproduced by permission.]

ITO, M., ISHIGAKI, A., NISHIKAWA, T. AND SAITO, T. (2001) 'Temporal variation in the wavelength of hummocky cross-stratification: implications for storm intensity through Mesozoic and Cenozoic', *Geology*, **29**, 87–89, Geological Society of America Inc. [**Figure 5.14.**]

JENKYNS, H. AND WILSON, P. A. (1999) 'Stratigraphy, paleoceanography, and evolution of Cretaceous Pacific guyots: relics from a greenhouse Earth', *American Journal of Science*, **299**, 341–392. [**Figure 8.12.**]

JENKYNS, H. C., GALE, A. S. AND CORFIELD, R. M. (1994) 'Carbon- and oxygen-isotope stratigraphy of the English Chalk and Italian Scaglia and its palaeoclimatic significance', *Geological Magazine*, **131**, 1–34, Cambridge University Press [**Figure 5.7.**]

JOHNSON, K. R. AND HICKEY, L. J. (1990) 'Megafloral change across the Cretaceous/Tertiary boundary in the northern Great Plains and Rocky Mountains, U.S.A.', *Geological Society of America Special Paper*, **247**. [**Figure 10.15.**]

KELLEY, S. P., SPICER, R. A. AND HERMAN, A. B. (1999) 'New ^{40}Ar/^{39}Ar dates for Cretaceous Chauna Group tephra, north-eastern Russia, and their implications for the geologic history and floral evolution of the North Pacific region', *Cretaceous Research*, **20**, 97–106. [**Figures 4.44, 4.50, 4.51, 4.52.**]

KENDALL, C. G. ST. C. AND LERCHE, I. (1988) 'The rise and fall of eustasy', pp. 3–17 in WILGUS, C. K., HASTINGS, B. S., KENDALL, C. G. ST. C., POSAMENTIER, H. W., ROSS, C. A. AND VAN WAGONER, J. C. (eds), *Sea-level changes: an integrated approach, Society of Economic Paleontologists and Mineralogists, Special Publ.*, **42**. [**Figure 3.2.**]

KENNEDY, W. J. AND COBBAN, W. A. (1976) 'Aspects of ammonite biology, biogeography, and biostratigraphy', *Special Papers in Palaeontology*, **17**, 94pp., Palaeontological Association [**Figure 5.15.**]

KERR, A. C., ITURRALDE-VINENT, M. A., SAUNDERS, A. D., BABBS, T. L. AND TARNEY, J. (1999) 'A new plate tectonic model of the Caribbean: Implications from a geochemical reconnaissance of Cuban Mesozoic volcanic rocks', *Geological Society of America Bulletin*, **111**, 1581–1599. [**Figure 2.14.**]

KOPPERS, A. P. A., STAUDIGEL, H., WIJBRANS, J. R. AND PRINGLE, M. S. (1998) 'The Magellan seamount trail: implications for Cretaceous hotspot volcanism and absolute Pacific plate motion', *Earth and Planetary Science Letters*, **163**, 53–68. [**Figure 7.20** copyright © 1998 by permission of Elsevier Science.]

LARSON, R. L. (1991) 'Latest pulse of Earth: evidence for a mid-Cretaceous superplume', *Geology*, **19**, 547–550, Geological Society of America [**Figure 7.17.**]

LARSON, R. L. (1995) 'The mid-Cretaceous superplume episode', *Scientific American*, February. [**Figure 2.17** by permission of Scientific American Inc., all rights reserved.]

MARSHALL, J. F. AND DAVIES, P. J. (1982) 'Internal structure and Holocene evolution of One Tree Reef, Southern Great Barrier Reef', *Coral Reefs*, **1**, 21–28. [**Figure 5.2.**]

MARTILL, D. M. (1993) *Fossils of the Santana and Crato Formations, Brazil*. Palaeontological Association, Field Guides to Fossils, No. **5**, 159 pp. [**Figure 2.11.**]

MATSUMOTO, T. (1980) 'Inter-regional correlation of transgressions and regressions in the Cretaceous Period', *Cretaceous Research*, **1**, 259–373. [**Figure 3.5** © 1980 Academic Press Inc. (London), modified with permission.]

MCCABE, P. AND PARRISH, J. T. (1992) 'Tectonic and climatic controls on the distribution and quality of Cretaceous coals', *Controls on the distribution and quality of Cretaceous coals, Geological Society Special Paper*, **267**, Geological Society of America. [**Figure 4.39.**]

MCCORMICK, M. P., THOMASON, L. W. AND TREPTE, C. R. (1995) 'Atmospheric effects of the Mt Pinatubo eruption', *Nature*, **373**, 399–404. [**Figure 7.8** copyright © 1995 Macmillan.]

METCALFE, I. (1988) 'Origin and assembly of South-East Asian continental terranes', pp. 101–118 in AUDLEY-CHARLES, M. G. AND HALLAM, A. (eds), *Gondwana and Tethys, Geological Society Special Publication*, **37**, Geological Society Publishing House. [**Figure 2.9b,c.**]

MILNER, A. C., MILNER, A. R. AND EVANS, S. E. (2000) 'Ch. 21, Amphibians, reptiles and birds: a biogeographical review', pp. 316–332 in CULVER, S. J. AND RAWSON, P. F. (eds), *Biotic response to global change — the last 145 million years*, Cambridge University Press. [**Figure 2.7.**]

MOLINA, E. *et al.* (1996) 'The Cretaceous/Tertiary boundary mass extinction in planktic foraminifera at Agost, Spain', *Revue de Micropaléontologie*, **39**, 225–243. [**Figure 10.4** by permission of the author.]

MONTANARI, A. AND KOEBERL, C. (2000) *Impact stratigraphy: The Italian Record*, Lecture Notes in Earth Sciences Series, **93**, Springer-Verlag GmbH and Co. KG. [**Figures 11.3, 11.7, 11.9, 11.13, 12.4** copyright © Springer-Verlag Berlin Heidelberg 2000.]

MULL, C. G., HOUSEKNECHT, D. W. AND BIRD, K. J. (2003) *Revised Cretaceous and Tertiary Stratigraphic Nomenclature in the Central and Western Colville Basin, Northern Alaska*, Alaska Department of Natural Resources/US Geological Survey. [**Figure 4.12** by permission of the author.]

NATIONAL GEOPHYSICAL DATA CENTER, Marine Geology and Geophysics Division Website: http://www.ngdc.noaa.gov/mgg/image/WorldCrustalAge.gif [**Figures 1.15, 7.14** data from MÜLLER, ROEST, ROYER, GAHAGAN AND SCHLATER, Digital Age Map of the Ocean Floor, Scripps Institution of Oceanography, Ref. Series No. 93-30.]

NEAL, C. R., MAHONEY, J. J., KROENKE, L. W., DUNCAN, R. A. AND PETTERSON, M. G. (1997) 'The Ontong-Java Plateau', in *Large Igneous Provinces: Continental, Oceanic, and Planetary Flood Volcanism*, **100**, American Geophysical Union. [**Figure 7.21** NAKANISHI, M., TAMAKI, K. AND KOBAYASHI, K. (1992) *Geophysical Journal International*, **109**, 701–719.]

ODP (Ocean Drilling Program) website: http://www-odp.tamu.edu/publications/171B_SR/chap_03/images/11_f01.gif [**Figure 10.3** copyright © Ocean Drilling Program.]

ODSN PLATE TECTONIC RECONSTRUCTION SERVICE: http://www.odsn.de/odsn/services/paleomap/paleomap.html, Ocean Drilling Stratigraphic Network, Research Center for Marine Geosciences, Kiel/University of Bremen [**Figures 1.21b–d, 4.8.**]

OGG, J. (1995) 'Magnetic polarity timescale of the Phanerozoic', *Global Earth Physics — a handbook of physical constants*, American Geophysical Union. [**Figure 1.27.**]

ORTH, C. J., GILMORE, J. S., KNIGHT, J. D., PILLMORE, C. L., TSCHUDY, R. H. AND FASSETT, J. E. (1981) 'An iridium abundance anomaly at the palynological Cretaceous–Tertiary boundary in northern New Mexico', *Science*, **214**, 1341–1343, American Association for the Advancement of Science. [**Figure 10.14.**]

PARRISH, J. T. AND BARRON, E. J. (1986) *Paleoclimates and Economic Geology*, Society of Economic Paleontologists and Mineralogists, Short Course No. 18. [**Figures 9.1, 9.2** © copyright SEPM 1986, all rights reserved.]

PERCH-NIELSEN, K., MCKENZIE, J. AND HE, Q. (1982) 'Biostratigraphy and isotope stratigraphy and the "catastrophic" extinction of calcareous nanoplankton at the Cretaceous/Tertiary boundary', *Geological Society of America Special Paper*, **190**, 353–371. [**Figure 10.11.**]

PINDELL, J. L. (1994) 'Evolution of the Gulf of Mexico and the Caribbean', pp. 13–39 in DONOVAN, S. K. AND JACKSON, T. A. (eds), *Caribbean Geology — An Introduction*, University of the West Indies Publishers' Association, Kingston, Jamaica. [**Figure 2.13** © the author.]

RAMPINO, M. R. AND SELF, S. (2000) 'Volcanism and biotic extinctions', in SIGURDSSON, H. (ed.), *Encyclopedia of Volcanoes*. [**Figure 6.10** copyright © 2000 Elsevier Science (USA), reproduced by permission.]

SELF, S., KESZTHELYI, L. AND THORDARSSON, T. (1998) 'The importance of pahoehoe', *Annual Review of Earth and Planetary Sciences*, **26**, 81–110. [**Figure 7.12** copyright © 1998 Annual Reviews (www.AnnualReviews.org).]

SEPKOSKI, J. J. JR. (1990a) 'Evolutionary faunas', in BRIGGS, D. E. G. AND CROWTHER, P. R. (eds), *Palaeobiology, A Synthesis*, Blackwell Science Ltd. [**Figure 1.24** adapted from SEPKOSKI, J. J. JR. (1984) 'A kinetic model of Phanerozoic taxonomic diversity, III, Post-Paleozoic families and mass extinctions', *Paleobiology*, **10**, 246–267.]

SEPKOSKI, J. J. JR. (1990b) 'The taxonomic structure of periodic extinction', *Geological Society of America Special Paper*, **247**, 33–44. [**Figure 10.2.**]

SHARPTON, V. L. *et al.* (1996) 'A model of the Chicxulub impact basin based on evaluation of geophysical data, well logs, and drill core samples', *Geological Society of America Special Paper*, **307**. [**Figure 12.8.**]

SHEEHAN, P. M. *et al.* (1996) 'Biotic selectivity during the K/T and Late Ordovician extinction events', *Geological Society of America Special Paper*, **307**. [**Figures 10.16, 10.17.**]

SMIT, J. (1999) 'The global stratigraphy of the Cretaceous–Tertiary boundary impact ejecta', *Annual Review of Earth and Planetary Sciences*, **27**, 75–113. [**Figure 13.4** copyright © 1999 by Annual Reviews www.AnnualReviews.org.]

SOHL, N. F. (1987) 'Cretaceous gastropods — contrast between Tethys and the temperate provinces', *Journal of Paleontology*, **61**, 1085–1111, Paleontological Society. [*Figure 1.2.*]

SPICER, R. A., AHLBERG, A., HERMAN, A. B., KELLEY, S. P., RAIKEVICH, M. I. AND REES, P. M. (2002) 'Palaeoenvironment and ecology of the middle Cretaceous Grebenka flora of northeastern Asia', *Palaeogeography, Palaeoclimatology, Palaeoecology*, **184**, 65–105. [**Figures 4.45, 4.47.**]

SPICER, R. A. AND HERMAN, A. B. (2001) 'The Albian–Cenomanian flora of the Kukpowruk River, western North Slope, Alaska: stratigraphy, palaeofloristics, and plant communities', *Cretaceous Research*, **22**, 1–40. [**Figures 4.14, 4.23.**]

SPICER, R. A. AND PARRISH, J. T. (1990) 'Late Cretaceous–early Tertiary palaeoclimates of northern high latitudes: a quantitative view', *Journal of the Geological Society, London*, **147**, 329–341. [**Figure 4.63.**]

STEUBER, T. (2002) 'Plate tectonic control on the evolution of Cretaceous platform-carbonate production', *Geology*, **30**, 259–262. [**Figure 8.13.**]

STEUBER, T. AND LÖSER, H. (2000) 'Species richness and abundance patterns of Tethyan Cretaceous rudist bivalves (Mollusca: Hippuritacea) in the central–eastern Mediterranean and Middle East, analysed from a palaeontological database', *Palaeogeography, Palaeoclimatology, Palaeoecology*, **162**, 75–104. [**Figure 2.16** copyright © 2000 by permission from Elsevier Science, modified from PHILIP, J. *et al.* (11 authors) (1993) 'Late Cenomanian (94–92 Ma)', pp. 153–178 in DERCOURT, J., RICOU, L. E. AND VRIELYNCK, B. (eds) *Atlas Tethys Palaeoenvironmental maps*, Gauthier-Villars, Paris.]

STOREY, B. C. (1995) 'The role of mantle plumes in continental breakup: case histories from Gondwanaland', *Nature*, **377**, 301–308, Macmillan [**Figure 7.23.**]

THOMAS, B. AND SPICER, R. A. (1987) 'Angiosperm radiation', *The Evolution and Palaeobiology of Land Plants*, Croom Helm Ltd. [**Figures 4.32, 4.33, 4.34, 4.35.**]

THORDARSSON, T. AND SELF, S. (1993) 'The Laki (Skaftár Fires) and Grímsvötn eruptions in 1783–85', *Bulletin of Volcanology*, **55**, 233–263. [**Figure 7.10.**]

UNIVERSITY OF CHICAGO 'PALEOGEOGRAPHIC ATLAS PROJECT': http://pgap.uchicago.edu/ [**Figures 1.21a, 2.1.**]

VAIL, P. R., MITCHUM, R. M. JR. AND THOMPSON, S. III (1977a) 'Seismic stratigraphy and global changes of sea level, Part 3: relative changes of sea level from coastal onlap', pp. 63–81 in PAYTON, C. E. (ed.), *Seismic stratigraphy — applications to hydrocarbon exploration, American Association of Petroleum Geologists, Memoir* **26**. [**Figure 3.8.**]

VAIL, P. R., MITCHUM, R. M. JR. AND THOMPSON, S. III (1977b) 'Seismic stratigraphy and global changes of sea level, Part 4: global cycles of relative changes of sea level', pp. 83–97 in PAYTON, C. E. (ed.), *Seismic stratigraphy — applications to hydrocarbon exploration, American Association of Petroleum Geologists, Memoir* **26**. [**Figure 3.9.**]

VALDES, P. J., SPICER, R. A., SELLWOOD, B. W. AND PALMER, D. C. (1999) *Understanding Past Climates — Modelling Ancient Weather*, Interactive CD-ROM. Gordon and Breach Science Publishers. [**Figures 4.57, 4.62** © P. J. Valdes, R. A. Spicer, B. W. Sellwood and D. C. Palmer.]

VAN DER HAAR, T. AND SUOMI, V. (1971) *Journal of Atmospheric Science*, **28**, American Meteorological Society. [**Figure 4.1.**]

VAUGHAN, A. P. M. (1995) 'Circum-Pacific mid-Cretaceous deformation and uplift: a superplume related event?', *Geology*, **23**, 491–494, Geological Society of America. [**Figure 7.19.**]

WARD, P. D. (1990) 'The Cretaceous/Tertiary extinctions in the marine realm; a 1990 perspective', *Geological Society of America Special Paper*, **247**. [**Figures 10.5, 10.6, 10.10.**]

WILSON, R. C. L. (1998) 'Sequence stratigraphy: a revolution without a cause?', pp. 303–314 in BLUNDELL, D. J. AND SCOTT, A. C. (eds), *Lyell, the Past is the Key to the Present. Geological Society, London, Special Publications*, **143**, Geological Society Publishing House. [**Figure 3.11** based on TUCKER, M. E. (1993) 'Carbonate diagenesis and sequence stratigraphy', pp. 51–72 in WRIGHT, V. P. (ed.), *Sedimentology Review/1*, Blackwell Scientific Publications.]

WINTERER, E. L. AND SAGER, W. W. (1995) 'Synthesis of drilling results from the Mid-Pacific Mountains: regional context and implications', pp. 497–535, in WINTERER, E. L., SAGER, W. W., FIRTH, J. V. AND SINTON, J. M. (eds), *Proceedings of the Ocean Drilling Program, Scientific Results*, **143**, Texas A & M University in cooperation with National Science Foundation and Joint Oceanographic Institutions, Inc. USA. [**Figure 1.17b.**]

WINTERER, E. L., VAN WAASBERGEN, R., MAMMERICKX, J. AND STUART, S. (1995) 'Karst morphology and diagenesis of the top of Albian limestone platforms, Mid-Pacific Mountains', pp. 433–470, in WINTERER, E. L., SAGER, W. W., FIRTH, J. V. AND SINTON, J. M. (eds), *Proceedings of the Ocean Drilling Program, Scientific Results*, **143**, Texas A & M University in cooperation with National Science Foundation and Joint Oceanographic Institutions, Inc. USA. [**Figure 1.17a.**]

YOUTCHEFF, J. S. *et al.* (1987) 'Variability in two Northwest Alaska coal deposits', *Alaskan North Slope Geology*, **1** (1 Oct.), 228–229, Pacific Section, Society for Economic Paleontologists and Mineralogists. [**Figures 4.37, 4.38.**]

ZIEGLER, P. A. (1990) *Geological Atlas of Western and Central Europe*, Geological Society Publishing House. [**Figure 2.9a.**]

Index

Note: page numbers refer to text, figures and tables.

palaeosol 134, 144, 157, 201–202
Palaeotethys (Ocean) 26–28
palaeotopography 61, 274
Palaeozoic (Era) 30–31, 60, 201
palladium 314
palmate 119, 120
palygorskite 145
palynoflora 146, 155
palynology 305
palynomorph 123, 145
palynostratigraphy 146
Pangaea 26–28, 30, 60, 180, 198, 229, 233, 270, 273
Panthalassa (Ocean) 26–28
Parana basalt 222
Parana–Etendeka Province flood basalts 233, 240, 247
parasequence 79, 82, 166–167, 182
Paratoxodium 110, 122, 124, 157, 159
parent (isotope) 39
Paris Basin 37–38, 283–284
partial melting 45, 60
partial pressure 261–263, 265
peat 122–126, 144, 190, 256–257
pebble lag 134
pedogenic 134
percent extinction 285
peridotite 52
permafrost 88–89, 193
Persian Gulf 263
Peruc (flora) 154–155
petal 114–115
petiole 126
Petriccio 315
petroleum 190, 194
pH 260
Phanerozoic (Eon) 26, 30–31
phenotype 149
phloem 127
Phoenicopsis 142
phosphate 172
phosphorus 123, 194, 205–207
    cycle 204
photic zone 11, 189
photosynthesis 35–36, 45, 187–189
physiognomy (leaf) 148–152, 155–156, 158–159
phytoplankton 189, 300, 307
Picea alba 88
Picea nigra 89–90
Pinatubo (Mount) 210, 213, 216–222, 247
pinnate 119
pioneer species 305
Pityophyllum 102–103, 105, 110, 122
planar deformation feature (PDF) 320
plankton 10, 19, 24, 32, 34–35, 41, 170, 172, 176, 189, 200, 202, 207, 264–266, 286–287, 289, 293–295, 297–300, 309, 313–315, 340
platanoid 121–122

Platanophyllum 121
Platanus 138, 140
plate margin
    (constructive) 46, 48
    (destructive) 46
    (passive) 50, 52, 63, 76, 78, 229, 257
plate tectonics 9, 21, 44, 48, 50, 68, 70–72, 226, 234, 311
plateau (volcanic) 47, 233, 235–240, 247
platinum 314
    group 314, 318, 325
Pleistocene (Epoch) 24, 78, 328
Plenus Marls 169–170
Pliocene (Epoch) 24
plume (mantle) 23, 50, 56, 72, 82, 230, 233, 235–236, 238, 241, 243, 247, 258
plume (volcanic/ash) 210, 213–214, 216, 220–221, 225, 245, 247, 280
pluton 94
Podocarpus 105
Podozamites 102–103, 105–106, 112, 122, 124
point bar 107
polar
    desert 91
    forest 86, 98, 100, 122, 125, 158, 161, 280–281, 343
    heat transport 85, 163, 176, 179, 276
    ice 29, 163, 186
    light regime 14, 90–91, 129
    winter 158
polarity (normal and reversed) 36, 231, 244
pollen 34, 97, 114, 117–119, 121, 123, 145–146, 157, 251, 254, 287, 304–305, 308, 310
Pompeii 215
Popigai crater 322
Populophyllum 120
Populus balsamifera 88
Populus tremuloides 89
positive feedback (loop) 191, 193–194, 206, 264–265, 280
positive forcing mechanism 192
potassium 123
potassium–argon (decay scheme) 39
Potomac Group 118–119, 121
Poty 341
precipitation 151–152, 157, 306
predator 160, 180–181, 183
Prince Creek Formation 97, 124–126, 129–131, 156, 159
protozoans 308
province (faunal or floral) 180
Pseudofrenelopsis 255
pteridophytes 199
pteridosperms 99
Pu'u O'o 222
Pula 15
pull of the recent 30
Pyrenees 28, 40, 59, 80
pyrite 95, 203
pyroclastic flow 215